INTERPRETATION OF ELECTRON DIFFRACTION PATTERNS

Interpretation of
Electron Diffraction Patterns

K. W. Andrews, D.Phil., D.Sc., F.I.M., F.Inst.P.
D. J. Dyson, B.Sc., A.Inst.P.
S. R. Keown, A.Met., A.I.M.
The United Steel Companies Ltd, Rotherham

Springer Science+Business Media, LLC

Library of Congress Catalog Card Number: 68–19540

Additional material to this book can be downloaded from http://extras.springer.com

ISBN 978-1-4899-6228-7 ISBN 978-1-4899-6475-5 (eBook)
DOI 10.1007/978-1-4899-6475-5

Originally published by Plenum US in 1967.
Softcover reprint of the hardcover 1st edition 1967
First published by
HILGER & WATTS LTD

PREFACE

The general purpose and scope of this manual is set out in the Introduction. It is hoped that the book will form part of the 'equipment' of laboratories and research workers using electron microscopes with diffraction. Although the book has primarily been written with this object in mind, and not as a text or reference book in the ordinary sense, we nevertheless hope that anyone seeking an elementary approach to the subject may find it useful to begin here. The explanatory and introductory material in Part 1 provides some, at least, of the background which users—particularly metallurgical users—of electron diffraction techniques will appreciate. The bias towards metallurgical applications is inevitable in this case but does not preclude applications in other spheres. We had begun to build up our own collection of tables, diagrams and relevant descriptive matter when the need for something like the present volume arose. We were encouraged to proceed further after consultations with the Committee (as constituted in 1965) of the Electron Microscopy and Analysis Group of the Institute of Physics and The Physical Society.

Some of the material presented here will naturally be familiar enough to X-ray crystallographers and some of it has appeared in books on the electron microscope. It is realized that the subject of electron diffraction is still developing, but some working basis is now needed from which research workers in this field can achieve their primary aim—to obtain and interpret as much information as they reasonably can from their diffraction patterns. Certain aspects of electron diffraction involve physical and mathematical concepts which are outside the scope of the book and not necessary to its objective. However, it has been thought useful to introduce in elementary terms those aspects which do in fact lead to practical consequences within the general aims of the book. In particular, certain 'geometrical effects' have been treated as simply as we believe possible, because it is desirable to show whether and to what extent any such effects can affect the accuracy to which distances or angles can be measured. Some of the other information given should help to explain and interpret the more subtle effects which can occur and for which more than one explanation can sometimes arise. Indeed, it is hardly to be expected that a full interpretation of all such effects can be made in every case at present. This situation is not a deterrent but an encouragement to further effort at the fundamental physical level and in translating the fruits of basic physical research into working concepts.

It is hoped that users of this volume will compile data of their own, and, if they think their data are of interest to others, submit them for possible inclusion in a later edition. It would be much appreciated if they would draw our attention to any errors or inconsistencies and suggest improvements within the limits of the stated aims.

We have pleasure in recording our thanks to the members of the Committee of the Electron Microscopy and Analysis Group for their support and helpful suggestions, and to Mr F. B. Pickering who made several useful proposals about the scope of the manuscript. Contributions of useful material have been collected from various authors and publishers to whom acknowledgement is given at the appropriate places in the text. This general acknowledgement here does not imply any less appreciation for their help.

We should however like to record our thanks specifically to Prof. J. Nutting, Dr P. M. Kelly, and their colleagues at Leeds University, for their contributions, to Prof. W. M. Hirthe and his colleagues at Marquette University, Wisconsin, for the use of their tables, and to Dame Kathleen Lonsdale, F.R.S. for permission to use material from the *International Tables for Crystallography*. Drs R. J. Peavler and J. L. Lenusky set a new standard and range for the table of angles between planes for cubic crystals. We have confirmed and corrected their table and added further information.

We are grateful to a number of our colleagues, in particular Mr D. W. Hogan and Mr B. R. Clarke, for discussion of certain points—and to Mr R. Day for his help with computing. Our thanks are also due to Mrs Jean Williams for her unfailing care and patience in typing the manuscript and to Miss Audrey Sherwood, and her staff, and Mr C. L. Wade, for their skill in preparing the illustrations. We also record our appreciation of the work of Mr N. Goodman, Mr D. B. Tomlinson, and staff at Hilger and Watts Ltd, for their help, encouragement and patience in preparing the finished article from our material. We are specially grateful to our wives for their assistance in proof-reading and encouragement throughout.

This volume has been published with the support of the authors' company and we record our thanks to Dr K. J. Irvine for his encouragement and to Dr F. H. Saniter, O.B.E., Director of Research, for permission to publish.

ROTHERHAM
March, 1967

K. W. ANDREWS
D. J. DYSON
S. R. KEOWN

CONTENTS

STEREOGRAPHIC PROJECTIONS

TABLES

IMPORTANT NOTES

1. We have generally given angles in degrees to two decimal places, though in some tables we have given them to three or in degrees and minutes to keep them in their original form.

2. In crystallographic work, stereographic projections have been referred to standard nets of three or even four different sizes. We strongly recommend the largest of the available sizes, viz. that with a 30-cm diameter. The stereographic projections reproduced in the book are 5 inches in diameter, which is the size of the smallest standard. The advantages of the larger size for actual work need no emphasis—the most obvious is that the increased size enables points to be plotted with much greater accuracy (e.g. $\pm\frac{1}{2}°$ or better).

3. In order to promote the use of 30-cm projections, we investigated the possibility of using Wulff nets since some of them had been available in the U.K., but learnt that no more were to be printed. We therefore borrowed a master copy on stable base from the original source, the U.S. Naval Oceanographic Office, Washington. We much appreciate the loan of this chart. The original copy was used to produce further master copies including some at the exact 30-cm size. These charts form the basis of a transparent Wulff net which is sold by Polaron Ltd of Finchley, London, N3. This company has also produced on a stable transparent base a polar net and standard copies of the principal projections given in this book. (It was not considered necessary to reproduce the Wulff or polar nets in this volume.) In addition, copies of most of the other projections are available printed on paper. The production and sale of these nets is regarded as part of a general service providing accessories to electron microscope users and others employing crystallographic techniques. The project is however particularly relevant to the purpose of this book, and we thank Dr G. Kaye (Polaron Ltd) for advancing this complementary project.

SYMBOLS AND ABBREVIATIONS

a Unit cell edge length—lattice parameter
b Unit cell edge length—lattice parameter
c Unit cell edge length—lattice parameter
d Interplanar spacing
e Electron charge
f Focal length; atom scattering factor
h Miller index; Plank's constant
i Miller-Bravais index
k Miller index
k kilo (as in kV)
l Miller index
m Space group symbol
m Electron mass; domain size; magnification
n An integer
p Multiplicity factor; Miller-Bravais direction index
q Miller-Bravais direction index
r Vector length; ring radius; Laue zone radius; Miller-Bravais direction index
s A small deviation from a reciprocal lattice point origin; Miller-Bravais direction index
t Thickness
u Crystal direction index
v Crystal direction index; electron velocity
w Crystal direction index
x Co-ordinate of atom position
y Co-ordinate of atom position
z Co-ordinate of atom position

A Structure type
Å Angstrom unit
A A number
B A number
C An integer; a constant; reciprocal lattice layer number
D Structure type
D Ring diameter
F Space group symbol
F Structure factor
H Hexagonal system
H Twin or habit plane index
I Space group symbol
I Crystal lattice vector; lens current; intensity of reflection
K Twin or habit plane index
L Structure type
L Twin or habit plane index; camera length; a number
M unspecified metal atom(s)
M Magnification; a number
N $= \Sigma h^2$
O Origin
P General point
Q General point
R Rhombohedral system
R A distance; distance between Kikuchi lines
U Index of twin plane normal
V Volts
V Index of twin plane normal
W Index of twin plane normal
X Crystallographic axis
Y Crystallographic axis
Z Crystallographic axis

α Interaxial angle; a fraction of
β Interaxial angle; a fraction of
γ Interaxial angle; a fraction of
δ A small quantity (e.g. $\delta\theta$)
θ Bragg angle
λ Wavelength; general angle
ν Frequency
ρ Angle between directions; general angle
ψ General angle
ϕ General angle; inner potential

a^* Reciprocal lattice parameter
b^* Reciprocal lattice parameter
c^* Reciprocal lattice parameter
d^* Interplanar spacing in reciprocal lattice

X* Reciprocal cell axis
Y* Reciprocal cell axis
Z* Reciprocal cell axis

Introduction

Modern electron microscopes are usually equipped with facilities for obtaining diffraction patterns and microscope images from the same area of a specimen. Normally the best results are gained if the fullest use is made of these combined facilities. Accurate information can be obtained about the identity of the phases present, their orientation relationships, the formation of zones, atomic ordering, twinning, faulting, growth characteristics, etc.

The tables, diagrams and other data presented here have been found useful when interpreting electron diffraction patterns obtained from instruments primarily designed as microscopes, but will also be of use to users of other electron diffraction facilities. Physical metallurgists in particular have appreciated the possibilities of diffraction, and although our data have a bias towards metallurgical applications, they also have a more general use; users can add to them their own standard data and also other tables as they become available. In this way a useful reference manual will be obtained.

Part 1 outlines some of the important, fundamental principles, and in particular gives a full, yet simple account of the methods for interpreting diffraction patterns and the crystallographic knowledge necessary to do this. It represents an approach which has been found useful in a metallurgical research laboratory, where it was possible to aim at only a measure of sufficiency rather than completeness. The coverage is therefore not such as would be justified in a rigorous textbook. There is, however, no account which approaches the subject in this way and, generally, users of the electron microscope will want no more detail than is given here. Reference can be made to the textbook by Pinsker[1] for the more fundamental aspects of electron diffraction and to numerous texts which cover the crystallographic aspects.[2-5]

Parts 2 and 3 mainly consist of tables of interplanar spacings for metals and compounds, together with certain standard stereographic projections. These interplanar spacings are for commonly occurring phases, have a slight bias towards ferrous metallurgy and embrace a variety of crystal structures. The projections included consist of single and double ones. The latter define orientation relationships which are commonly found between two given crystal structures. The single projections have a variety of uses, some of which are included in the succeeding text.

The successful pursuit of structural studies in connection with the development of all kinds of materials requires the use of all the information gained from the use of several techniques. Just as optical and electron microscopy cover a large range of magnifications, one extending and complementing the range of the other, so is it with X-ray and electron diffraction. In the authors' experience X-ray diffraction has proved particularly useful for recognizing many metallurgical constituents and accuracy is such that lattice parameter differences in the third and fourth decimal place become meaningful. The same accuracy cannot be achieved when using electron diffraction analysis. With X-rays, however, under standard experimental conditions, more or less continuous diffraction rings are obtained from samples containing particles within an order of magnitude on either side of 1 μ (10^4 Å). On the other hand, a typical upper limit for the crystal size of particles giving ring patterns by electron diffraction is approximately 10^{-5} to 10^{-6} cm (10^3 to 10^2 Å).

With crystallites intermediate between these two limits, X-rays give broad diffuse ring patterns similar to the ever-present carbon rings of diffraction patterns from carbon replicas, while electrons give spot patterns. Thus, in this range, electron diffraction identification, either by using carbon extraction replica methods or thin foils, extends the range of X-ray analysis and may reveal the presence of unsuspected intermediate or metastable phases and ordering effects which occur prior to the formation of the larger sized particles. The relative merits of these two methods of electron diffraction analysis have been discussed.[6] As the (small) resolving power of an electron diffraction camera is proportional to the wavelength of the radiation, less ambiguous solutions of a diffraction ring pattern are likely to be obtained from X-ray photographs where the resolution is much higher. In a sample where two phases with the same crystal structure and virtually the same lattice parameter are present, it becomes increasingly difficult to distinguish between them in the electron diffraction camera, especially as the structures become more complex. The distinction between such phases on X-ray powder patterns is considerably more certain.

Selected-area diffraction patterns from thin metal foils are, however, frequently derived from only one crystal, or a small number of crystals, and are virtually single-crystal spot patterns. These are equivalent to X-ray single-crystal spot patterns and a similar degree of accuracy can be achieved from both types of pattern. These patterns provide both interplanar angle and spacing measurements for identification. Again ambiguities can arise with similar crystal structures.

Part 1

OUTLINE OF PRINCIPLES

1 Fundamentals of electron diffraction

1.1 ELECTRONS AS WAVES

The diffraction of electrons is possible because of their wave-mechanical behaviour and their conformity to the relationship

$$\lambda = \frac{h}{mv} \tag{1}$$

where λ is the wavelength of the electrons, h Planck's constant and mv the momentum of the particle.

Electrons accelerated by a potential difference of V volts have a kinetic energy of $\frac{1}{2}mv^2$, where

$$\tfrac{1}{2}mv^2 = Ve$$

e being the electronic charge.

Eliminating v by using equation (1) gives

$$\lambda = \frac{h}{\sqrt{(2meV)}} \text{ Å}$$

$$= \frac{12{\cdot}236}{\sqrt{V}} \text{ Å} \tag{2}$$

This is the wavelength associated with an electron of mass m grams travelling with a velocity of v cm/sec. A relativity correction is needed for actual conditions involving the voltages used, so that the actual formula is slightly more complicated than equation (2). An electron microscope operating at 100 kV has electrons of wavelength 0·037 Å as the source of radiation.

1.2 THE BRAGG LAW

When a suitably focused beam of electrons passes through a crystal, diffraction will occur if the three Laue conditions are simultaneously fulfilled. These are

$$\left.\begin{aligned} a(\cos\alpha_1 - \cos\alpha_2) &= n_1\lambda \\ b(\cos\beta_1 - \cos\beta_2) &= n_2\lambda \\ c(\cos\gamma_1 - \cos\gamma_2) &= n_3\lambda \end{aligned}\right\} \tag{3}$$

where a, b and c are the crystal lattice parameters and n_1, n_2 and n_3 are the Laue orders of diffraction. The cosines of the angles define the directions of the incident and diffracted beams. The combined conditions for diffraction are more generally represented by the Bragg Law, which may be said to *contain* the Laue conditions. This is illustrated in

Fig. 1. In the circular inset, a parallel beam of electron waves impinges on a crystal and makes an angle of incidence θ with a set of crystallographic planes of interplanar spacing d and Miller indices hkl. By the construction shown, diffraction occurs when the ray paths via successive planes in the system differ from each other by an exact number of wavelengths. The diffracted ray then leaves the plane at an angle θ (or 2θ with the

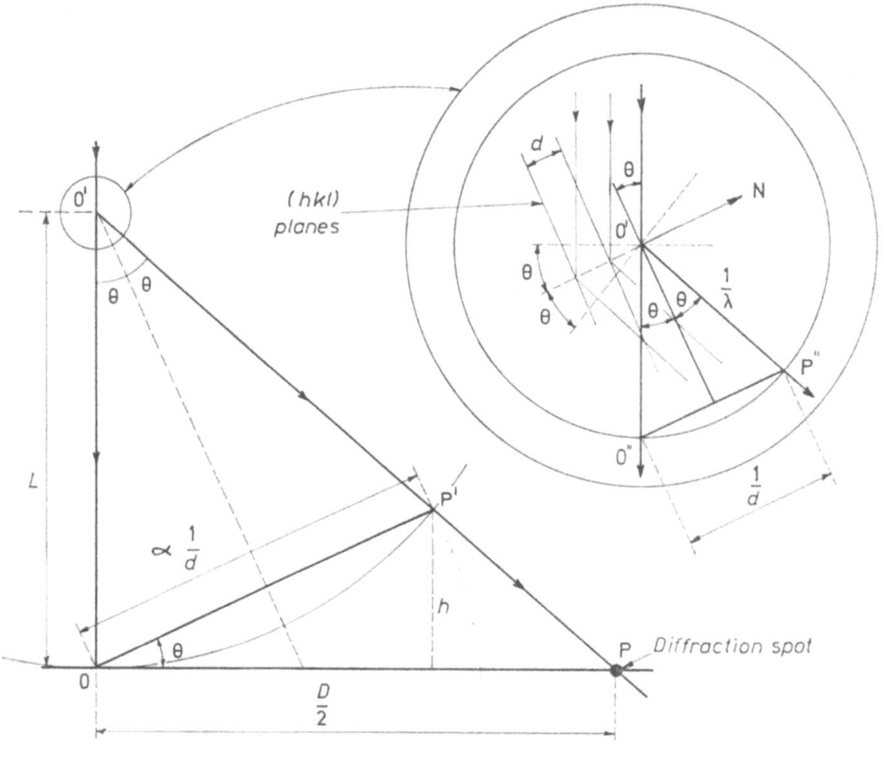

FIG. 1. The basic geometry of diffraction. For electron diffraction the angle θ is small and P′ very near to P.

incident beam). When this occurs the path difference, which is $n\lambda$, must equal the extra lengths along successive ray paths and is given by $2d \sin \theta$, i.e.

$$n\lambda = 2d \sin \theta \tag{4}$$

In X-ray and electron diffraction work it is common practice to make no distinction between the nth order diffraction from a plane of spacing d and the first order from a parallel plane of spacing d/n. All d-spacings are recorded as for $n = 1$, but the Miller indices may then contain a common factor.

1.3 FORMATION OF DIFFRACTION PATTERNS

An alternative form of equation (4) with *n* equal to unity is

$$\frac{1}{d} = 2\left(\frac{1}{\lambda}\right) \sin \theta \qquad (5)$$

In the inset in Fig. 1, a *sphere* of radius $1/\lambda$ is constructed with its centre at O′. This intersects the direct beam at O″ and the diffracted beam at P″. The distance O″P″ is seen to be $2 \sin \theta/\lambda$ which is therefore equal to $1/d$. Furthermore, the direction of O″P″ is parallel to O′N which is perpendicular to (*hkl*). The construction of the 'sphere of reflection' or 'Ewald sphere' (as in Fig. 1) is a well-known device for interpreting diffraction patterns and will be referred to later.

Equation (5) is thus a useful form for interpreting the diffraction pattern from a single crystal in the electron beam. The pattern is recorded on a flat film or plate, which is usually normal to the incident beam direction O′O in Fig. 1. The diffracted ray proceeds along O′P′ and makes a diffraction spot on the film at P.

Let OO′ equal *L*, the effective camera length, and make O′P′ also equal to *L*. Then the triangle O′OP′ is similar to O′O″P″ (inset). Therefore, make *L* equal $M(1/\lambda)$, where *M* is a scale or 'magnification' factor. Therefore

$$OP' = M\left(\frac{1}{d}\right) = M.d^*$$

The use of d^* for $1/d$ is general. The vector $\mathbf{d^*}$ is the reciprocal lattice vector (§2.1.) perpendicular to the plane. Hence all such distances as OP′ are parallel in direction and proportional to reciprocal lattice vectors, $\mathbf{d^*}$. In Fig. 1

$$\left.\begin{aligned} OP &= L \tan 2\theta \\ OP' &= 2L \sin \theta \end{aligned}\right\} \qquad (6)$$

For electron diffraction patterns the wavelength is small, and so θ is also small ($< 3°$). Therefore P′ is very near to P and the distances OP′ and OP approximate to $2L\theta$. (The scale is deliberately enlarged in Fig. 1.)

It follows that if the photographic plate were tilted through the angle θ, the points P and P′ would actually coincide, but the distance of other diffraction spots from the centre would be increased. If the incident beam is *exactly perpendicular* to a reciprocal lattice layer, a pattern of spots will form because the distance of the layer from the sphere at any point—*h* in Fig. 1—is less than the finite size of the diffraction spot. The smaller the wavelength, the easier it is to fulfil this condition.

In the symmetrical (single-crystal type) pattern there is thus a tendency for the pattern to fade out with increasing θ. Spots may be formed away from the central region if the sphere cuts a second (or third) layer of the reciprocal lattice (see §5.3), or if there are slightly misoriented fragments or sub-grain boundaries in the crystal. Completely separate grains may give superimposed patterns right from the centre or only in fringe regions. Strictly symmetrical diffraction spot patterns are not necessarily obtained without special effort.

Polycrystalline ring patterns arise because grains of many orientations have planes in the right positions for diffraction and the diffracted rays then lie anywhere on a cone such as would result from rotating O′P′ about O′O in Fig. 1.

2 Relation between crystal lattice, stereographic projection and reciprocal lattice

The full interpretation of patterns is materially advanced by the use of the reciprocal lattice and stereographic projection.

2.1 RECIPROCAL LATTICE

The reciprocal length $1/d$ (or the vector $\mathbf{d}^*$), which has already appeared in equation (5), is further understood in relation to the general reciprocal lattice concept.

It is evident that the single-crystal electron diffraction spot pattern is closely similar to, and with decreasing wavelength approximates more closely to, a network of points that are at distances from the centre of the diffraction pattern (O in Fig. 1) proportional to $1/d$ or d^*. Furthermore, the line joining O to any point is also parallel to the normal to the diffracting plane and so represents the *vector* $\mathbf{d}^*$ on a suitable scale. The array of points therefore determines vectors from the origin O. Because it is a regular network it is regarded as a layer of a reciprocal lattice. The reciprocal lattice is a regular three-dimensional pattern of points which is related to the three-dimensional crystal lattice by the requirement that the vector distances in the former, viz. $\mathbf{d}^*_{hkl}$, are perpendicular to (hkl) planes in the space lattice and have lengths d^*_{hkl} inversely proportional to the corresponding interplanar spacings d_{hkl}.

The connection between the two lattices must be such that the vector $\mathbf{d}^*_{100}$ must be perpendicular to (100) and ($\bar{1}00$) which form two opposite faces of the unit cell parallelepiped. The length d^*_{100} is inversely proportional to d_{100}, which is not necessarily equal to the cell edge a. The plane (100) does, however, contain the two other cell edges, [010] and [001], so that d^*_{100} is perpendicular to OY and OZ in the crystal lattice. This condition is sufficient to provide an appreciation of the complete relationship for all cases except the most general (the triclinic) by reference to Fig. 2.

Fig. 2 covers two-dimensional lattices with rectangular axes (*a*) and oblique axes (*b*). Three-dimensional lattices (except triclinic) follow by adding Z axes normal to the plane of the diagram and extending the two dimensional formulae analogously. In both diagrams ON is perpendicular to the plane (hk) or $(hk0)$ in the crystal lattice. The reciprocal unit cell edges are then made to satisfy the above conditions for d^*_{100}, and d^*_{010}, i.e. $a^* = d^*_{100}$, $b^* = d^*_{010}$; for example OX* is perpendicular to OY and OY* to OX.

The interplanar spacing formula for Fig. 2(*a*), i.e. rectangular axes, is given in the literature[2–5] (orthorhombic case for $l = 0$) or is easily deduced. It can be written

$$\frac{1}{d^2} = \frac{h^2}{a^2} + \frac{k^2}{b^2} \tag{7}$$

This is precisely the result obtained in the reciprocal lattice, since $d^* = 1/d$, $a^* = 1/a$, $b^* = 1/b$, so that

$$(\mathrm{OP'})^2 = (d^*)^2 = h^2(a^*)^2 + k^2(b^*)^2 \tag{8}$$

(Pythagoras' theorem). The lattice is thus built up of points at distances, a^*, $2a^*$, $3a^*$, etc., along the OX* axis and b^*, $2b^*$, etc., along the OY* axis. In Fig. 2(*a*) the vector $\mathbf{d^*}$, length OP′, is illustrated for $h = 2$ and $k = 3$.

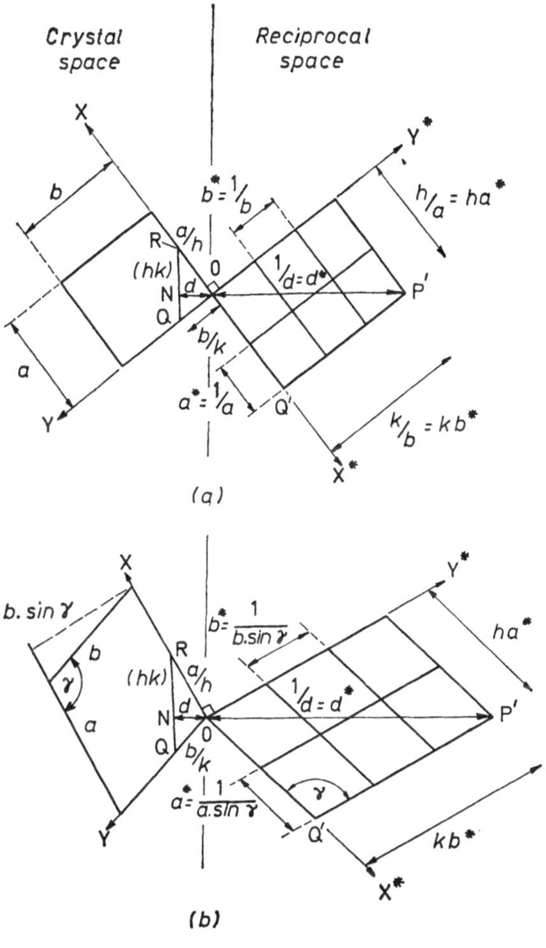

FIG. 2. Summary of reciprocal lattice relationships for rectangular- and oblique-plane lattices. (These relationships lead to all three-dimensional cases except the triclinic.) (*a*) Rectangular axes. (*b*) Oblique axes.

The three-dimensional orthorhombic case is found by adding $l^2(c^*)^2$ ($= l^2/c^2$) to equation (7) and is a simple Pythagorean sum in three dimensions. For the tetragonal case, $a = b$, $a^* = b^*$ and for the cubic case, $a = b = c$, $a^* = b^* = c^*$. It is a matter of simple geometry involving similar triangles to prove that $\mathrm{R\hat{O}N} = \mathrm{P'\hat{O}Q'}$ so that OP′ is collinear with ON.

In Fig. 2(*b*) the formula for the interplanar spacing (crystal lattice) is

$$\frac{1}{d^2} = \frac{h^2}{(a \sin \gamma)^2} + \frac{k^2}{(b \sin \gamma)^2} - \frac{2h}{(a \sin \gamma)} \cdot \frac{k}{(b \sin \gamma)} \cos \gamma \tag{9}$$

which is equivalent to the usual monoclinic formula[2] with $l = 0$, apart from the convention that generally makes β the angle $\neq 90°$, i.e. equation (9) interchanges b and c for convenience.

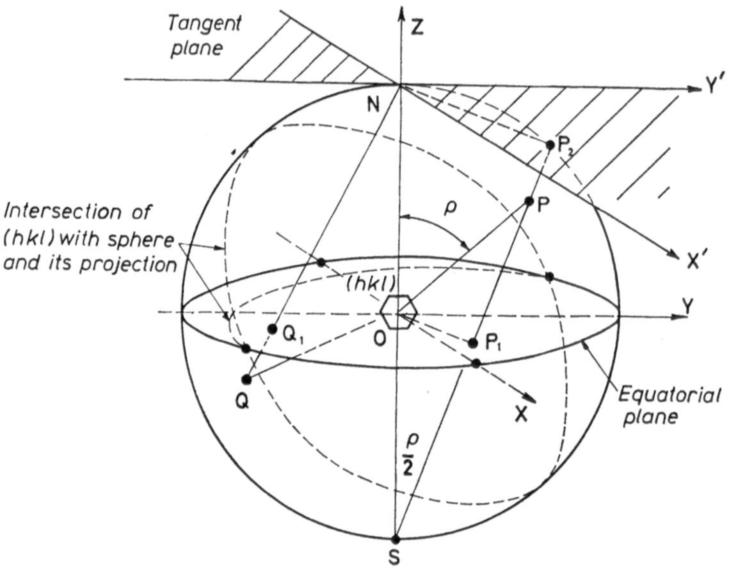

FIG. 3. Basic construction for stereographic projection.

It is required that $a^* = d^*_{100} = 1/(a \sin \gamma)$; similarly $b^* = d^*_{010} = 1/(b \sin \gamma)$. Putting $h = 1$, $k = 0$ and $h = 0$, $k = 1$ in (9) shows at once that the corresponding reciprocal formula must be

$$(d^*)^2 = h^2(a^*)^2 + k^2(b^*)^2 - 2hka^*b^* \cos \gamma \tag{10a}$$

or

$$(d^*)^2 = h^2(a^*)^2 + k^2(b^*)^2 + 2hka^*b^* \cos \gamma^* \tag{10b}$$

where $\gamma^* = X^*\hat{O}Y^* = 180° - \gamma$.

Reciprocal lattice points using a^*, b^* as cell edges therefore give the pattern shown on the right of Fig. 2(*b*) and a typical case OP′ is shown for $h = 2$, $k = 3$. Equation (10a) follows directly from the cosine formula in the triangle P′OQ′ which endorses the correctness of the construction. Again triangles P′OQ′ and OQR are similar and it is easy to show that OP′ is collinear with ON.

The addition of OZ or OZ* perpendicular to this diagram gives a monoclinic lattice, and if $\gamma = 120°$, a hexagonal lattice results. All rectangular cases would equally well follow by putting $\gamma = 90°$.

2.2 STEREOGRAPHIC PROJECTION

The stereographic projection or stereogram (see Figs. 3 and 4) is essentially a device for condensing three-dimensional angular relationships onto a plane (comparable with the production of a circular map from the terrestrial globe).

In Fig. 3 a crystal is supposed to be at the centre of a sphere and a chosen plane (*hkl*) is indicated. This plane can be represented in one of two ways:

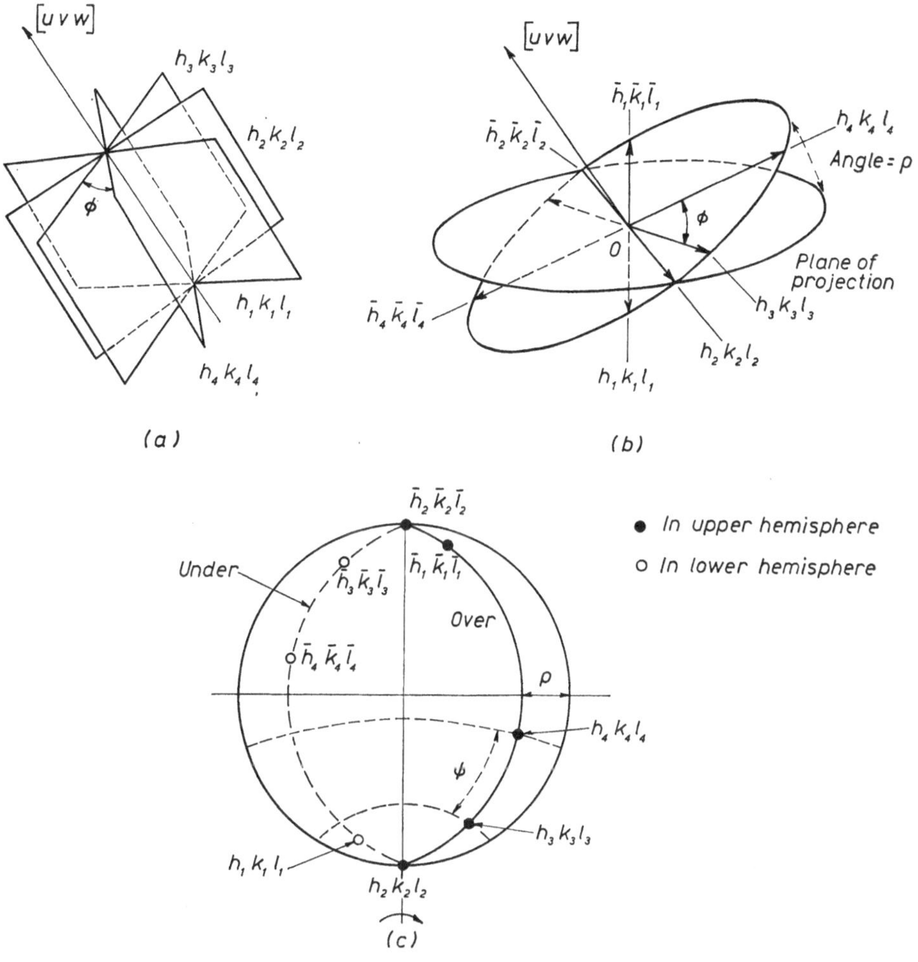

FIG. 4. Stereographic projection in relation to planes and zones. (*a*) Planes in a zone. (*b*) Perpendiculars to planes as radii of a great circle of a sphere. (*c*) Projection of the circle on equatorial plane.

(*a*) By representing the chosen plane by its normal OP. The required projection of P is found by joining P to S, in which case the corresponding point in the stereographic projection is P_1. Here the equatorial plane has been taken as the plane of projection. P_1 is thus a projection of P, the *pole* of the plane (*hkl*).

(*b*) By extending the chosen plane so that as a diametral plane it cuts the sphere in a great circle. This representation of the plane, which is not required further here, can be obtained by extending the plane *hkl* until it cuts the sphere in a circle and then projecting points on this circle on the plane of projection. The construction is indicated by broken curves in Fig. 3. The projection circle in this diagram is perpendicular to OZ and, in crystals other than triclinic, would therefore represent the (001) plane monoclinic with $\gamma \neq 90$).

In some cases, the plane X′Y′ tangential to the sphere at N is taken instead of the equatorial plane, but the same result is obtained, only the scale being altered. Thus P projects to P_2, but similar triangles, such as SOP_1, SNP_2, maintain proportionality, making $NP_2 = 2OP_1$.

In Fig. 4(*a*) some of the planes *hkl* which lie in a zone [*uvw*] are shown. The normals to these planes are clearly perpendicular to [*uvw*] and radiate from it like spokes of a wheel [Fig. 4(*b*)]. Hence, analogous with the projection of (*hkl*), the zone [*uvw*] and the planes in it can also be represented by a great circle or by its pole [*uvw*]. The projection of the great circle on the plane of projection is shown in Fig. 4(*c*). It follows that the projection can be used to represent (*a*) a plane by a great circle *and/or* by its 'pole' at 90° from it or (*b*) a zone by its pole *and/or* by a great circle at 90° from it. All planes in a zone have poles which lie on the great circle. Similarly, the planar great circle contains points which represent the poles of directions lying in the plane. In *either case* the equation

$$hu + kv + lw = 0 \tag{11}$$

must be obeyed.

2.3 SUMMARY OF INTER-RELATIONSHIPS

It is useful to appreciate the inter-relation of the two lattices and the projection, as in the table below, for which equation (11) is fundamental. Thus a plane (*hkl*) in the crystal is represented by the great circle or by the projection of its pole. The normal to the plane is parallel to **d*** in reciprocal space and this vector joins the origin to a point representing *hkl*. The reverse applies to the zone axis [*uvw*], where the direction is in real space and the plane in reciprocal space, but the stereographic projection can represent either or both.

Summary of relationships for the equation $hu+kv+lw = 0$

Symbols or relationship	Crystal space		Stereographic projection		Reciprocal space
(*hkl*)	Plane	↔	Great circle *or* Point = projection of its pole	↔	Lattice point or, more exactly, the vector joining the origin to this point.
[*uvw*]	Zone axis	↔	A pole *or* Great circle	↔	Plane
$hu+kv+lw = 0$ (*a*) [*uvw*] constant	Equation defines planes in the zone.	↔	Planes represented by points on the great circle for the zone.	↔	Equation defines lattice points which lie in the reciprocal plane.
(*b*) (*hkl*) constant	Equation defines zone axes contained in the plane.	↔	Points representing these zones lie on the great circle for the *plane.*	↔	Equation defines the planes which contain the reciprocal vector.

Equation (11) represents (hkl) planes which lie in zone [uvw]. These planes are usually represented by points on the great circle for the zone. This great circle represents a plane parallel to the plane in reciprocal space which contains all the reciprocal points, hkl, satisfying the equation. Alternatively, the equation represents the directions in crystal space which are contained in a given (hkl) plane, or the planes in reciprocal space which intersect along the reciprocal vector for (hkl).

2.4. EXTENSION OF ZONE EQUATION

Finally equation (11) is a special case of

$$hu+kv+lw = \pm C \tag{12}$$

where C is an integer. This equation represents:

(*a*) *In crystal space, keeping h, k, l constant*: The co-ordinates u, v, w of lattice points which lie in the Cth parallel plane from the origin, the indices for which are hkl.

(*b*) *In reciprocal space, keeping u, v, w constant*: The co-ordinates h, k, l of reciprocal lattice points which lie on the Cth parallel plane from the origin, the indices of which are uvw.

Relationship (*b*) is of more immediate value in interpreting electron diffraction patterns.

3 Indexing polycrystalline and single-crystal patterns

3.1 GENERAL

The type of transmission diffraction pattern obtained in an electron diffraction camera depends largely on the crystal size of the diffracting medium and to a somewhat lesser extent on the size of the limiting aperture used. These two factors determine the number of crystals which contribute to the observed pattern.

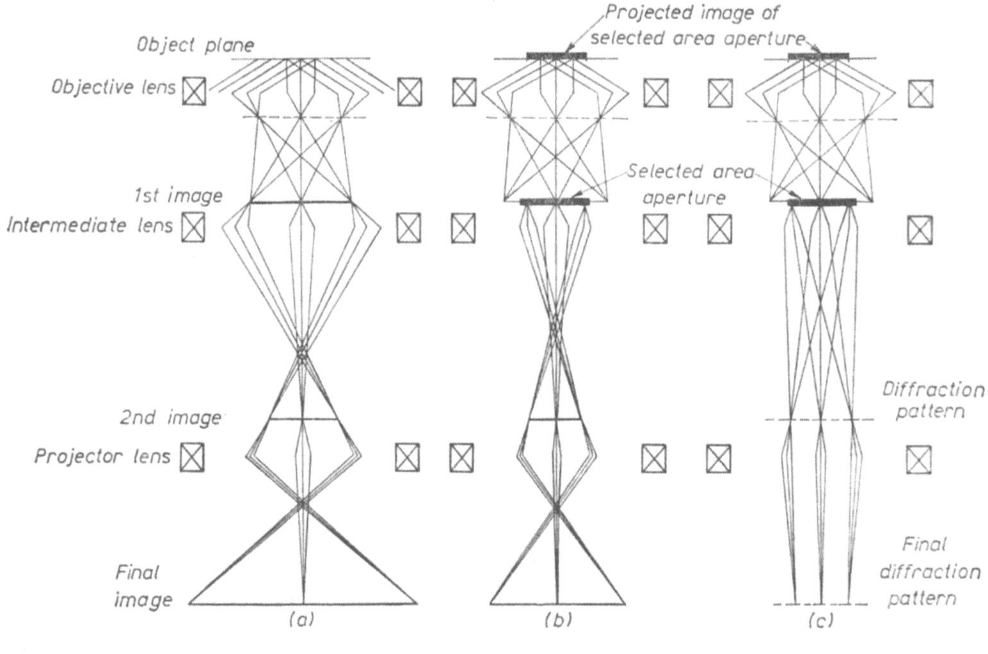

FIG. 5. Image formation in the electron microscope. (*a*) Normal image formation. (*b*) Selected-area image formation. (*c*) Selected-area diffraction pattern.

Modern electron microscopes enable electron diffraction patterns to be obtained from selected areas of the sample. Defocusing the intermediate lens to collect the diffracted image from the objective lens, enables a magnified image of the diffraction pattern to be obtained on the fluorescent screen (see Fig. 5). This pattern can be observed directly and also photographed if required. An aperture may be inserted in the object plane of the intermediate lens to select a restricted region of the microscopic image. There is a diffraction pattern at the 'cross-over' plane between the intermediate and projector lenses. If the intermediate lens is defocused, this pattern extends downwards until it coincides with the object plane of the projector lens and is thus focused by the projector lens on the fluorescent screen or a photographic plate.

If the selector aperture defines a large number of randomly orientated crystallites, as would be analogous to conditions in X-ray powder diffraction, then Debye–Scherrer (ring) patterns will be obtained. Complete rings are produced when sufficient crystals are present

(*a*) To satisfy the diffraction conditions for each set of $\{hkl\}$ planes.

(*b*) To give random reflections for definite values of θ, all diffracted rays with the same value of θ combining to form a cone of rays which intersects the fluorescent screen in a circle.

With fewer crystallites present in the selector aperture, condition (*a*) will again be fulfilled, but (*b*) will not. The 'rings' lose their continuity about 360° and a spotty ring pattern results.

A single crystal in the electron beam will give rise to a more or less regular and ideally symmetrical arrangement of spots, each of which has a different value of *hkl*. Spots which are equidistant from the centre and diametrically opposite, are associated with the same set of crystal planes and are indexed as *hkl* and $\bar{h}\bar{k}\bar{l}$. Their simultaneous appearance is due to the effects of small wavelength, divergence of the electron beam, and other instrumental factors. Otherwise the curvature of the reflecting sphere tends to favour diffraction from slightly different zones as θ increases away from the centre of the pattern and on either side. Many $\{hkl\}$ planes will not satisfy condition (*a*) and therefore, as one tends towards a single-crystal pattern, so certain rings disappear. These are the single-crystal type spot patterns already considered in §1.3.

The solution of these patterns leads to the identification of the selected area of the sample via its crystal structure. Patterns are solved by relating distances and angles in the crystal lattice. From Fig. 1 and equation (6)

$$\frac{D/2}{L} = \tan 2\theta \qquad (13)$$

where D is the ring diameter, $D/2$ the distance of any spot from the centre of the pattern, and L the effective camera length. Eliminating θ from equation (13) by using the Bragg relationship and assuming θ to be small so that

$$\tan 2\theta \simeq 2 \sin \theta \simeq 2\theta$$

then

$$D = 2L.2\theta$$

$$= 4L.\frac{\lambda}{2d}$$

Therefore

$$\frac{D.d}{2} = L\lambda = \text{Camera constant} \qquad (14)$$

The camera constant may be found by calibration with a sample of known lattice parameter that gives sharp ring or spot patterns. The former is preferable because it gives a continuous calibration around the centre. This procedure eliminates the difficulty of the exact determination of L. Instrumental variations and differences in the camera length generally prevent electron diffraction in the electron microscope from being a high-accuracy technique. However, methods of achieving the best possible accuracy will now be discussed.

3.2 ACCURACY

Inaccuracies in the interpretation of electron diffraction patterns, even if the patterns are not complicated by double diffraction, streaking, etc., can arise if two important factors are not fully appreciated. These factors are:

(*a*) The precision with which the diffraction pattern is actually located in relation to the area of the image from which it is desired.

(*b*) The accuracy of determining interplanar distances owing to variations in the camera constant λL.

3.2.1 *Accuracy of selection: procedure*

Agar[7] has outlined the correct procedure for obtaining a diffraction pattern from a selected area of the specimen. This procedure involves the following stages:

(*a*) Adjust the intermediate lens current to focus the selector aperture on the screen. In most commercial electron microscopes, the aperture is focused at a magnification of about ×20 000 to ×60 000. The plane of the aperture now coincides with the object plane of the intermediate lens.

(*b*) Adjust the objective lens current to focus the specimen image. The specimen image is then co-planar with the selector aperture.

(*c*) Re-adjust (decrease) the intermediate lens current to produce the diffraction pattern on the screen. Some commercial instruments facilitate this operation by step switching from an 'imaging' position to a 'diffraction' position followed by a fine adjustment of the lens current. This adjustment can be judged by making the central spot as small as possible or by focusing the rings of a polycrystalline pattern. The diffraction image from the back focal plane of the objective lens is thus imaged in the object plane of the intermediate lens and consequently appears magnified on the viewing screen.

(*d*) Adjust (defocus) the condenser lens to illuminate the specimen with an almost parallel beam of electrons. This decreases the screen intensity of the diffraction pattern and often photographic exposures of the order of ½ to 2 minutes are required to record the pattern.

Agar has shown that the second stage (the objective lens adjustment) of the procedure produces the most serious errors if it is not carefully carried out, but obviously the whole procedure must be stringently adhered to for maximum accuracy. Phillips[8] has suggested that stages *c* and *d* of the procedure should be reversed so that the final setting of the intermediate lens can be more easily carried out.

3.2.2 *Accuracy of camera constant*

In equation (14) it has been assumed that $2\lambda L$, the camera constant, is in fact a constant. However, small variations in its value will be obtained because of fluctuations in the voltage supply to the instrument, incorrect procedure in obtaining diffraction patterns, changes in specimen height, errors of measurement, etc. If the errors associated with these various deviations are minimized or corrected as shown, it is possible that the total error in the camera constant will enable *d*-spacings to be measured to an accuracy approaching 0·1 per cent. While this is not as good as the accuracy which can be obtained by X-ray diffraction, it should enable distinction to be made between phases of like crystal structure and somewhat different lattice parameter. It must, however, be emphasized that this accuracy can only be approached by careful measurement and correction.

Effect of instrumental variations on $2\lambda L$

λL will change if the high voltage supply to the electron microscope fluctuates appreciably. Stabilization ratios of 5000 : 1 can now be approached, which cut these fluctuations to a minimum. Equation (2) shows that the wavelength λ of the electrons is inversely proportional to the square root of the accelerating voltage. With 100-kV electrons, a stabilization ratio of 1000 : 1 (a minimum value) and a camera length of 400 mm, the camera constant will vary by approximately 1 part in 2500. This is negligible. However, we have assumed that the voltmeter is calibrated and can be accurately set to give electrons of energy corresponding to any value of accelerating voltage. This is not always so, but can easily be checked.

Faulty lens settings will alter $2\lambda L$ if the correct procedure for obtaining diffraction patterns is not followed. Phillips[8] has shown that

$$\frac{\Delta L}{L} = \frac{\Delta f_0}{f_0} + \frac{\Delta m}{m} + \frac{\Delta M}{M} \tag{15}$$

where $\Delta L/L$ is the proportional error in the camera length L, $\Delta f_0/f_0$ the proportional error in the focal length f_0 of the objective lens, $\Delta m/m$ the proportional error in the magnification of the immediate lens, and $\Delta M/M$ the proportional error in the magnification of the projector lens.

In some instruments the strength of the projector lens is fixed so that $\Delta M/M = 0$. However, in other instruments the projector lens current is variable and the accuracy of the setting obviously has a marked effect on $2\lambda L$.[8] $\Delta m/m$ is the error in re-adjusting the intermediate lens strength to image the diffraction pattern from the back focal plane of the objective lens [stage (*c*) of the recommended procedure]. This adjustment can be made with considerable accuracy, so $\Delta m/m$ is usually very small.

As indicated above, an error Δf_0 in f_0 can arise when focusing the image in the same plane as the selector aperture [stage (*b*) of the recommended procedure] or from variations in specimen height. Phillips[8] states that variations in λL of about 3 per cent can arise from the $\Delta f_0/f_0$ component by varying the objective lens current by 9 mA in a total of 500 mA ($\sim$ 1·4 per cent).

Since the effective camera length is defined as

$$L = f_0 \times m \times M$$

where m is the magnification of the intermediate lens and M the magnification of the projector lens, the focal length of the objective lens, f_0, is the most sensitive factor which affects the reproducibility of L. In turn, f_0 depends on the plane of the object. The factors which control this are specimen holders of different length, support grids that are not flat, thick and buckled specimens, and the effects due to tilting the specimen.

The change in L due to plate position should, in an efficient microscope, be small. It is determined by the amount of 'play' in the plate-holder.

Fig. 6 (due to Armitage and MacConaill[9]) shows a typical variation of camera constant with objective lens current. The curve covers a range of specimen heights of about 0·8 mm. It shows that $2\lambda L$ varies from 3·4 to 4·5 cm. Å. This represents a 25 per cent variation. Such errors should therefore be minimized. If the objective lens is thin, the proportionate error in the camera length is then related to the focal length of the lens and the lens current[10] by

$$\frac{\Delta L}{L}=\frac{\Delta f_0}{f_0}=\frac{2\Delta I}{I} \tag{16}$$

so that changes in camera constant should be approximately twice those in the lens current. This is, however, not found to correspond to the observed facts, and reasons for this have been put forward.[10,11]

The A.E.I. EM.6 electron microscope normally operates with a constant projector lens current so that the size of the diffraction pattern and the camera constant are fixed.

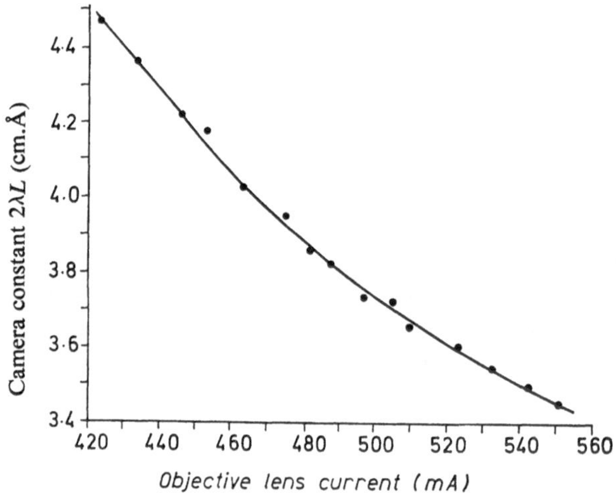

FIG. 6. Variation of camera constant with objective lens current. (After Armitage and MacConaill.[9])

However, there is a facility in the microscope which enables the size of the diffraction pattern to be continuously varied with the projector lens current. If the camera constant is plotted against the projector lens current, a linear relationship is obtained (Fig. 7), and this graph can be used to determine the camera constant for diffraction patterns of a wide range of sizes.

Effects of some physical factors

Electrons are scattered both elastically and inelastically by the crystalline diffracting medium. An electron is considered to have undergone an elastic collision when it suffers a change in momentum and direction, this change being balanced by a recoil of particles within the diffracting medium as a whole. At the Bragg angle this scattering is coherent and the scattered electrons are in phase in the direction of the diffracted beam. The elastic or coherent scattering is mainly responsible for the intensity of the observed reflections. Any 'halo' effect or background intensity will then be an indication of the presence of inelastic scattering,[12] even though elastic scattering from imperfections may contribute to this background level between reflections. With thin crystals, very little background is observed, the majority of the scattering being elastic. With thicker crystals, inelastic scattering becomes more important. Inelastically scattered electrons suffer a change in momentum and energy which may cause a change in the internal electronic structure of the scattering atoms. Inelastic scattering is mainly confined to small angles and has little effect on the diffraction patterns or image contrast.

Pinsker[1] has dealt fully with the elastic interaction of the electron beam with the specimen, particularly in respect to the effects of refraction and inner potential. The

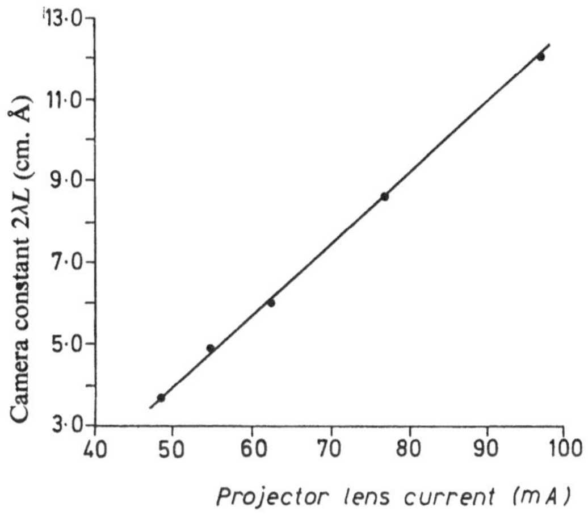

FIG. 7. Relationship between camera constant and projector lens current.

potential field within a lattice will cause the electrons to receive an additional acceleration which will result in a change in wavelength according to the relation

$$\lambda' = \frac{h}{\sqrt{[2me(\phi+V)]}} \tag{17}$$

where ϕ is the inner potential of the lattice [cf. equation (2)]. As a result the Bragg angle for any system of planes will be altered since $n\lambda' = 2d \sin \theta'$ [equation (4)].

It follows that, although the incident radiation is the same for the standard and the specimen under examination (see next section), a difference in the value of ϕ for the two phases will lead to a different emergent wavelength and a possible error in the calculation of the spacings. Fortunately, the inner potentials are of the order of 10 volts and so complete neglect of ϕ makes an error of $\sim$ 0·01 per cent in d.

For fast electrons normal to the entrance and exit surfaces of the crystal the effect of refraction is negligible. Under favourable conditions the effects of refraction can be observed for angles of incidence near 45°.[13] Thus the actual magnitude of the effect in transmission studies depends mainly on the change in electron path and to a lesser extent on the change in wavelength. For normal transmission microscopy both effects can be neglected for the kind of practical application to which the information presented here is relevant.

Variations in 2λL due to inaccuracies in measurement of patterns

Many of the errors in determining λ or L can only be estimated and not accurately obtained. A second method is thus needed for calculating the camera constant. This is done by measuring the ring diameters of the diffraction pattern obtained from a polycrystalline specimen of a standard material of known lattice parameter. It is preferable for this ring pattern to be superimposed on the unknown spot pattern. Plate 1 provides a typical example. It has been assumed that the standard gives a ring pattern and the unknown a spot pattern. If, however, both the unknown and the standard give a ring pattern, obvious complications will arise. In such a case a standard which gives a sharp spot pattern, e.g. thallium chloride, can be used to calibrate the ring pattern. Obviously this standard cannot define the constant along any radius from the centre but it can still be used to give the same accuracy.

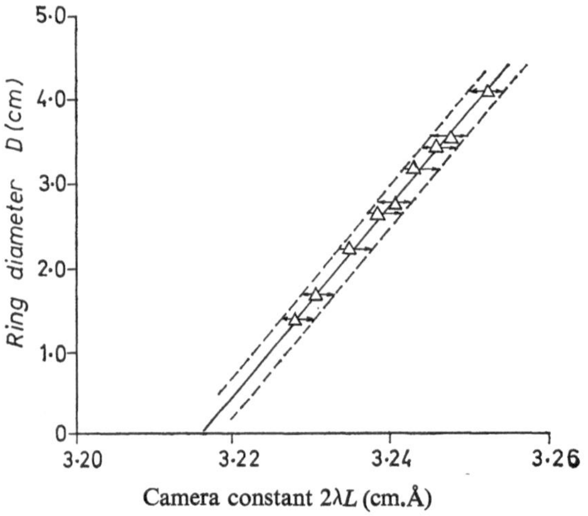

FIG. 8. Variation of camera constant with ring diameter.

The errors that arise in the measurement of rings and spots may now be considered. The breadth of the diffracted ring depends on the finite width of the diffracted beams, the grain size of the specimen, the state of strain in the specimen, the thickness of the specimen, any voltage drift during the exposure, and the grain size of the emulsion on the photographic plate. Ideally, the smallest line breadth is necessary but the authors have found that with a polycrystalline aluminium specimen the rings can be measured to an accuracy of $\pm$ 0·01 mm when using a vernier microscope or microdensitometer. The accuracy of the camera constant is determined by this error combined with the errors in the previous section due to variation of the inner potentials of the specimen and standard and to variation in camera length because more than one plate is used.

In some microscopes, lens aberrations may lead to distortion of the rings into ellipses. Thus, since D is not the same for each point on a 'ring', $2\lambda L$ will not be a constant around this ring. The extent to which this variation occurs can be seen in Fig. 8, where the broken lines represent the maximum and minimum values of $2\lambda L$ for a given ring. This figure also shows that the camera 'constant' increases approximately linearly with ring diameter, i.e. with distance from the centre of the plate. It is necessary, therefore, to construct such a figure in order to use the appropriate value of the camera constant with D values from the diffraction pattern of the unknown specimen. The measurement of the diffraction pattern from the unknown is subject to the same errors also. These errors will combine with that for $2\lambda L$ to give a resultant error in d.

A more rigorous proof, showing in more detail the reason for the increase of $2\lambda L$ with D, will now be given. It has already been indicated that the increase is due to the decreasing validity of the approximation for small angles

$$\tan 2\theta \simeq 2 \sin \theta \simeq 2\theta$$

The geometry of Fig. 1 shows that the difference OP − OP′ progressively increases with increasing D. A correction can thus be made by using equation (18).

Using the Bragg equation with $n = 1$, and equation (13), and setting $x = \tan 2\theta = D/2L$, we get

$$\frac{\lambda}{2d} = \sin\theta = \left(\frac{1-\cos 2\theta}{2}\right)^{\frac{1}{2}}$$

$$= \left\{\frac{1}{2}\left[1-\left(\frac{1}{1+x^2}\right)^{\frac{1}{2}}\right]\right\}^{\frac{1}{2}}$$

$$= \frac{1}{\sqrt{2}}[1-(1+x^2)^{-\frac{1}{2}}]^{\frac{1}{2}}$$

which on expanding the expression $(1+x^2)^{-\frac{1}{2}}$ first and then the whole function as a power series gives

$$\frac{\lambda}{2d} = \tfrac{1}{2}x(1-\tfrac{3}{8}x^2 + \text{terms in } x^4, x^6, \text{etc.})$$

so that to a close approximation

$$d = \frac{2\lambda L}{D}\left(1+\frac{3D^2}{32L^2}\right) \tag{18}$$

When D is 5 cm and L 40 cm, the correction to d is ~ 0·15 per cent. This amount is subtracted.

Variations in apparent ring diameters due to spot shape

In specimens which contain stacking faults or have very small dimensions in one direction (platelets etc.) the effect in reciprocal space is to elongate the diffraction spot in a direction perpendicular to the fault or plate. The general interpretation of the streaks so formed is considered later (§4.3), but it is necessary to point out here some very elementary deductions about the way in which the presence of such streaks, as seen in Plate 2, might effect the actual measurement of spacings. This part anticipates somewhat the general analysis of spot patterns (§3.4).

In Fig. 9(a) a series of such spots is shown in the zero layer through the origin of the reciprocal lattice and with elongations normal to the plane. Here the effect of the elongation is simply to extend the appearance of the diffraction pattern further outwards from the centre since the Ewald sphere can be intersected by the streaks from such points. In this condition the interplanar spacing is only slightly affected. In Fig. 9(b) the streaks are shown in a row of spots in a non-zero layer [$C \neq 0$ in equation (12)]. There could be a condition where the zero row of spots is absent or unextended. In this case too, spots appear from this layer nearer to the origin than they would otherwise be and also extend further from it. This condition was noted by Bendler.[14] A point such as A

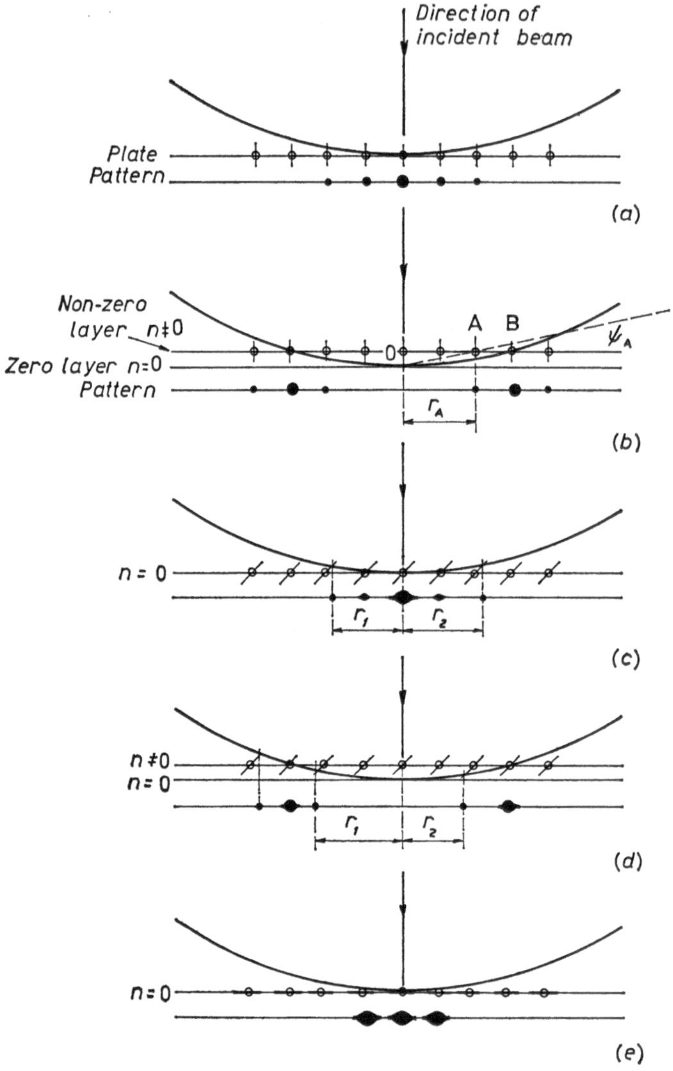

FIG. 9. Effect of reciprocal lattice point shape on the diffraction pattern.

would, if it were centred exactly in the sphere, appear to lie in a zone at an angle ψ with the plane of projection. If the *true* distance $OA = D_A/2$, then the measured distance $r_A = (D_A/2) \cos \psi_A$. Similarly $r_B = (D_B/2) \cos \psi_B$. Hence the row of spots will show a systematic apparent change in λL which is actually due to a real change in r.

Figs. 9(*c*) and (*d*) show the corresponding cases for streaks at an angle to the plane of the pattern. In this case the streaks are visible but shortened or lengthened in projection according to the angle with respect to the electron beam. Also the apparent position will differ increasingly with distance from the centre and in general $r_1 \neq r_2$. For case (*c*) in general, however, $r_1 + r_2 = D$ (approx.). Fig. 9(*e*) shows that streaks which are more or less parallel to the plane of the pattern simply make the measurement r from O, or the distance between corresponding spots D, less accurate.

In Figs. 9(*a*), (*c*) and (*e*) the effect is shown for a row of spots through O. Figs. 9(*b*) and (*d*) are more complicated for the spots are not in a row through the origin. The following treatment has been used by the authors in interpretating such patterns from cubic metals. Fig. 10 shows the construction for two points P_1 and P_2, deriving from planes which have indices $(h_1k_1l_1)$ and $(h_2k_2l_2)$ respectively. It is assumed that stacking faults (or the thin dimension of a platelet) are parallel to planes of indices $(h_3k_3l_3)$. The plane of the diffraction pattern does not pass through P_1, P_2, but through points such as

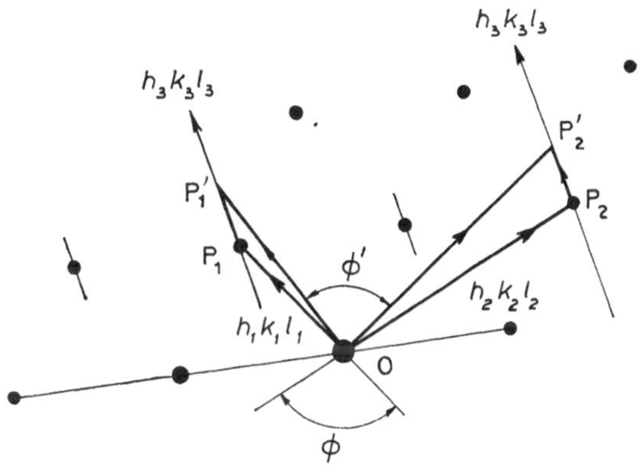

FIG. 10. Apparent changes or errors in spacings and angles due to streaks in reciprocal space.

P_1', P_2' according to the way it is tilted. Such points are represented in reciprocal space by the vectors whose components are

for P_1': $h_1 \pm \alpha h_3,\ k_1 \pm \alpha k_3,\ l_1 \pm \alpha l_3$

for P_2': $h_2 \pm \beta h_3,\ k_2 \pm \beta k_3,\ l_2 \pm \beta l_3$

where α and β are fractions. This is simply the process of adding indices vectorially as in the indexing of single-crystal patterns (§3.4).

In the general case

$$|\, d^* \,|^2 = \Sigma h^2 a^{*2} + \Sigma 2klb^*c^* \cos \alpha^* \tag{19}$$

and

$$\cos \phi = \frac{(\Sigma h_1 h_2 a^{*2} + \Sigma(k_1 l_2 + l_1 k_2) b^* c^* \cos \alpha^*)}{|\, d_1^* \,|\,|\, d_2^* \,|} \tag{20}$$

[Equation (19) is the general form of (10).] In the cubic case, (20) reduces to the familiar formula

$$\cos\phi = \frac{h_1h_2+k_1k_2+l_1l_2}{\sqrt{(h_1^2+k_1^2+l_1^2)}\sqrt{(h_2^2+k_2^2+l_2^2)}} \tag{21}$$

Also, (19) becomes

$$|d^*|^2 = |a^*|^2 (h^2+k^2+l^2) \tag{22}$$

If now the reciprocal sphere cuts such a point at P_1' rather than at its centre,

$$|d_1^*|^2 = |a^*|^2 (\Sigma h_1^2 \pm 2\alpha\Sigma h_1h_3 + \alpha^2\Sigma h_3^2)$$

from which we obtain

$$\frac{\Delta|d_1^*|^2}{|d_1^*|^2} = \frac{\alpha^2\Sigma h_3^2 \pm 2\alpha\Sigma h_1h_3}{\Sigma h_1^2} \tag{23}$$

For small values of α this approximates to

$$\frac{2\Delta d_1^*}{d_1^*} = -2\frac{\Delta d}{d}$$

Similarly,

$$\cos\phi_1 = \frac{\Sigma(h_1\pm\alpha h_3)(h_2\pm\beta h_3)}{\sqrt{[\Sigma(h_1\pm\alpha h_3)^2]}\sqrt{[\Sigma(h_2\pm\beta h_3)^2]}} \tag{24}$$

The table below illustrates the magnitude of these effects for {111} reflections in a face-centred cubic metal with [110] as the zone of the diagram, and stacking faults in (111), i.e. $(h_1k_1l_1) = (\bar{1}11)$, $(h_2k_2l_2) = (1\bar{1}1)$, $(h_3k_3l_3) = (111)$. The effects of increasing α from 0 to $\frac{1}{6}$ and β from 0 to $\frac{1}{6}$ are indicated for a lattice parameter of ~ 3·6 Å.

α	β	$h_1 \pm \alpha h_3$	d_1'	$h_2 \pm \beta h_3$	d_2'	ϕ or ϕ'
0	0	$\bar{1}11$	2·079	$1\bar{1}1$	2·079	70°32′
$+\frac{1}{6}$	$+\frac{1}{6}$	$\frac{\bar{5}}{6}\frac{7}{6}\frac{7}{6}$	1·948	$\frac{7}{6}\frac{\bar{5}}{6}\frac{7}{6}$	1·948	80°10′
$+\frac{1}{6}$	$-\frac{1}{6}$	$\frac{\bar{5}}{6}\frac{7}{6}\frac{7}{6}$	1·948	$\frac{5}{6}\frac{\bar{7}}{6}\frac{5}{6}$	2·171	69°18′
$-\frac{1}{6}$	$-\frac{1}{6}$	$\frac{\bar{7}}{6}\frac{5}{6}\frac{5}{6}$	2·171	$\frac{5}{6}\frac{\bar{7}}{6}\frac{5}{6}$	2·171	62°54′

It follows that the effects can be quite appreciable, and care should be taken when dealing with patterns from crystals which give streaks. The effect could be sufficient to confuse the identification, but this is unlikely if patterns are properly inspected and, when possible, the tilting stage is used to examine the effects of small tilts.

3.2.3 *Practical calibration of $2\lambda L$*

To calibrate $2\lambda L$ accurately it is necessary to obtain a diffraction pattern from a standard known substance such as aluminium, gold, thallium chloride, etc. $2\lambda L$ is then simply obtained from $Dd = 2\lambda L$.

The best technique is to superimpose the standard substance on the specimen by evaporation so that both materials are practically co-planar and both patterns are obtained on the same plate. An example is shown in Plate 1. However, sometimes small particles are masked by such an evaporated layer and fail to diffract. An alternative method then is to obtain the standard in some adjacent region and to photograph both regions on successive photographic plates, taking care not to alter any of the microscope controls. It is particularly important not to re-focus the image.

Two different specimen techniques are used in the authors' laboratory to obtain standards on only part of the specimen grid:

(*a*) The standard substance is evaporated onto a glass slide and the resulting layer is floated off in water. The specimen grid is then simply dipped into the water and a small piece of the standard substance is caught on the grid in a suitable position alongside the specimen.

(*b*) The standard substance is evaporated onto only part of the sample. This involves masking off regions of the sample, and a simple way of doing this is to tilt the supporting grid so that the evaporated metal is deposited at a grazing angle. The grid bars shelter part of the replica and the standard substance is obtained in only small regions.

To allow for the small variation of $2\lambda L$ with D, as illustrated in Fig. 8, it is advisable to derive unknown d values using a $2\lambda L$ value obtained from a standard ring or spot with approximately the same D value.

With thin foils, the matrix metal itself acts as a 'built-in' standard, but, owing to the high intensity and consequent 'spreading' of some of the matrix spots, it is difficult to obtain $2\lambda L$ to a high degree of accuracy.

Another calibration technique exists but it is doubtful whether high accuracies can be guaranteed.[9] This technique has already been indicated in connection with Fig. 6, where the variation in $2\lambda L$ is plotted against the corresponding changes in objective lens current necessary to focus specimens at various heights. The technique produces a useful quick method of obtaining the camera constant, but for really accurate work the use of standards is recommended.

3.2.4 *Summary of procedure for obtaining accurate results*

Arrange for the unknown substance to be co-planar with the standard in the specimen holder.

Follow closely the rules set out by Agar to obtain a diffraction pattern from the selected area. This should preferably show the diffraction patterns from both standard and material under investigation, superimposed on one another.

Obtain the variation of $2\lambda L$ with D along several diameters of the 'rings' of the standard pattern. It is convenient to choose diameters that correspond with prominent rows of reflections on the pattern of the unknown.

Thus a value of $2\lambda L$ for many points on the plate is known. This can be corrected for the effect of the different inner potentials of standard and unknown.

Measure each spot or ring on the unknown diffraction pattern as accurately as possible. The error associated with this measurement, combined with the error in $2\lambda L$ from the above procedure, determines the final error in d.

The effect of small crystal size can be minimized by obtaining ring or spot profiles to determine the peak position more accurately.

3.3 POLYCRYSTALLINE SPECIMENS

The polycrystalline specimen is assumed to have a large number of individual, randomly oriented grains or crystallites. Some of these will have orientations that satisfy the conditions of Fig. 1, and a diffraction spot appears at a distance $R \tan 2\theta$ from O. All possible diffracted rays satisfying this condition therefore lie on a cone of semi-vertical angle, 2θ, and thus come from planes with the same d-spacing, which would be obtained by rotating O′P about O′O. This cone will intersect the plane of the photographic plate in a circle. It is equivalent to rotating the reciprocal vector about O and the diffraction condition for this cone does not limit the position of the reciprocal lattice or the corresponding crystal orientation in any other way. Furthermore, owing to the symmetry of the crystal there may be several combinations of the same indices giving the same d-spacing and therefore the same d^*. This is the multiplicity factor p. As an example consider the spacing for tetragonal {711} in relation to the formula

$$\frac{1}{d^2} = \frac{h^2+k^2}{a^2} + \frac{l^2}{c^2} \tag{25}$$

The first two indices and the + or − signs can be interchanged, i.e.

$$\pm h \ \pm k \ \pm l \text{ and } \pm k \ \pm h \ \pm l$$

which makes $p = 16$. It is noted, however, that {551} also gives the same spacing since $5^2+5^2 = 7^2+1^2$. In this tetragonal case, interchange of h and k, with $h = k$, now makes no difference and so p is reduced to $2^3 = 8$ (± signs only). In the corresponding cubic case, however, $a = b = c$ and all three indices can be interchanged and so for both sets of planes $p = 24$. Therefore, in the tetragonal case $16+8 = 24$ planes and in the cubic case $24+24 = 48$ planes can contribute the same diffraction ring.

One feature sometimes found in ring patterns is a pronounced arcing of the rings and is representative of a preferred orientation in the sample. In rolled-metal specimens or vacuum-deposited films, it is commonly found that a certain direction in each crystallite will orientate itself so that it becomes normal to the specimen surface. This is almost equivalent to decreasing the number of crystallites, since some 'randomness' has been lost. Some of the rings will disappear and the lengths of the arcs will depend on the degree of ordering of the crystallites.

The possibility of separating lines of closely similar spacings is conditioned by the small wavelengths used and thus the small values of θ. The *resolution* is found from equation (14).

$$\Delta d = d_2 - d_1 = 2\lambda L \left(\frac{1}{D_2} - \frac{1}{D_1}\right) = \frac{2\lambda L}{D_2 D_1} \Delta D$$

for two spacings d_1, d_2 from two ring diameters D_1, D_2. A typical camera constant $2\lambda L$ is 3·2 cm. Å. For a ring diameter of 1·6 cm ($d = 2{\cdot}0$ Å)

$$\Delta d = \frac{\Delta D}{8} \text{ (Å)}$$

Hence, if the accuracy of measurement is 0·1 mm (which is better than general), the resolution is 0·025 Å. This is at least an order of resolution lower than is obtainable with normal X-ray procedures. It is clear that, as long as rings can be resolved from each other, the sequence of θ or d values can be related to a given lattice, e.g. cubic or some other lattice with the appropriate axial ratio(s). Absolute identification by accurate d-spacings is, however, hampered by the above limitation. In some cases, for example, relatively small differences in lattice parameter may be important (as with cubic carbides or nitrides), and supplementary X-ray diffraction analysis may be required to establish the lattice parameters or even simply to confirm the existence of two phases of similar structure and lattice parameters.

3.3.1 *Simple procedure for indexing ring patterns*

(*a*) *When identity of substance is known*:
Measure the diameters (D) of the rings.
Convert these distances into interplanar (d) values via the camera constant.
Compare the measured d values with standard ones for the particular substance (see Part 3). Each ring can then be indexed.
Confirmatory X-ray work where possible or necessary.

(*b*) *When identity of substance is unknown*:
Measure the diameters (D) of the rings.
Convert these distances into interplanar (d) values via the camera constant.
The arrangement of diffraction rings may indicate a certain recognizable crystal structure or type.
Use *A.S.T.M. Index*[15] (see §3.5.1).
Use sample for confirmatory X-ray diffraction where possible or necessary.

3.4 SINGLE-CRYSTAL PATTERNS

3.4.1 *General*

Having only one reciprocal lattice present means that the Ewald sphere will only cut through a certain number of regularly spaced reciprocal lattice points. Thus a spot pattern results which is an enlarged image of a certain reciprocal lattice plane already envisaged in the construction of Fig. 1.

Fig. 11 represents diagrammatically a pattern of spot locations in a single plane through O, the centre of the diagram which represents the 'straight-through', unscattered electrons. If two directions such as OA, OB are taken, spots will lie at equal intervals, a and b respectively, along these directions. Other spots occupy corners of parallelograms formed in the manner shown. A pattern is analysed by measuring

(*a*) The distances of spots from O, i.e. $D_1/2$, $D_2/2$, $D_3/2$, from which corresponding d-spacings are calculated, and

(*b*) The angles such as ϕ_1, ϕ_2, ϕ_3 between the reciprocal vectors and hence between the planes giving rise to the diffraction spots which subtend these angles at O.

Assuming that the identity of the diffracting medium can only be certain phases, then from the values of d and ϕ it is, in principle, possible to proceed:

(*a*) To identify the phase giving the pattern by comparison of measured and standard sets of *d*-spacings, although this identification may not be absolute without the provision of patterns from more than one orientation.

(*b*) To establish or confirm the identity of the phase known to be, or suspected of being, present by comparing the calculated and measured interplanar angles.

This process is, however, complicated by the existence of the multiplicity of planes with the same *d* owing to permutations of the same indices. Account must also be taken of those cases where two basically different values of *d* happen to coincide, as in the *cubic* system for

$$(h_1^2+k_1^2+l_1^2) = (h_2^2+k_2^2+l_2^2)$$

e.g. (557)/(771) or (300)/(221). Also the multiplicity of planes in a set allows for a number of possible directions for the vector of length $1/d_1 = \mathbf{d}_1^*$ and likewise for any other

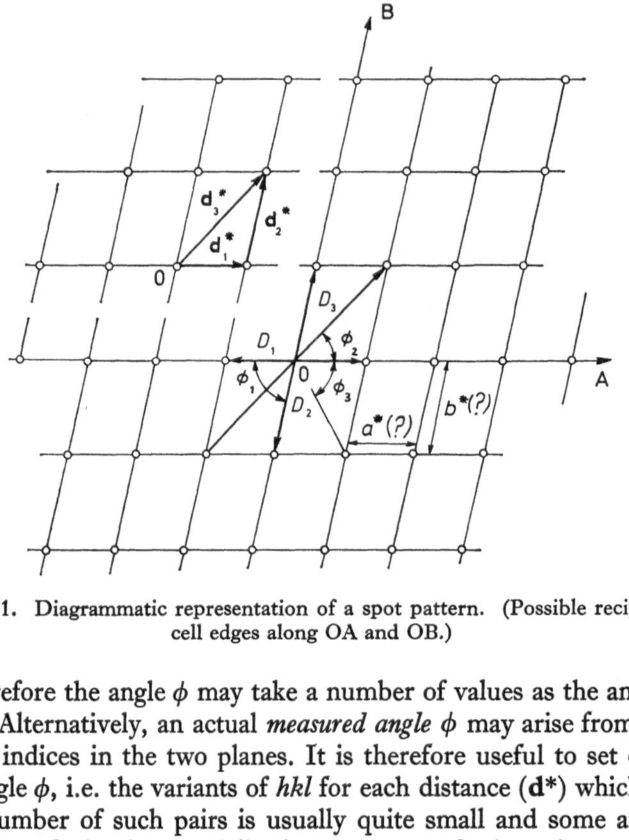

FIG. 11. Diagrammatic representation of a spot pattern. (Possible reciprocal cell edges along OA and OB.)

vector $\mathbf{d}_2^*$. Therefore the angle ϕ may take a number of values as the angle between the two directions. Alternatively, an actual *measured angle* ϕ may arise from more than one combination of indices in the two planes. It is therefore useful to set out the possible pairs for the angle ϕ, i.e. the variants of *hkl* for each distance ($\mathbf{d}^*$) which could give the angle ϕ. The number of such pairs is usually quite small and some are equivalent to choosing a variant of what is essentially the same type of orientation.

From any pair of indices $(h_1k_1l_1)$, $(h_2k_2l_2)$ a zone axis $[uvw]$ for the direction normal to the patterns is obtained from the equations

$$\left.\begin{aligned} u &= k_1l_2 - k_2l_1 \\ v &= l_1h_2 - l_2h_1 \\ w &= h_1k_2 - h_2k_1 \end{aligned}\right\} \tag{26}$$

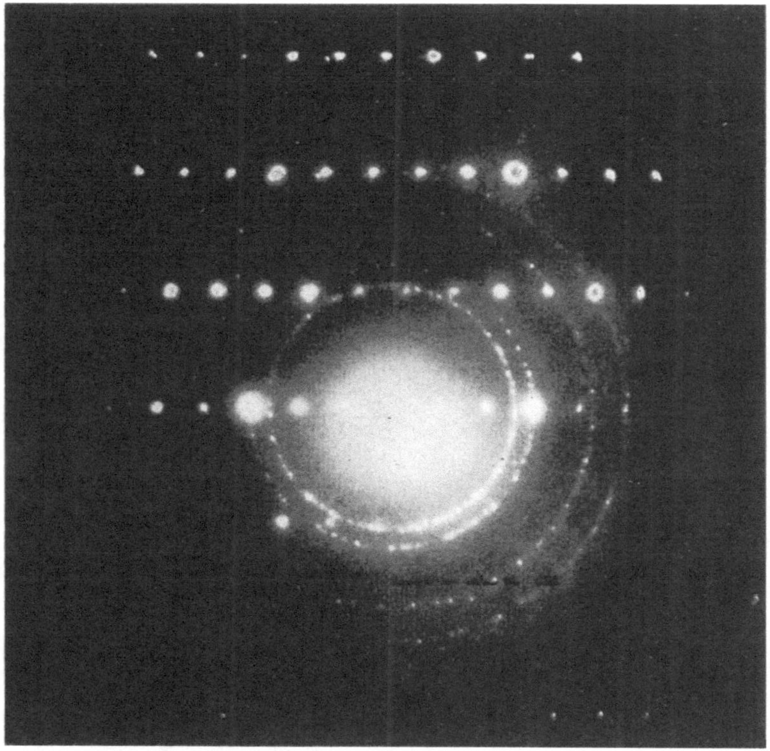

PLATE 1. Aluminium standard (ring) pattern superimposed on carbide (spot) pattern.

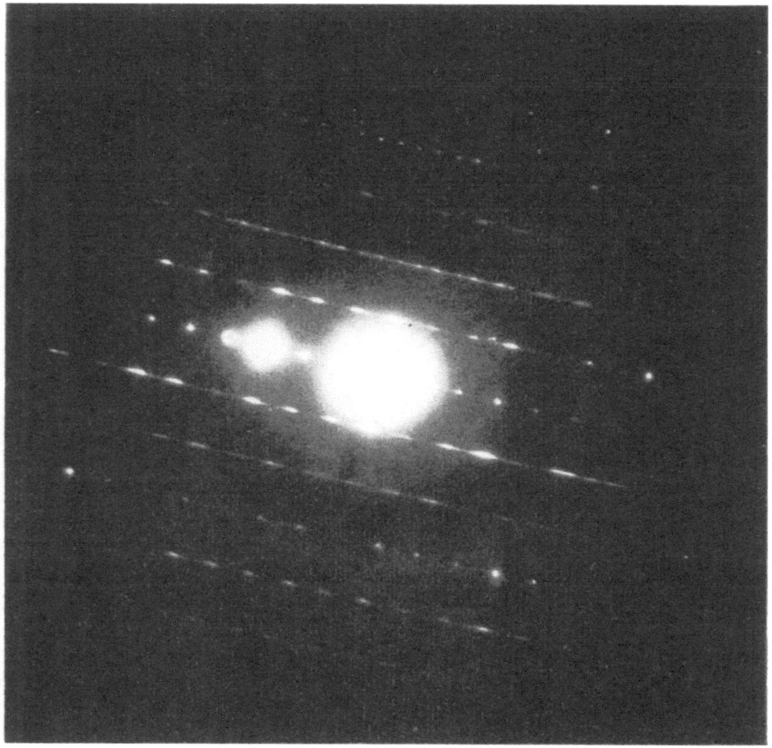

PLATE 2. Streaks on an electron diffraction pattern due to a faulted crystal lattice.

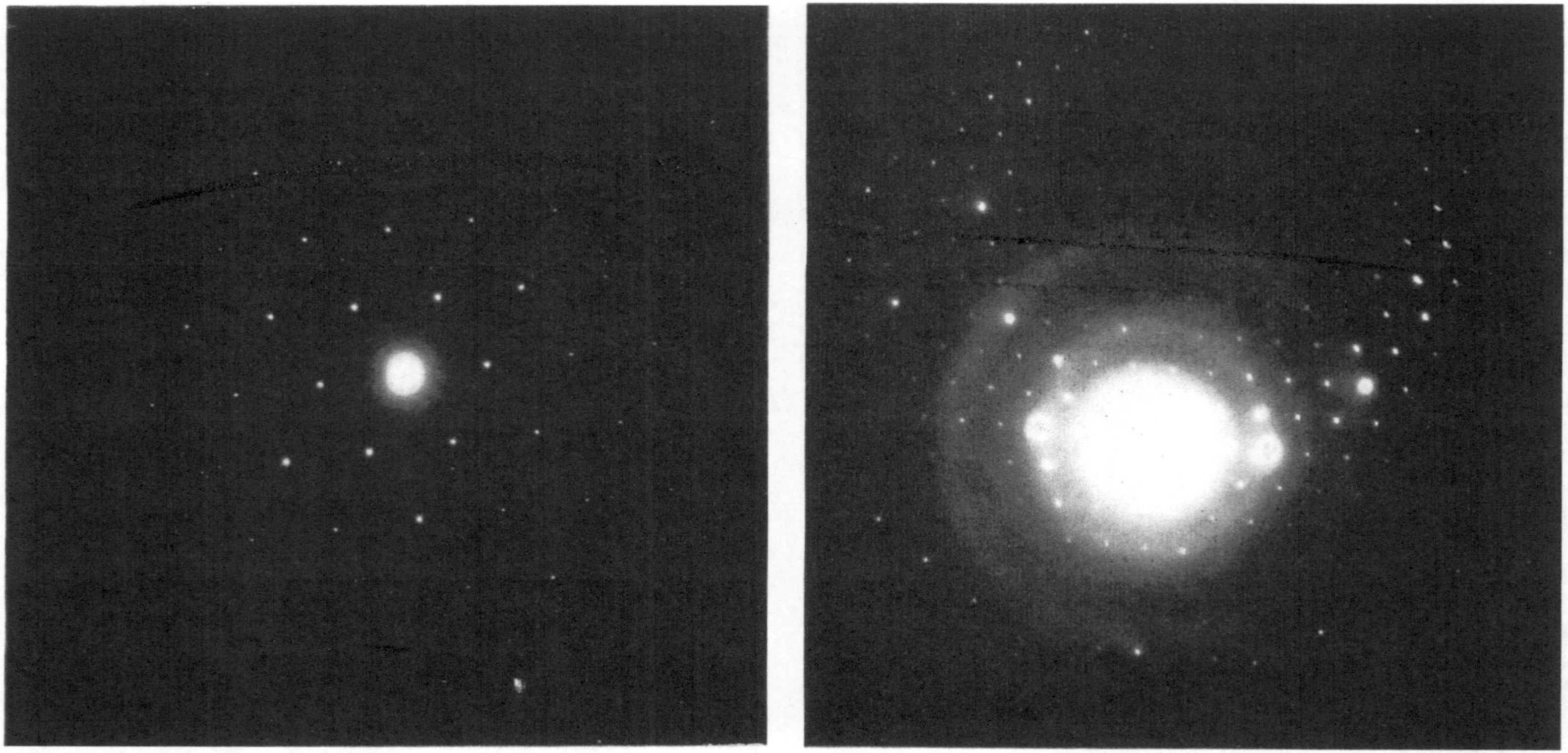

(a) (b)

(c)

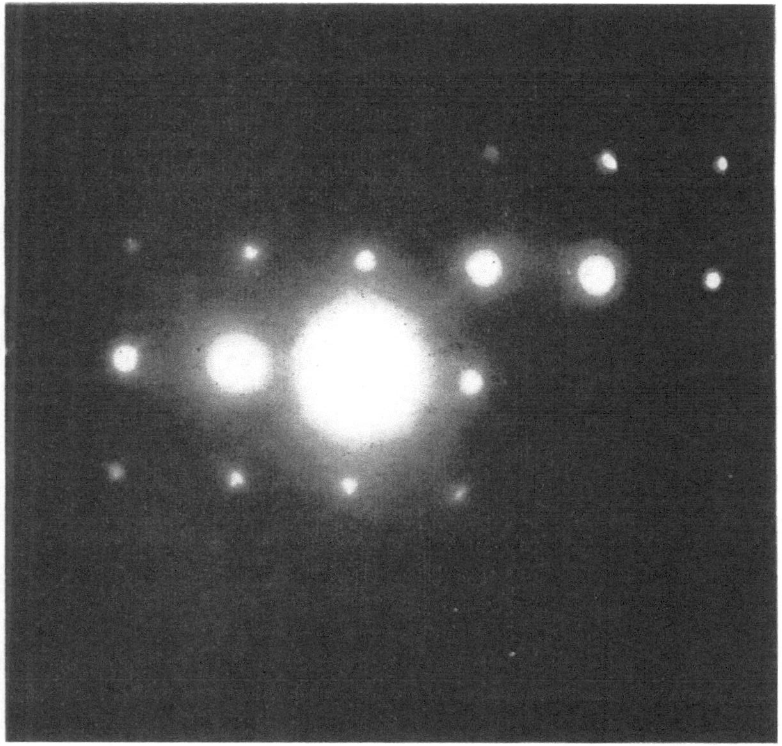

(d)

PLATE 3. Diffraction patterns illustrating the effects of tilt from a prominent zone axis and the formation of Laue zones. (*a*) Negligible tilt, (*b*) $\sim 1\frac{1}{2}°$ tilt, (*c*) $\sim 5°$ tilt, and (*d*) $\sim 6°$ tilt.

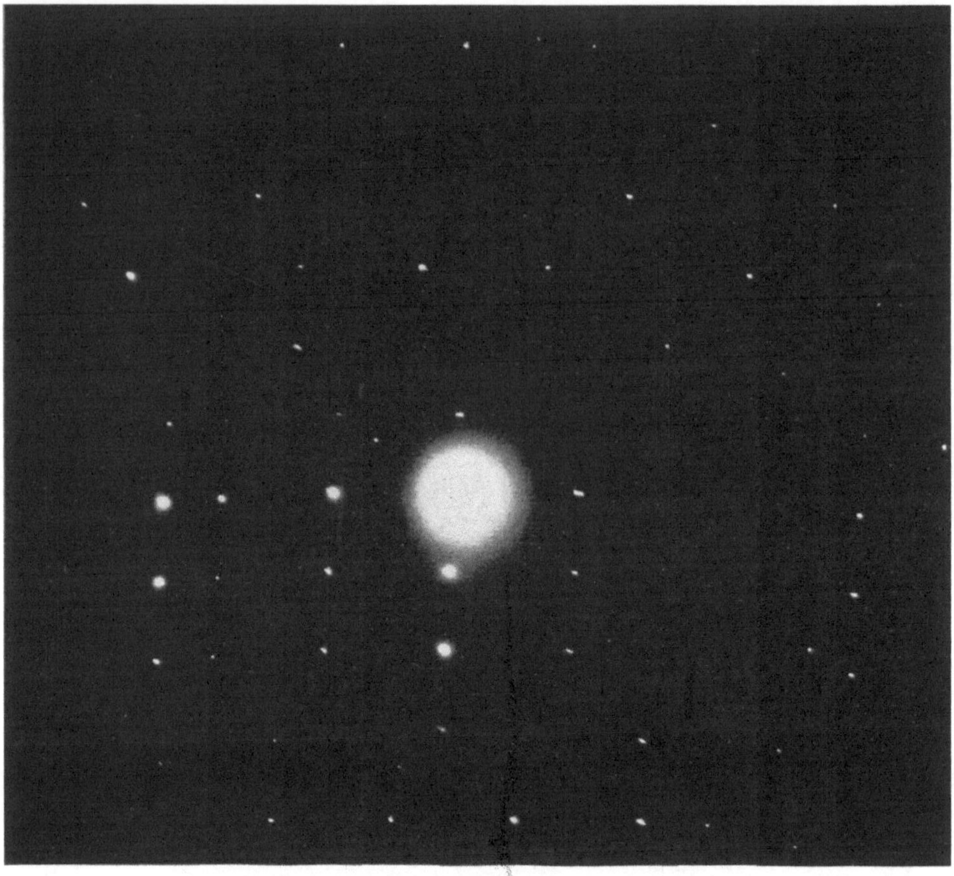

PLATE 4. Electron diffraction pattern from M_6C.

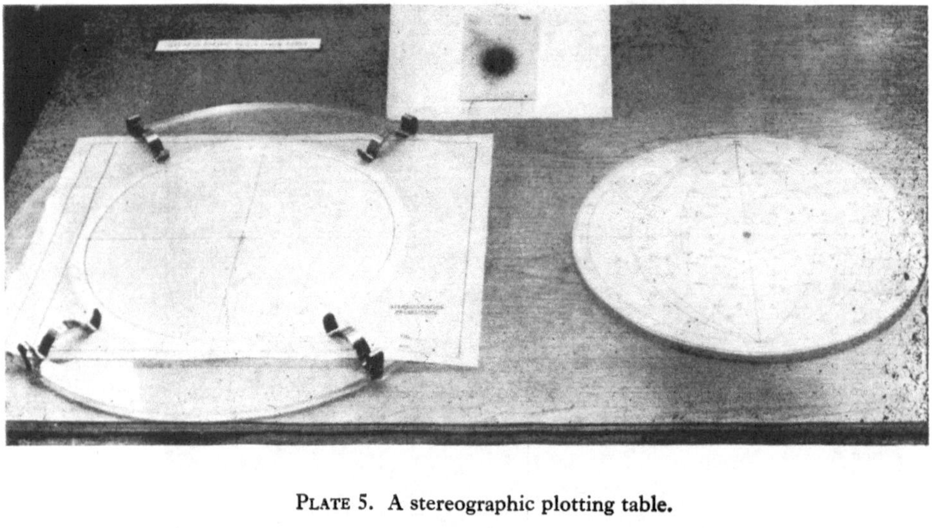

PLATE 5. A stereographic plotting table.

This direction is the one normal to the diffraction pattern, and for all cubic directions and certain directions in other crystal systems the plane of the pattern has the same indices as this direction.

It is then necessary to confirm that other spots which form part of an unbroken sequence about the origin also satisfy the corresponding zone [equation (11)], viz.

$$hu+kv+lw = 0$$

The indexing of the other spots may be assisted by adding the indices vectorially, e.g. in Fig. 11

$$\mathbf{d}_3^* = \mathbf{d}_1^* + \mathbf{d}_2^*$$

and
$$h_3 = h_1+h_2, \quad k_3 = k_1+k_2, \quad l_3 = l_1+l_2$$

In general all spots in a zone satisfying equation (11) can be inter-related by combinations of indices such as

$$\left.\begin{aligned} h_3 &= mh_1+nh_2 \\ k_3 &= mk_1+nk_2 \\ l_3 &= ml_1+nl_2 \end{aligned}\right\} \qquad (27)$$

where m and n are $\pm$ integers.

The interpretation may be assisted considerably by reference to a stereographic projection. This will be

(*a*) The standard projection for the phase or structure present if that is known (or suspected). This includes permanent standard projections for cubic phases, or

(*b*) A projection drawn from the photograph itself by the procedure indicated below.

In case (*a*), the standard projection can be used to find which zones contain planes belonging to the sets indicated by the d values obtained from the photograph. The points on such zones will always lie on (projected) great circles. Alternative possibilities can be found and checked on the same diagram.

In case (*b*), the procedure indicated in Fig. 12 can be followed. A projection circle of the right diameter is drawn on a piece of tracing paper and then placed over the film so that the centre of the circle lies over the centre of the diffraction pattern. Lines joining the centre and the spots are then drawn in and continued to the circumference of the circle. The projection circle then corresponds directly to the zone [*uvw*], and the points where the radii cut the circumference are the projected poles of the planes giving the diffraction spots. Angles ϕ may then be measured round the circumference and should correspond to the angles in a standard projection.

It may be useful to use a standard projection and a Wülff net to locate the poles of principle planes or directions within the circle, or alternatively to establish rotations which would tilt the standard projections into the [*uvw*] projection or vice versa.[16] Such angular movements are indicated in Fig. 4(*c*).

Also from Fig. 12, it is clear that points with a common factor always lie on the same line through O, and that equations (11) and (27) apply and can be used even when spots or vectors are more distant from the origin and not so easily indexed in the first instance by direct reference to d and ϕ.

3.4.2 *Indexing of imperfect patterns*

Many patterns are not simply related to one zone, and, especially with increasing distance from O (increasing θ), groups of spots are found which appear to belong to slightly different planes in the reciprocal lattice. This effect could be due to slight curvature of the specimen, but, in the first instance, should be assumed to be due to the finite wavelength effect, which gives a finite radius to the Ewald sphere. As a result, points such as P in Fig. 1 arise from reciprocal lattice points as P′, but points further from O require the angle P′OP to increase progressively in order for P′ to remain on the sphere and satisfy the reciprocal condition for diffraction.

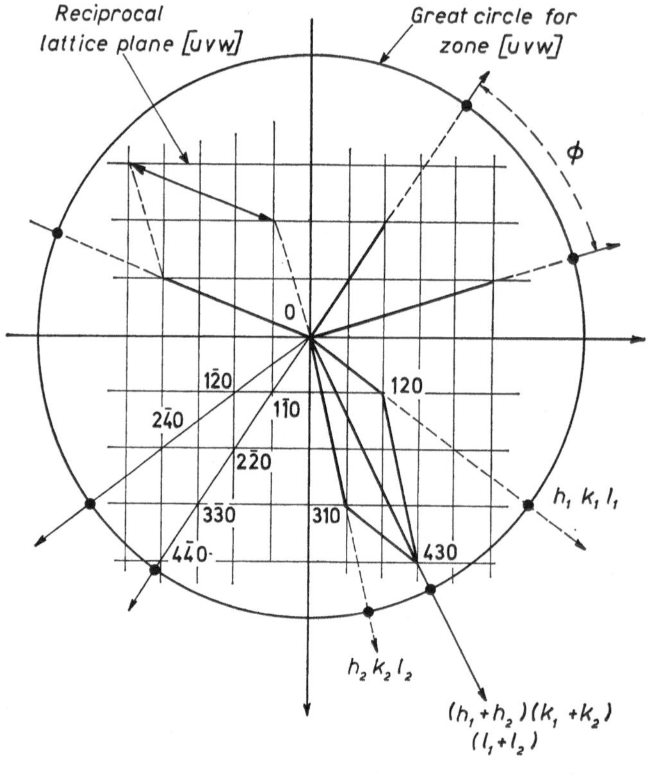

FIG. 12. Relation between stereographic projection and reciprocal lattice or diffraction pattern. The figure shows (*a*) measurement of angles along the great circle, (*b*) equivalence of adding indices and vectors (here illustrated for (120)+(310) = (430) with [*uvw*] = [001]), and (*c*) common factors.

Procedure for indexing imperfect patterns

Start from O and index outwards initially assuming one zone (Figs. 11 and 12).

To index successive spots it is useful to use the addition of indices [equations (11) and (27) and Figs. 11 and 12] proceeding outwards.

When the pattern departs from its primary zone, the addition of indices will still lead to correct indexing of further spots, but *new* reciprocal vectors, not previously used, will be required. The new vectors are added vectorially to the reciprocal lattice vectors in the primary zone, with which they make an angle, in order to give the indices of the next set of spots away from the centre in the direction determined by the sum of the new and old vectors.

Such spots should be immediately tested in equation (12), viz.

$$hu+kv+lw = C = \pm 1, \pm 2, \pm 3, \text{ etc.}$$

which is the extension of (11) to include layers in the reciprocal lattice parallel to (11) but not passing through O. In general, larger values of C occur in the pattern further from O and in some cases only even values of C are found. (See §3.5.3 which deals with centred lattices.)

Alternatively, the new vectors and points lie in a reciprocal plane through O tilted slightly from the planes of the original indexing. If $[u_0v_0w_0]$ indicates the primary zone and $[u_1v_1w_1]$ a secondary zone, it is useful to find the angle between $[u_0v_0w_0]$ and $[u_1v_1w_1]$. If this is of the same order of magnitude as θ, it is likely that only the same reciprocal lattice is involved and the points should equally well fit equation (12). If, however, the angle differs appreciably from θ, crystal curvature, or even a subgrain structure, may be present. It is clearly desirable that equation (11) or (12) should be used only for one zone as far as possible.

3.4.3 *A note on Laue zones*

The procedure of §3.4.2 is considered to be the simplest for most purposes because the indexing of the diffraction pattern enables zone axes which are closely parallel to the electron beam to be calculated. This is particularly so if the pattern is practically symmetrical. A typical example is shown in Plate 3(*a*). An example in which the spots further from the centre can be accounted for by using equation (12) with $C > 0$, is given in Plate 4 and Fig. 20 (see also §5.3.1). Although it is not clear from the diffraction pattern (Plate 4) alone, this effect is probably to be regarded as a special case of the so-called Laue zones. An alternative method of determining the exact position of a zone axis depends on the appearance of these Laue zones. The word 'zone' here refers to bands of diffraction spots such as those that appear in Plates 3(*b*), (*c*) and (*d*).

A Laue zone arises when a prominent crystallographic zone axis is tilted slightly with respect to the direction of the electron beam. For example, Plate 3(*b*) shows a displaced band of diffraction spots which is *elliptical* in general form and includes spots near the central region. The tilt here corresponds to approximately $1\frac{1}{2}°$. With larger tilts, patterns such as Plate 3(*c*) are obtained. The curves on these plates are arcs of ellipses of fairly low eccentricity. A second Laue zone is present some distance away from the first and there is a blank region in between the two. The band of reflections through the centre is said to be the zero-order Laue zone and all reflections correspond to points on a plane through the origin of the reciprocal lattice. The second band of reflections corresponds to the first-order Laue zone and arises when the Ewald sphere cuts the next plane above in the reciprocal lattice. The important consideration is that in this particular orientation the three Laue conditions [equations (3)] are satisfied only in the zones and not in the regions between them. It has already been shown that diffraction is more likely to occur if the wavelength is shorter. Diffraction will also be more likely if the crystal is thin in one direction, particularly if the normal to this direction is collinear with the incident electron beam, as in thin foil microscopy. The reason for this is clear from equation (32) and will be described more fully later. The intensity distribution about a reciprocal lattice point depends on the crystal dimensions t_1, t_2 and

t_3. If the crystal is sufficiently thin in one direction, i.e. t_3 is small, then the reciprocal lattice points are extended in this direction. This has the effect of extending the range of intersection of the Ewald sphere with the reciprocal lattice points, and this tends to reduce the size of the blank regions so that they may even disappear. Some reflections which correspond to an intersection of the Ewald sphere with the extension of the reciprocal lattice points will show below maximum intensity, and will satisfy equation (4) because of the relaxation of the Laue conditions. Only when the intersection of sphere and lattice point occurs at the origin of the lattice point will maximum intensity be observed. Such points, generally few in number, are said to be in the exact Bragg condition.

Another important factor which decides whether well-defined Laue zones appear is the indexing of the zone axis that is approximately normal to the plane of the diffraction pattern. In the cubic crystal system the distance between lattice points in the direction [*uvw*] is

$$I_{uvw} = a\sqrt{(u^2+v^2+w^2)}$$

a formula which can be obtained from the equation in the right-hand lower corner of Table 1. The reciprocal of this distance is the spacing in reciprocal space for (reciprocal) planes of these indices. Hence the planes are closer together if the lattice parameter is higher or if the indices are high. Alternatively, Laue zones are more likely to occur with simpler substances and simpler indices. For this reason, some simple cubic substances show the zones clearly when the orientation is near to that of a cube face in the plane of the diagram, e.g. Plate 3(*d*).

The following table illustrates these points.

Plate number	Compound	Lattice parameter (Å)	[*uvw*] for the Laue zones	I_{uvw} (Å)†	$\frac{1}{I_{uvw}}$ (Å^{-1})	Number of zones on pattern
3(*a*)	AlN	a_0 = 3·110 c_0 = 4·975	[00.1]	4·975	0·201	1
3(*b*)	$M_{23}C_6$	a_0 = 10·621	[110]	15·021	0·067	2
3(*c*)	M_3C	a_0 = 16·276 b_0 = 10·034 c_0 = 11·323	[24$\bar{1}$]	52·903	0·019	3
3(*d*)	α-Fe	2·866	[100]	2·866	0·349	1
4	M_6C	11·082	[$\bar{1}$38]	95·331	0·010	4

† From Table 1 (facing p. 72) using the appropriate modification of formula according to the system.

Patterns such as Plates 3(*b*), (*c*), (*d*) can be used to determine the orientation of the crystal by measuring the angle ρ between the tilted zone axis and the vertical. This requires an approximate estimate of the position of the centre of the ellipse (which to a good approximation can be considered to be a circle) and can be obtained by completing the arc through the stronger reflections. This can be done with greater accuracy for Plate 3(*b*) than for Plate 3(*c*) or (*d*). If the distance from the centre of the pattern to the centre of this circle is r cm and L is the camera length in cm, then

$$\tan \rho = \frac{r}{L}.$$

If r^* is the reciprocal distance derived from r, equation (14) gives

$$\frac{r}{r^*} = L\lambda = \text{Camera constant}$$

and $\tan \rho = \lambda . r^*$ if r^* is obtained directly from the pattern instead of from r. At values of $\rho < 5°$ the approximation $\tan \rho \sim \rho$ radians can be used.

It must be appreciated that if a zone of simple indices like [100] is tilted in this way with respect to the beam, appreciable values of ρ can be reached before another zone axis of small indices can be found normal to the plane of the diagram.

The importance of determining ρ will again appear in the section on the determination of orientation relationships (§4.1).

3.4.4 *The problem of unknown structures*

The above approach implies that the identity of the diffracting medium is known or can be determined by a process of elimination after taking note of all the possibilities that can occur. Sometimes, however, any *a priori* assumptions as to the composition or structure of the specimen or phase are not always justified. A good example of this is in the precipitation processes in a multi-alloyed steel. In such cases a more detailed procedure has been found necessary, but does not always guarantee a result.

Several diffraction patterns from the same region of the sample are obtained, each with a different zone axis. These can be obtained from a single particle if a tilting stage is available. These patterns will provide a large number of interplanar spacings and angles by applying the appropriate equations as with crystals of known structure. These values themselves, together with the observed symmetries of the diffraction patterns, may give some idea of the crystal structure, e.g. an exactly hexagonal (60°) array of spots will almost certainly come from a hexagonal or cubic structure. It is also advisable to obtain diffraction patterns with rows of closely spaced spots. These will define a direction [*uvw*] in the crystal lattice where u, v and w have *small values* almost certainly $\leqq 2$. These spots will also define a set of crystal planes (*hkl*) (perpendicular to [*uvw*]) which have a large *d-spacing*. This will give some indication of the probable dimensions of the crystal unit cell. It is thus easier to index a high d-spacing spot (or ring) in any structure, since it indicates a high lattice parameter, and less ambiguity is likely. Another useful technique for solving unknown patterns is based on the existence of relationships between $1/d^2$ values. These relationships enable one to ascertain whether the crystal is cubic, tetragonal, hexagonal or orthorhombic, providing a ring pattern or sufficient different spot patterns have been obtained. The method is as follows.

(*a*) *Cubic system*: From

$$\frac{1}{d^2} = \frac{h^2+k^2+l^2}{a^2} = \frac{N}{a^2} = N.a^{*2} \tag{28}$$

where N can have all integral values except those which cannot be expressed as the sum of three squares, i.e.

$$N \neq (8m+7)4^n$$

where m and n are integers. When a body-centred cubic (b.c.c.) lattice is under investigation, N can only take even values, and with a face-centred cubic (f.c.c.) lattice, N can only equal $4n$ or $8n+3$, since h, k and l must be all odd or all even.

Assuming it is possible to get a large enough number of d-spacings from the diffraction data, it should not be too difficult to find the value of a^{*2} and hence a_0, since a common difference between certain pairs of $1/d^2$ values will become obvious. The values of $1/d^2$ for $d > 1{\cdot}00$ Å have been tabulated.[17]

A useful procedure, which may also be used, particularly when cubic symmetry is present, is the ratio method. From equation (14), and the above relationship,

$$D^2\left(\frac{1}{2\lambda L}\right)^2 = \frac{1}{d^2} = N.a^{*2}$$

and for two points

$$\frac{d_2^2}{d_1^2} = \frac{D_1^2}{D_2^2} = \frac{N_1}{N_2} \tag{29}$$

Therefore, if ring diameters or spot distances are compared in pairs by placing the two values opposite each other on the C and D scales of a slide rule, somewhere on the same scales the $\sqrt{N}$ values will also be opposite each other. Therefore the two integers N_1 and N_2 will also be found opposite on the A and B scales. For spot patterns, however, it is necessary to check the angles between planes with these N values to see if they agree with the correct tabulated values. If several d-spacings have been obtained, the slide can be turned upside down so that the B scale (inverted) is in contact with the D scale. If d values are marked off along the latter scale, the B scale can be adjusted until integers fall opposite these d values. These integers are the N values corresponding to the respective interplanar spacings.

(*b*) *Tetragonal system*: Here

$$\frac{1}{d^2} = \frac{h^2+k^2}{a^2}+\frac{l^2}{c^2} \tag{25}$$

$$= (h^2+k^2)\,a^{*2}+l^2c^{*2}$$

$$= Ma^{*2}+l^2c^{*2}$$

where M can take the values

1, 2, 4, 5, 8, 9, 10, 13, 16, 17, 18, 20, etc.

If any M value is doubled, a second permissible M value is obtained. Thus, if a factor of 2 is prominent among many of the $1/d^2$ values, tetragonal symmetry will probably be present, and the lattice constants can be calculated using the above equation, since the reflections involved are all of the form $\{hk0\}$. Other $\{hkl\}$ reflections are of course necessary to determine c_0 and can mask this ratio effect.

The following relationships thus exist:

(i) Different $1/d^2$ values which have the same l value. Then

$$\frac{1}{d_1^2}-\frac{1}{d_2^2} = (M_1-M_2)\,a^{*2}$$

Hence subtracting $1/d^2$ values from one another can result in values for a^{*2} when $M_1-M_2 = 1$ or multiples of this quantity.

(ii) Different $1/d^2$ values which have the same M value.

$$\frac{1}{d_1^2}-\frac{1}{d_2^2} = (l_1^2-l_2^2)\,c^{*2}$$

Hence subtracting $1/d^2$ values from each other can also reveal c^{*2} or multiples of this value, and $(l_1^2-l_2^2)$ can take the values

1, 3, 4, 5, 7, 8, 9, 11, 12, 13, 15, 16, 17, 19, 20, 21, etc.

(iii) Different $1/d^2$ values with different l and M values. Then

$$\frac{1}{d_1^2}-\frac{1}{d_2^2} = (M_1-M_2)\,a^{*2}+(l_1^2-l_2^2)\,c^{*2}$$

Thus, if a list of differences between all the pairs of $1/d^2$ values are calculated, a^* and c^* should be distinguishable and a_0 and c_0 can be calculated. It is obvious that the factor $(M_1-M_2)a^{*2}$ can be found in more ways than $(l_1^2-l_2^2)c^{*2}$ and therefore a^{*2} will appear more often than c^{*2}.

(*c*) *Hexagonal system*: Here

$$\frac{1}{d^2} = \frac{4}{3}\frac{(h^2+hk+k^2)}{a^2}+\frac{l^2}{c^2} \tag{30}$$

$$= \frac{4}{3}La^{*2}+l^2c^{*2}$$

where L can take the values

1, 3, 4, 7, 9, 12, 13, 16, 19, 21, etc.

Thus this system differs from the previous one by the factor which precedes a^{*2}. It is more convenient to take the 4/3 with the a^{*2} and therefore write the equation for $1/d_2$ as

$$\frac{1}{d^2} = La'^{*2}+l^2c^{*2}$$

If any L value is trebled, a second permissible L value is obtained. This 3 : 1 ratio distinguishes the hexagonal system, but since it relies on the appearance of $\{hko\}$ reflections, it can be masked as before.

Subtraction between pairs of $1/d^2$ values will, as with the tetragonal system, lead to values of a'^{*2} and c^{*2}, $(l_1^2-l_2^2)$ taking the same permissible values and (L_1-L_2) taking any value obtainable by subtracting any L value from any other.

(*d*) *Orthorhombic system*: Here

$$\frac{1}{d^2} = \frac{h^2}{a^2}+\frac{k^2}{b^2}+\frac{l^2}{c^2} \tag{31}$$

$$=h^2a^{*2}+k^2b^{*2}+l^2c^{*2}$$

The solution is time consuming since all the variables are independent. Subtraction between pairs of $1/d^2$ values will of course give a^{*2}, b^{*2} and c^{*2} when certain reflections are used, but these will tend to be masked by the multitude of other varieties that can occur. It is better to obtain supplementary X-ray diffraction data if possible. Assistance can be obtained by using some of the procedures indicated in some of the general textbooks.[17,18]

(*e*) *General case*: Ito[19] has provided a treatment for the most general case (triclinic) which could provide correct solutions. It is limited by the fact that some of the planes of low indices required to initiate a solution may not give diffraction spots (or rings) owing to systematic or incidental zero structure factors.

If sufficient sample is available, there is a clear advantage in being able to obtain X-ray diffraction patterns. The spacings from either source can also be examined to see whether they fit any of the compounds in the *A.S.T.M. Index*[15] or other tabulations of data.

The procedure will, if successful, provide the identification of a phase by its single-crystal spot pattern(s) or at least enable a unit cell to be determined. If the latter is the case, more information may be obtained by preparing larger quantities of the material and submitting it to chemical analysis. The electron-probe microanalyser might also be of use in this connection if the particles are sufficiently large or can be specially prepared.

3.4.5 *Summary of simple practical procedures for indexing single-crystal patterns*

(*a*) *When identity of crystal is known*:

Measure the distance (D or r) of three diffraction spots which form a parallelogram with the origin.

Convert these distances into interplanar (d) values via the camera constant.

Using a table of interplanar spacings for the particular substance (see Part 3), correlate the measured d values with actual d values and thus obtain the indices hkl for each spot.

Measure the angles (ϕ) between the chosen diffraction spots.

Check the indexing by comparing the measured ϕ values with calculated ϕ values. (See Parts 2 and 3.)

Define the exact (hkl) indices (so far we have only determined the family of indices $\{hkl\}$) by deciding whether the values of hkl are positive or negative. This is done by summation of vectors. For example, if $\mathbf{h}_1\mathbf{k}_1\mathbf{l}_1+\mathbf{h}_2\mathbf{k}_2\mathbf{l}_2 = \mathbf{h}_3\mathbf{k}_3\mathbf{l}_3$, then $h_1+h_2 = h_3$, $k_1+k_2 = k_3$ and $l_1+l_2 = l_3$.

Completely index the remaining spots by (*a*) continuing a sequence of indices along a prominent direction, and (*b*) adding two sets of indices of prominent directions to index the spots between prominent directions.

Obtain the zone axis $[uvw]$ by cross multiplying the indices of any two reflections as follows:

$$\begin{aligned} u &= k_1l_2-k_2l_1 \\ v &= l_1h_2-l_2h_1 \\ w &= h_1k_2-k_1h_2 \end{aligned} \qquad \begin{array}{c|cccc|c} h_1 & k_1 & l_1 & h_1 & k_1 & l_1 \\ & & \times & \times & \times & \\ h_2 & k_2 & l_2 & h_2 & k_2 & l_2 \end{array} \tag{26}$$

All the indices should satisfy the condition

$$hu+kv+lw = 0 \tag{11}$$

near the centre of the pattern, but on moving outwards it may be necessary to extend the equation to

$$hu+kv+lw = C = \pm 1, \pm 2, \ldots$$

to include all the spots.

The indexing can also be checked by reference to stereographic projections. All the spots should lie on the circumference of a stereogram of the zone, as plane of projection, or on a great circle for the zone in a stereogram with standard orientation. Of course, ϕ values can also be checked from stereograms, or by calculation, or from tables if available.

(*b*) *When identity of crystal is unknown*:

As for the known crystal, choose three spots which form a parallelogram with the origin and measure D or r.

Using the camera constant, obtain interplanar spacings from the measured D or r values.

Measure the angles (ϕ) between the diffraction spots.

The symmetry of the diffraction pattern can give some indication of the zone and the crystal structure. The patterns in Figs. 32 and 33 illustrate some common patterns in two simple systems.

Certain angles can indicate to a person with some experience that certain planes are present, e.g. the common cubic angles between planes such as {100}, {110}, {111}, {112}.

It is advisable to obtain data from several orientations of the same specimen and to examine the *d*-spacings and angles for any information they may give about the symmetry or possible crystal system. Attention should be given to the details indicated in §3.4.4.

3.5 ADDITIONAL NOTES

3.5.1 *General*

The *A.S.T.M. Index* was originally compiled by workers in the field of X-ray diffraction. Many users have, however, made their own collections of the *d*-spacings of phases which commonly occur in their work. The following comments may be useful in relation to the use of compilations in conjunction with electron diffraction work.

(*a*) The tabulations of the *d*-spacings from X-ray measurements are in order of magnitude of d and in general are to an accuracy greater than can be achieved by electron diffraction. These tabulations do not usually go below $d = 1$ Å, but computer programmes are available for quick calculations of *d*-spacings.

(*b*) While *d*-spacings below 1 Å are not generally needed, it is an advantage to have them available, especially from small simple unit cells. Contrary to this is the large number of spacings which are possible from even simple large unit cells (see Part 3).

(*c*) For the more frequently occurring complicated structures, it is useful to have tabulations of the *d*-spacings listed under zones or layers of the reciprocal lattice.

(*d*) The X-ray tabulations also miss out diffractions which (i) will not occur because of symmetry considerations, and (ii) have an extremely weak intensity such that they are barely, if at all, visible. These may occur in single-crystal diffraction patterns and tend to 'fill in' parallel rows of spots in a zone.

(*e*) Tables of angles should be available when solving spot patterns, whereas ring patterns are independent of angular relationships. These angles are in some cases standard and a selection is included in this compilation. In the more complicated structures, the angles must be calculated; information is given on this aspect.

(*f*) As the solution to the ring patterns is independent of angular relationships, it is therefore solely dependent on *d*-spacing measurements. The accuracy obtainable, as discussed, will determine whether or not it is possible to discriminate between two phases with similar lattice parameters.

(*g*) A single spot pattern can never justifiably be accepted as evidence of a phase being present whereas a ring pattern can.

3.5.2 *Relaxation of the Laue conditions, cross-grating patterns and foil thickness*

In textbooks dealing with electron diffraction it may be stated that the pattern is able to form for a stationary sample because one of the Laue conditions is considerably relaxed. If the case of transmission through a thin film is considered, the crystallographic axes may be imagined to be transformed so that *a* and *b* are in the plane and *c* normal to it (*c* is assumed to be orthogonal to *a* and *b*). The first two equations in equations (3) are then obeyed exactly, but the third is represented by a broadened peak [third term of equation (32) in §4.3]. The circle, which would normally not pass through the intersections of the first two sets of curves represented by equations (3) (these are intersecting hyperbolae in the plane of the pattern), is therefore broadened into a band which does pass through the intersections, and so, in effect, all three conditions of equations (3) are fulfilled. The smaller the *c* dimension the more nearly the pattern approaches that which would be obtained from a true two-dimensional lattice or net, i.e. cross-grating pattern. The thicker the specimen the sharper the *c* peak, so the more sharply defined the circle corresponding to this condition. Hence the pattern is less likely to form unless the specimen is tilted one way or another. In this case the spots which do appear will lie in one or more arc-shaped areas known as Laue zones (see §3.4). In many metallurgical applications, especially with thin foils (100–1000 Å thick), it is likely that true three-dimensional diffraction is occurring except where thin plate-like precipitates are present. The number of unit cells in the *c* direction is thus sufficient to give a fairly well defined third Laue condition.

On the other hand, it is possible to study thicker foils with higher voltage electrons and the tendency is for the wavelength to decrease according to equation (2). This reduces the curvature of the Ewald sphere and thus there is a great probability that it will pass through several reciprocal points (of finite 'size') in a single reciprocal lattice plane. Hence this geometrical effect favours the formation of a sufficiently extended diffraction pattern and tends to counteract the tendency for the pattern to disappear with thickness—a tendency which is more likely to be important in the range up to 100 Å or so in many materials.

3.5.3 *Centred lattices*

A face-centred lattice (three faces centred) gives diffraction spots in reciprocal space which represent a body-centred lattice, the extinctions corresponding to reciprocal points of zero intensity. The converse applies to a body-centred lattice in real space. The following notes illustrate and extend these observations.

(*a*) For diffraction from a face-centred lattice, h, k and l are all odd or all even. Consider the equation $hu+kv+lw = 0$.

(i) All odd points hkl can only make this true if two out of three u, v or w (no common factor) are odd. This makes $u+v+w$ even, which is the *reciprocal equivalent* of the condition in real space for diffraction from a body-centred lattice.

(ii) Even hkl points are either even multiples of three odd indices which clearly lie on the same planes through the origin as the series given in (i), or even multiples of mixed sets of odd and even indices, the points corresponding to the prime indices having zero intensity.

Hence *all points* of finite intensity can be found on planes with $u+v+w$ even. This may be appreciated by reference to an example. 314 does not occur since hkl are not all odd or all even but 628 does. In reciprocal space 628 lies in a zonal plane containing 111 and 517 into which it can be resolved vectorially and which has indices [$\bar{3}$12] obeying the rule $u+v+w$ even.

(*b*) A face-centred lattice may, however, be oriented with *any* zone axis [uvw] normal to the plane of diffraction. The following cases then arise.

(i) One index, u, v or w, is even, the other two are odd. $hu+kv+lw = 0 =$ even is satisfied for all possible indices hkl. Therefore any other points not on this zero plane obey $hu+kv+lw = C = \pm 2, \pm 4, \pm 6, \ldots$, i.e. only 'even layers' occur (see Plate 4 and Fig. 20).

(ii) Two of u, v, w are even, the third is odd. This combination means that for $C = 0, \pm 2, \pm 4, \pm 6, \ldots$, only even points can lie on the planes. But $C = \pm 1, \pm 3, \pm 5, \ldots$ contain only odd points.

(iii) With u, v, w all odd. The same applies here as for (ii).

(*c*) Similarly, for a body-centred lattice, $(h+k+l)$ must be even for non-zero intensity. Hence all three indices are even or two are odd. $hu+kv+lw = 0$ is satisfied for any one h, k or l even, if uvw are all odd or all even. All points can be accommodated on reciprocal planes which satisfy this rule.

(*d*) A body-centred lattice may be oriented.

(i) With u, v, w all odd. This zone contains points $hu+kv+lw = C = 0$ and for $C \neq 0$ only even layers.

(ii) With one of u, v, w even, the equation is then satisfied with hkl all even and for some points with one even and two odd. Similar points giving an even sum thus lie on planes with $C = 0, \pm 2, \pm 4, \pm 6, \ldots$ It is, however, possible to make $hu+kv+lw$ odd with other permissible values of hkl satisfying the primary condition $h+k+l$ even, i.e. $C = \pm 1, \pm 3, \pm 5, \ldots$, e.g. $[uvw] = [213]$, $C = 0$ for ($\bar{2}$11), $C = 1$ for (121).

(iii) With two of [uvw] even and one odd. This case is similar to (ii), e.g. [uvw) = [243], $C = 0$ for [11$\bar{2}$], $C = 1$ for (12$\bar{3}$).

(*e*) For structures with only one face centred, or those with systematic absences due to glide planes, similar considerations apply for certain orientations. Generally 'systematic absences' due to centring are not affected, but others may be replaced by finite diffraction intensities due to double reflection.

(*f*) It is noted that at least two instances have been reported where the electron diffraction pattern for a phase indicates absences due to a centred lattice which is not otherwise indicated by X-ray diffraction or even by electron diffraction in other orientations of the same phase.[20,21]

4 Orientation relationships, trace analysis and some other structural or diffraction phenomena

This chapter describes some of the useful information that can be obtained from a single-crystal diffraction pattern, particularly that relating to the atomic structure of the diffracting medium. It provides some information on how to interpret extra spots and streaks.

4.1 ORIENTATION RELATIONSHIPS

Indexing a spot pattern as described in §3.4 enables the orientation of the diffracting medium with respect to the incident beam or photographic plate to be calculated. The zone axis calculated from any pair of sets of Miller indices represents the direction in the crystal which is collinear or almost so with the incident beam. A stereogram constructed on the plane perpendicular to this zone axis is obtained as in §3.4. The cubic system is the only one where the (hkl) plane and the normal $[uvw]$ direction have the same indices for all directions. The hexagonal system in particular must be handled with some care and this is dealt with more fully in §9.2.

The mutual orientation of any two or more phases can be determined from a diffraction pattern if the patterns from each phase appear superimposed after a single exposure. All that is necessary is to superimpose the stereographic projection of each phase in its relative orientation with respect to the others. It is found more often than not that the zone axis of one or even both phases does not have simple values of u, v and w and that similarly the planes which have satisfied the Bragg condition and contain this zone axis are not the closest-packed planes. As it is these planes which are often significant in a transformation, it is necessary to locate them on the stereogram and examine their proximity to similar planes in the second phase. This process is assisted by having standard stereographic projections of each of the phases available so that a quick rotation from the standard stereograms can be made to locate these poles. A stereographic plotting table is very useful for this kind of procedure[16] (see Appendix).

The multiplicity of planes for a given set $\{hkl\}$ is important when determining these relationships. This is particularly so with the cubic system and to a lesser extent with the others. Consider, for example, a $\langle 115 \rangle$ zone in a cubic crystal. The direction $[11\bar{5}]$ is contained in the $(\bar{1}10)$ (231) and (321) planes. However, the equivalent direction $[115]$ is contained in the $(\bar{1}10)$ $(32\bar{1})$ and $(23\bar{1})$ planes. These two directions are separated by an angle of $31°35'$. If the diffraction pattern comes from such a zone, either set of indices is a possible choice as both are correct. To draw the stereogram for such a pattern involves rotating the standard (001) cubic projection through an angle of $15°48'$ either clockwise or anti-clockwise about $[\bar{1}10]$. By doing this it is easy to verify that for any fixed point on a second superimposed projection which is stationary while the great circle of this cubic phase is being rotated, there will be a different environment of poles depending on whether $[11\bar{5}]$ or $[115]$ is chosen for this zone. Any stationary pole on one stereogram will become coincident with different poles on the other stereogram, depending on the rotation. Thus from any two superimposed patterns two or more orientations are possible. Any two or more may be complementary, which therefore lessens the number of different ones. Therefore one or even two such diffraction patterns are insufficient for uniquely defining an orientation relationship. When several patterns, all of which preferably have different zone axes normal to the plate, have been solved, one orientation relationship which consistently recurs will become obvious. If there is no such obvious orientation, it may be that the two phases grow independently and are in fact not related.

Needless to say, patterns from each phase which are symmetrical about the centre are usually chosen since they generally have prominent rows of spots. Using a tilting stage, the specimen can be rotated about a given row of spots for one phase so that different zones which are perpendicular to this direction are brought into positions where they satisfy the diffraction conditions. If it is possible to measure the rotations accurately, an unambiguous result can be obtained for the orientation of this phase.

4.1.1 *Summary of procedure for obtaining orientation relationships*

Solve fully each independent pattern.

Determine the zone axis of each pattern.

Construct and superimpose stereographic projections for the two phases as orientated about these zone axes.

Check that the ambiguity caused by the multiplicity of planes does not give a further relationship. If it does, determine it.

Locate poles of close-packed rows and other important planes.

Allow up to $\sim 1°$ or $2°$ deviation of the zone axis of either pattern from the pole position.

4.2 TRACE ANALYSIS

In addition to the crystallographic information obtained by the procedures so far outlined, further information can be obtained from the diffraction pattern when it is considered in conjunction with the selected-area electron micrograph. Thus, besides observing the shapes of precipitates (needles, platelets, etc.) and calculating their orientation with respect to the matrix, the growth directions and habit planes of the precipitates can also be determined. In particular, the directions of needles or the lines seen on a metal surface or in projection on the electron image of a foil can be studied in relation to the known orientation of matrix and precipitate. The analysis of these traces may indeed contribute to the solution of an orientation problem, and it does at least provide a means of determining habit planes and growth directions. The principles are similar to those already used in optical and X-ray work and reference may, for example, be made to Barrett.[2]

Important: In applying the following principles it is assumed that proper account has been taken of the fact that the structural image is rotated in relation to the diffraction pattern. This angle must be determined experimentally, for example, by the use of a small crystal of known structure and morphology; MoO_3 is often used.

The problem is summarized in Fig. 13(*a*) and concerns a platelet of a precipitated phase embedded in a thin foil. It is easy to consider the effects for a needle if the situation is described first for this more general case. The platelet is considered to cut the plane of the foil at an angle. The geometry in Fig. 13(*a*) shows how the situation in the plane of the foil can be completely defined by its indices (*hkl*) and the direction A which is the *apparent growth direction* in this plane. It is assumed that (*hkl*), or the indices of the zone axis M, i.e. [*uvw*], have been determined by the methods of the previous section (4.1). (In the cubic case, $u = h$, $v = k$, $l = w$.) If A is $[u_1v_1w_1]$, then it is easy to find the direction $B = [u_2v_2w_2]$ also contained in the plane (*hkl*). The

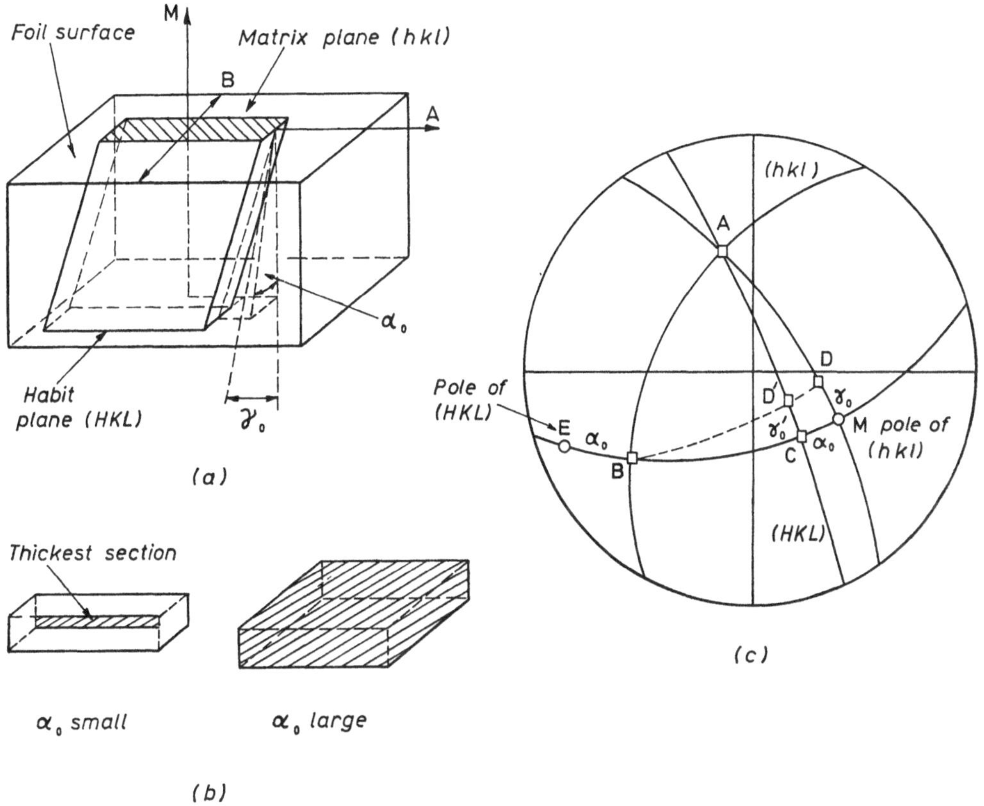

FIG. 13. Basis for trace analysis. (*a*) Geometry, (*b*) projected outline of plate (idealized), and (*c*) corresponding stereographic projection.

actual *habit plane* (*HKL*) is therefore assumed to have been reached from the vertical by turning through an angle α_0 about A (or $90-\alpha_0$ from the plane of the foil); both planes (*hkl*) and (*HKL*) are thus in the zone $[u_1v_1w_1]$. On the other hand, the sloping edge of this plate has a direction defined by the angle γ_0 being rotated about B from the vertical M = [*uvw*], through this angle. The *actual* (*maximum*) *growth direction* lies in this plane and for a square or disk-shaped precipitate is not closely defined or may be random. If, however, the plate is narrow in one dimension or even becomes needle-like, the direction becomes quite sharply defined, but the plane itself may then become more difficult to determine.

Fig. 13(*b*) shows idealized projected images of two such platelets. In practice the outlines are unlikely to be sharp and may be confused either by Moiré fringe effects if the two lattices are very similar or by other interference effects due to the wedge-shaped crystal on either side of the sloping interface.

Fig. 13(*c*) shows a stereographic projection of the situation represented by Fig. 13(*a*). In this diagram M is the pole of the plane (*hkl*) for which the great circle is shown. The directions A, $[u_1v_1w_1]$, and B, $[u_2v_2w_2]$, are represented by points along this great circle. A great circle through B and M is therefore at 90° to A, and so the rotation α_0 takes place along this great circle in the manner shown. The *direction* represented by the point C is thus in the plane (*HKL*) which also contains the pole A $[u_1v_1w_1]$. Likewise, the angular movement γ_0 is along the arc MD on the great circle MA. It follows that the plane (*HKL*) contains both A and C and is located by the great circle through these two directions and its pole at E. E is necessarily 90° from A and so lies along MB. Also, BE = MC = α_0. The three points AMB and the three points AEC each form a right-angled spherical triangle.

The slanting edge of the platelet which may be required as a prominent or significant growth direction is contained in a plane which contains the rotation axis B and the off-vertical direction D. This is represented by the great circle DB. The growth direction must also lie in (*HKL*). It is therefore given by D′ which is γ_0' from C. For the platelet nearly perpendicular to the foil, as shown here, $\gamma_0' \simeq \gamma_0$, but this is not generally the case.

The information required from the diffraction pattern and micrograph is thus the plane (*hkl*), the direction A = $[u_1v_1w_1]$, and so B = $[u_2v_2w_2]$. All that this provides is the knowledge that the habit plane contains A. Determining a second trace direction in another area would perhaps enable an absolute determination of (*HKL*) Neither of the angles α_0 and γ_0 can be found absolutely from a single area. In this kind of study, however, the appearance of well-formed flat plates nearly parallel to the surface in some cases and nearly perpendicular to it in others, can often make interpretation easy. In the latter case, B is nearly perpendicular to the plate and C is near to M (α_0 small). Hence the construction of the traces of planes perpendicular to B and containing A from several patterns will give a series of great circles spread about on either side of, or near to, AC, which is the true habit plane circle. On the other hand, for needle-like precipitates, the direction D is more important and lies in the great circle AM. If the longest needles are chosen, they will be nearly parallel to the foil surface and so to A ($\gamma_0 \rightarrow 90°$). A few such determinations should give a clear indication of which direction (of simple indices) corresponds to the long-growth direction.

The above approach is therefore *inexact* because complete information cannot readily be obtained. A study of the crystallography may, however, considerably limit the choices. Something may already be inferred about possible habit planes from the structure. For example, in a cubic case, platelets at 90° give an immediate indication, and in systems involving hexagonal phases, the basal plane is often, but not always, significant. In certain cases it should be possible to make approximate estimates of either α_0 or γ_0 or of both from a study of the shape in projection and a knowledge of foil thickness.

4.2.1 *Summary of procedure for trace analysis*

Obtain a structural image and an electron diffraction pattern from the same area bearing in mind the precautions outlined in §3.2.

To compensate for the amount of rotation produced by the microscope lenses, rotate the structural image until it orientates exactly with the diffraction pattern. This image rotation can be calibrated for a range of magnifications as mentioned previously.

To determine the plane (*hkl*) and its normal M, index the diffraction pattern.

Choose a prominent row of reflections on the diffraction pattern as a reference direction, and measure the angles which the two structural trace directions (A and B) make with the reference direction.

Draw a standard stereographic projection and indicate prominent planes and directions. Draw on the standard stereogram the zone axis (M) of the diffraction pattern and a great circle corresponding to the plane of the pattern (*hkl*).

On this great circle mark the reference direction from the diffraction pattern and plot the two structural trace directions (A and B) by marking off the previously measured number of degrees from the reference direction.

Draw a great circle through A and the zone axis (M) to define the trace direction.

Draw a great circle through B and the zone axis (M) to define the trace normal to the direction trace, and since B will be $90°$ from A, this defines the habit plane.

Carry out a similar procedure for each of the pairs of micrographs and diffraction patterns available.

By superimposing all the direction trace great circles on one stereogram, it should be found that a significant crystallographic pole is indicated by the intersection of the great circles.

Similarly the intersection of the normal traces will define the habit plane.

Notes

(*a*) With rod-shaped precipitates, only the A direction can be detected. As the needle does not have a habit plane, there is no B direction, and hence normal trace analysis using a projection is not necessary.

(*b*) It may be found that when a few directions such as A have been determined from different specimens, it is only necessary to establish that these lie in a plane of the same indices or type. If the points corresponding to direction A are plotted on a stereogram and a single great circle passes through them, this is (*HKL*). Alternatively, the points may indicate two or more planes of the same form {*HKL*}. If the directions A lie near to directions of simple indices, it may be easy to ascribe approximate indices to them using the vector addition rule [equation (27)]. The zone multiplication law, i.e. equation (26), can then be used (inversely) to find (*HKL*) from two directions. The remaining directions can then be checked by equation (11). This is an analytical method which does not require the stereographic projection.

4.3 THE INTERPRETATION OF STREAKS IN DIFFRACTION PATTERNS

The possible effects of streaks on measurements of spacings or angles have already been considered.

The elementary kinematic theory of diffraction enables the intensity of diffracted waves at any point to be calculated. From a three-dimensional array of lattice points the total scattered intensity at a distance R is given as

$$I = \frac{f^2}{R^2} \cdot \frac{\sin^2(\pi t_1 s_1)}{\sin^2(\pi s_1)} \cdot \frac{\sin^2(\pi t_2 s_2)}{\sin^2(\pi s_2)} \cdot \frac{\sin^2(\pi t_3 s_3)}{\sin^2(\pi s_3)} \tag{32}$$

where s_1, s_2 and s_3 are vector components which define a small deviation from the origin of a reciprocal lattice point, and t_1, t_2 and t_3 are the lengths of the parallelepiped along the same axes and define the volume of the sample which is being irradiated. Equation (32), apart from the first term, is called the interference function and its principal maximum occurs when the denominator is zero, i.e. when $s_1 = s_2 = s_3 = 0$ at the centre of the reciprocal lattice point. The second maxima (pair) occur when $s = \pm 3/2t$. At the exact lattice points the peak intensity becomes

$$I = \frac{f^2}{R^2} \cdot t_1^2 \cdot t_2^2 \cdot t_3^2$$

In any of the directions between the maxima at $s = 0$ and $s = \pm 3/2t$ there are points at which the function is zero, i.e. when $s = \pm 1/t$. Hence the centre peak in reciprocal space has a total width of $2/t$ and a half-peak width of $1/t$, and is the same for all reciprocal lattice points on a line in this direction.

If any value of t is large, then the fall-off in intensity from the central maximum with s is rapid in the corresponding direction. With small t values the fall-off with s is gradual and so the spot becomes elongated in the corresponding direction. This is equivalent to the appearance of a streak through a spot on the diffraction pattern if the specimen is in the correct orientation. If each value of t is small, i.e. with crystallites which are of a very small size, the intensity of the central maximum which is proportional to the square of the product of the t values, is lowered and the fall-off extends over a larger distance in each direction. Thus the reciprocal lattice points become enlarged and their 'reflecting power' decreases. This is equivalent to the line broadening observed in X-ray diffraction patterns from small crystallites. With very small crystallites the reciprocal lattice points will become connected if t approaches the value of an interplanar spacing.

In foils where, say, t_1 and t_2 are large and t_3 is small, the reciprocal lattice points will be extended in one direction giving cigar-shaped spots. Only if t_3 is nearly perpendicular to the incident beam will the effect of these be clearly observed in the diffraction pattern, since the streaks extend parallel to the normal to the planes. With needle-shaped precipitates where two t values are small, the reciprocal lattice will have disk-shaped points. Depending on how the Ewald sphere cuts these, enlarged spots or streaks may be obtained. A comprehensive account of the effect of crystal shape on single-crystal X-ray diffraction patterns has been given by Geisler and Hill.[22]

Stacking faults in a face-centred cube can lead to regions of a close-packed hexagonal structure, in an otherwise face-centred cubic structure, which are only a few layers thick. These regions are equivalent in shape to thin crystals and therefore give rise to streaking. Stacking faults in any structure can lead to streaking in diffraction patterns[23] (see Plate 2). Thin ordered domains (initially Guinier–Preston zones) may also produce an elongation of the reciprocal lattice spots. These domains have the same crystal structure as the matrix. If the domain has m unit cells along an edge, the lattice points have spikes extending to positions whose co-ordinates are $\pm 1/m$ away from the centre. Streaks can similarly arise from thin precipitates which are coherent with the matrix.[24] The important point is that the streaking associated with reciprocal lattice points is related to a small dimension in the crystal lattice and will only be observed for certain orientations of the crystal. The fuller interpretation of this small dimension can often be assisted by reference to the corresponding micrograph.

Recently Honjo, Kodera and Kitamura[25] have obtained diffuse streak patterns from single crystals which have a cubic structure. They have attempted to explain these in terms of linear chain scattering with a small number of the nearest neighbour atoms moved from their proper positions. However, all the features of the diffraction pattern cannot be explained in this manner. The streak pattern is broadened by an amount which corresponds to the angular spread of the primary beam. These streaks should not be confused with Kikuchi lines which have a totally different origin and are not broadened (see §4.7).

Fujime, Watanabe and Ogawa[26] have also observed unexpected streaking in electron diffraction patterns from hexagonal cobalt. They suggest that these are caused by the dynamical interaction of streaks (which occur because of stacking faults in hexagonal cobalt) with normal reflections.

4.4 TWINNING AND RELATED PHENOMENA

4.4.1 *General principles*

The appearance of extra spots at positions which agree with simple *fractional indices* (assuming the main spot pattern has been integrally indexed), is sometimes an indication of the presence of twinned crystals.

In general one can regard the relationship between the parent and twinned crystals as arising from reflection through a plane (HKL) or a rotation of 180° about a direction $[UVW]$ normal to such a plane. Other rotations need not be considered here. A plane $(h_1k_1l_1)$ will thus appear in the position $(h_2k_2l_2)$ where

$$h_2 = h_1 - \frac{2H(Uh_1+Vk_1+Wl_1)}{(HU+KV+LW)} \tag{33}$$

with similar expressions for k_2 and l_2.[27]

In the cubic case

$$h_2 = h_1 - \frac{2H(Hh_1+Kk_1+Ll_1)}{(H^2+K^2+L^2)} \tag{34}$$

Equations (33) and (34) can be used to convert indices in either direction although common factors may arise in h_2, k_2 and l_2.

4.4.2 *Some relationships for cubic crystals*

In the cubic case the length of the reciprocal vector joining O to the point $h_2k_2l_2$ will clearly equal the length of the integral vector $h_1k_1l_1$. Hence, although $h_2k_2l_2$ involves fractions,

$$h_1^2+k_1^2+l_1^2 = h_2^2+k_2^2+l_2^2 \tag{35}$$

It is better to retain the fractional indices for the general interpretation of patterns, but on squaring and adding the three expressions (34) above for a cubic case and multiplying out to remove fractions

$$h_2'^2+k_2'^2+l_2'^2 = (h_1^2+k_1^2+l_1^2)(H^2+K^2+L^2)^2 \quad (36)$$

Hence the fractional reciprocal vector $h_2k_2l_2$ will pass through the integral point $h_2'k_2'l_2'$. It may be useful to employ this equation to find possible indices for planes. For example, consider {111} twins in face-centred cubic metals. $(H^2+K^2+L^2)^2 = 9$. Therefore {111} planes, other than the twin plane itself, will become parallel to planes given by $\Sigma h_2'^2 = 9\times3 = 27$, which must therefore be of the form {511} (since (333) = 3(111) is not admissible). The fractional indices are thus (5/3, 1/3, 1/3). Similarly (100) becomes parallel to a plane of the form {221} and the reciprocal vector length is given by (2/3, 2/3, 1/3) for which $\Sigma h_2^2 = \Sigma h_1^2 = 1$. In the body-centred case with {211} twin planes, $(H^2+K^2+L^2)^2 = 36$. Hence (100) in a parent grain is parallel to (442) or (221), which, as before, must correspond to a reciprocal vector (2/3, 2/3, 1/3). It is interesting to note that, by the same procedure, (111) becomes parallel to (10,2,2) or (511) as before (vector (5/3, 1/3, 1/3)).

These examples incidentally illustrate another point of great value in interpreting twinned *cubic* metals. This is the principle described by Kelly[28] in which face-centred twins on {111} are shown to produce the same effects as body-centred twins on {211}. A twofold ⟨110⟩ axis lies along the intersection of these planes. Both twin systems are therefore represented on the same stereographic projection. The interpretation of cubic twin patterns is therefore much assisted by this diagram (Fig. 35) and other information included in Part 2. Two analytical methods for calculating the positions of matrix and twin reflections in electron diffraction patterns of twinned cubic crystals are those of Meieran and Richman[29] for twinning on a {112} plane in a body-centred cubic structure and of Johari and Thomas[30] who consider a more general case.

4.4.3 *Practical procedures*

A typical practical procedure for indexing patterns is as follows:

Index main points by usual method.

Allocate fractional indices to twin points.

Insert these in the formula for d or $1/d$. This will give a spacing equal to one already known for integral indices. Hence the forms of $\{h_1k_1l_1\}$ and $\{h_2k_2l_2\}$ are easily found.

The magnitudes of the fractions give some indication of the twin plane, since from equation (33) the denominators are of the form $(HU+KV+LW)$ or in the cubic case $(H^2+K^2+L^2)$. For example, in the face-centred cubic system with (111) twins the fractions are all thirds, whereas in the body-centred cubic system with (211) twins they should be sixths, but since the 2 cancels they are again thirds. In non-cubic cases it is more difficult to decide exact values for the fractions since the axial ratio affects the magnitudes of U, V and W relative to HKL.[27]

Further interpretation at this stage follows from the fact that the *equal* reciprocal vectors $(h_1k_1l_1)$, $(h_2k_2l_2)$, if properly chosen, add up to (HKL) or a multiple fraction of it. Some trial and error will soon make it clear which planes are likely to be twin planes. (For example, $(11\bar{1})$ in the cubic case when twinned on (111) becomes parallel to (115). The vector length is given by (1/3, 1/3, 5/3), which when added to $(11\bar{1})$ gives (4/3, 4/3, 4/3) = 4/3(111)).

A standard stereographic projection can be used to assist the interpretation. Use is then made of the principle that the actual assignment of signs or permutations to the possible indices must conform to the condition that the poles $(h_1k_1l_1)$, (HKL), $(h_2k_2l_2)$ lie on a single great circle with equal angles between the first pair and the second.

4.4.4 *Related phenomena*

The treatment of streaks in §4.3 will have made it clear that, if twin planes are sufficiently frequent, the structure will behave as though it is composed of parent crystals and twinned crystals which are small in a direction normal to the twin plane. If this dimension is small enough, streaks will be formed on either side of the principal and twin spots. The direction of these streaks is thus a direct indication of the twin plane itself. These are stacking faults, but narrow stacking faults and wider twin bands can occur together.

In addition to these effects face-centred cubic or close-packed hexagonal phases may pass into each other as a result perhaps of plastic deformation. Alternating regions of cubic and hexagonal phases may thus be found together. Some possible combinations of diffraction patterns are indicated in Parts 2 and 3.

4.5 ORDERED PHASES: SUPERLATTICES

4.5.1 *General*

The occurrence of ordered structures in metallurgy is well known and has been the subject of frequent X-ray studies. In the first instance it is sufficient to note that the occurrence of order can have simple geometrical effects on a diffraction pattern:

(*a*) Given the unit cell of the disordered pattern, with systematic absences of reflections according to the usual rules, then ordering gives reflections, generally fainter, in the forbidden positions.

(*b*) The ordering may involve more than one original unit cell, in which case additional diffraction spots appear at positions corresponding to fractional indices. These indices can be made integral by doubling if halves are involved and trebling if thirds appear, etc., i.e. the unit cell in real space is twice or three times as large in these two cases. In addition to these effects, the existence of small regions of order extending over several unit cells in a given direction has a corresponding effect on the diffraction in the region of lattice points in reciprocal space.

(*c*) Ordered domains (initially Guinier–Preston zones) produce elongation of reciprocal spots related to domain size. If the domain has m unit cells in say a cell-edge direction, the lattice points will have spikes extending to positions whose coordinates are $\pm 1/m$ away from the centre (see §4.3).

In certain circumstances where a pattern is not obtained from a suitable zone, this could lead to the apparent shift of individual spots. An example of this phenomenon in copper–aluminium alloys has been given by Marcinkowski and Zwell.[31]

(*d*) Alternatively, instead of showing streaks, discrete satellite spots may occur at fractional distances and these correspond to the effect (*b*), although other large unit cells may be implied. Thus it is possible for a single domain to represent an elongated or extended unit cell comprising several of the original unit cells.[32]

4.5.2 *Reasons for 'extra' reflections*

It is useful to appreciate the reasons for the appearance of 'extra' reflections in the ordinary case (*a*).

The intensity of a diffraction spot depends on a quantity known as the 'structure factor'. This is obtained by adding together the amplitudes of waves scattered by atoms at different points (*xyz*) in the unit cell and finding the resultant, necessarily taking account of atomic separations and hence of the extent to which waves scattered by any two atoms are out of phase with each other. The resultant intensity I depends on F^2, where F is the structure factor. It is found that

$$\begin{aligned} I \alpha |F|^2 &= [\Sigma f \cos 2\pi(hx+ky+lz)]^2 + [\Sigma f \sin 2\pi(hx+ky+lz)]^2 \\ &= A^2+B^2 = (A+iB)(A-iB) \end{aligned} \qquad (37)$$

where $i = \sqrt{-1}$. Hence it is much easier to put $F = A+iB$ and $F^* = A-iB$; then $|F|^2 = FF^* = A^2+B^2$

$$\left.\begin{array}{lr} \text{and since} & A = \Sigma f \cos 2\pi(hx+ky+lz) \\ \text{and} & B = \Sigma f \sin 2\pi(hx+ky+lz) \\ \text{therefore} & F = \Sigma f \exp 2\pi i(hx+ky+lz) \end{array}\right\} \qquad (38)$$

The summation includes each atom in the cell and f is the 'atomic scattering factor', i.e. the relative strength of each atom contributing to F.

In equations (37) and (38) (*xyz*) represents the (fractional) co-ordinates of the atoms. In a body-centred (cubic) lattice identical atoms are at 000 and $\frac{1}{2}\frac{1}{2}\frac{1}{2}$.

$$\begin{aligned} F &= f \exp[2\pi i 0] + f \exp\left[2\pi i\left(\frac{h}{2}+\frac{k}{2}+\frac{l}{2}\right)\right] \\ &= f[1+\exp \pi i(h+k+l)] \\ &= f[1+\cos \pi(h+k+l)] \\ &= 2f \cos^2 \frac{\pi}{2}(h+k+l) \end{aligned}$$

As $\cos \pi/2$, $\cos 3\pi/2$, ..., are all zero,

$$F = 0 \text{ when } (h+k+l) \text{ is odd and}$$

$$F = 2f \text{ when } (h+k+l) \text{ is even}$$

This is the basis of the extinction rules for all body-centred lattices.

In an alloy of composition AB say, the mean value of f is $\frac{1}{2}(f_A+f_B)$ and providing the alloy is *disordered*, i.e. the distribution is random, the 'odd' reflections will still be missing. If ordering occurs, then

$$\begin{aligned} F &= f_A + f_B \exp 2\pi i\left(\frac{h+k+l}{2}\right) \\ &= f_A+f_B \text{ for } h+k+l \text{ even} \\ &= f_A-f_B \text{ for } h+k+l \text{ odd} \end{aligned}$$

Hence faint reflections now occur in the other positions with intensities $\propto (f_A - f_B)^2$, depending on the difference in scattering between the two atoms.

As all reflections now occur, the lattice is no longer disordered and contains *one* unit AB (molecule?) per unit cell, i.e. the space group Im3m disordered becomes Pm3m ordered. The structure is then the same as certain ionic compounds such as CsCl.

Hence the effect of ordering is, in the first instance, to prevent the effects of identical atoms from cancelling each other out, by regular substitution of two types of atoms so that, although for certain reflections they 'oppose' each other, the net result is not zero.

4.5.3 *Reasons for double or multiple cells*

In the above case, order can be established in one unit cell. The same tendency to order, often associated with two different sizes of atoms, cannot always be accommodated so easily. For example, it is not possible for an ordered A_3B superlattice to develop from a single body-centred cubic cell since there are only 2 atoms/unit cell. A typical case is Fe_3Al. A larger cubic cell of double edge contains eight original cells and so sixteen atoms. Ordering is recognized, firstly, by the appearance of diffraction spots which are absent in the disordered alloy, and, secondly, by the appearance of spots halfway in between. Doubling the cell makes these spots correspond to integral values of *hkl*.

In this double body-centred cell the cube corners are all A (Fe) and half the centres must also be A (Fe), the other half B (Al). If these alternate regularly, these atoms alone behave like a face-centred sodium chloride lattice and the other A atoms like a simple cube. The net result is diffraction similar to that from a face-centred lattice, i.e. *hkl* are all odd or all even for the double cell. That this should be so is evident from the structure factor. Putting the origin on a B atom for convenience, we have

(*a*) 4B at 000, $0\frac{1}{2}\frac{1}{2}$, $\frac{1}{2}0\frac{1}{2}$, $\frac{1}{2}\frac{1}{2}0$	Face-centred cubic positions
(*b*) 4A at $\frac{1}{2}\frac{1}{2}\frac{1}{2}$, $\frac{1}{2}00$, $0\frac{1}{2}0$, $00\frac{1}{2}$	Intermediate face-centred cubic positions
(*c*) 8A at $\frac{1}{4}\frac{1}{4}\frac{1}{4}$, $\frac{1}{4}\frac{3}{4}\frac{3}{4}$, $\frac{3}{4}\frac{1}{4}\frac{3}{4}$, $\frac{3}{4}\frac{3}{4}\frac{1}{4}$, $\frac{3}{4}\frac{3}{4}\frac{3}{4}$, $\frac{3}{4}\frac{1}{4}\frac{1}{4}$, $\frac{1}{4}\frac{3}{4}\frac{1}{4}$, $\frac{1}{4}\frac{1}{4}\frac{3}{4}$	Mid-points diagonally between first two sets

The structure factor can be set out in full but the expressions for all space groups are already available in *International Tables for X-ray Crystallography.*[5] On page 517 of Vol. 1, for the group Fm3m is found the full expression for 192 atoms (4×48) in the most general position (xyz) ($x \neq y \neq z \neq 0$). By putting in the special co-ordinates and scaling to the actual numbers of atoms, the following sequence of terms is obtained:

$$F = 4\cos^2 2\pi\left(\frac{h+k}{4}\right)\cos^2 2\pi\left(\frac{k+l}{4}\right)\times$$

$$\left[\begin{array}{ll} f_B & \text{(a)} \\ +f_A \cos\pi h\cos\pi k\cos\pi l & \text{(b)} \\ +f_A\cos\frac{\pi h}{2}\cos\frac{\pi k}{2}\cos\frac{\pi l}{2}+f_A\cos\frac{3\pi h}{2}\cos\frac{3\pi k}{2}\cos\frac{3\pi l}{2} & \text{(c)} \end{array}\right]$$

Note:

$$\cos\frac{3\pi h}{2} = \cos\left(\frac{4\pi h}{2}-\frac{\pi h}{2}\right) = \cos\frac{\pi h}{2}$$

The first two terms require *hkl* all odd or all even. For all odd reflections

$$F = 4(f_B - f_A)$$

For all even reflections we have

(*a*) $F = 4(f_B + f_A) + 8f_A$
providing $h = 4h'$, $k = 4k'$, $l = 4l'$ or *one* only a multiple of 4.

(*b*) $F = 4(f_B + f_A) - 8f_A$
$= 4(f_B - f_A)$
providing $h = 2h'$, $k = 2k'$, $l = 2l'$ with $h'k'l'$ *all* odd or *one* odd.

Hence, intensities depend on the squares of $(f_B - f_A)$ or $(f_B + 3f_A)$ according to the indexing, but with the double cell the face-centred cubic condition is in any case necessary.

There is another type of superlattice found in steels, which is of considerable interest. The disordered alloy is body-centred cubic, and the ordered has the formula Ni_2AlTi. Other simple body-centred cubic ordered phases can form, but this one has atomic sites like Fe_3Al and so on ordering becomes face-centred cubic as above, except that only two out of the four sets of positions are identical, e.g.

$$F = 4\cos^2 2\pi\left(\frac{h+k}{4}\right)\cos^2 2\pi\left(\frac{k+l}{4}\right) \times \left(f_{Ti} + f_{Al}\cos \pi h \cos \pi k \cos \pi l + 2f_{Ni}\cos\frac{\pi h}{2}\cos\frac{\pi k}{2}\cos\frac{\pi l}{2}\right)$$

which immediately indicates the possible intensity relationships. The occurrence of these super lattice phases in certain alloy steels is summarized in the following table.

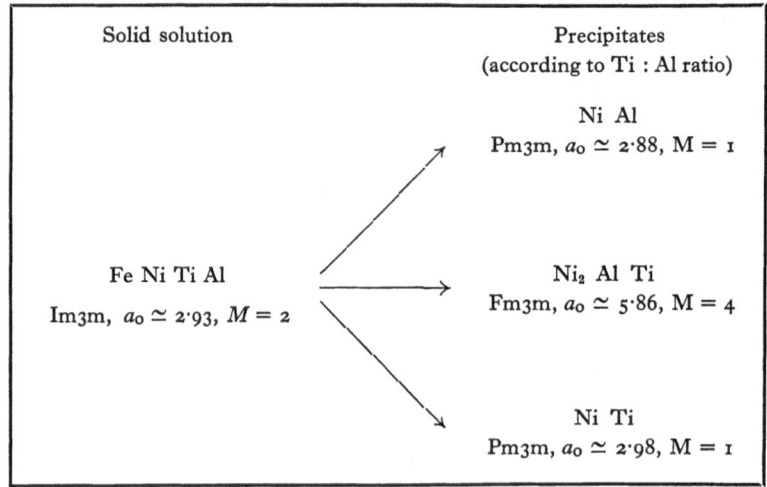

Solid solution		Precipitates (according to Ti : Al ratio)
	↗	Ni Al Pm3m, $a_0 \simeq 2{\cdot}88$, M = 1
Fe Ni Ti Al Im3m, $a_0 \simeq 2{\cdot}93$, $M = 2$	→	Ni_2 Al Ti Fm3m, $a_0 \simeq 5{\cdot}86$, M = 4
	↘	Ni Ti Pm3m, $a_0 \simeq 2{\cdot}98$, M = 1

4.6 DOUBLE REFLECTION

4.6.1 *General*

The phenomena described so far have concerned the appearance of extra diffraction spots which appear either on lattice points which would otherwise have zero intensity or at points in between. The phenomenon of double reflection can also lead to the appearance of finite intensity at integral points on single-crystal spot patterns which would otherwise have zero intensity.

The phenomenon is illustrated in Fig. 14. It simply arises when a ray diffracted from one set of planes $(h_1k_1l_1)$ (spacing d_1) is further diffracted by another set $(h_2k_2l_2)$ (spacing d_2). The paths of the rays in crystal space are necessarily *first* in a plane perpendicular to $(h_1k_1l_1)$ containing d_1^*. Then, *second*, the ray between the two sets of planes,

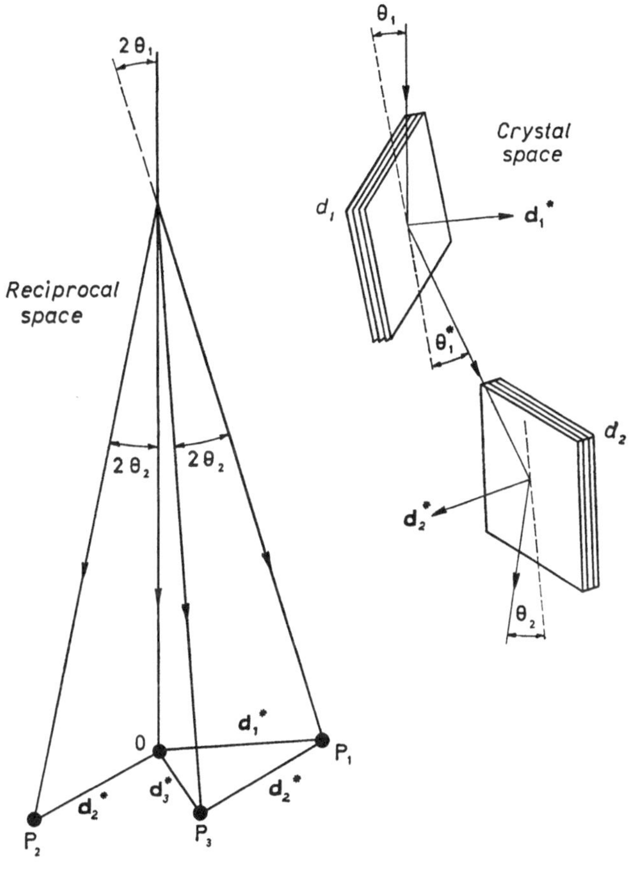

FIG. 14. Elements of double reflection.

d_2^*, and the re-diffracted ray must also lie in a second plane not necessarily coincident with the first. In reciprocal space this gives rise to a point P_1, $\mathbf{d}_1^*$ from O, for the singly diffracted ray and then a point P_3, $\mathbf{d}_2^*$ from P_1. The angles involved are quite small, $\sim 2°$, and so it is quite conceivable that diffraction could also have occurred from the planes $(h_2k_2l_2)$ giving rise to a point P_2 where OP_2 is the vector $\mathbf{d}_2^*$, P_2 being a point on the reciprocal lattice of non-zero diffracted intensity. As two reciprocal lattice vectors necessarily add up to a third (since it is a regular network), $OP_3 = \mathbf{d}_3^*$ is also a reciprocal lattice vector. It follows at once for all such cases that:

(a) $h_3 = h_1 \pm h_2$
$k_3 = k_1 \pm k_2$
$l_3 = l_1 \pm l_2$
(all plus or all minus signs)

(b) The intensity at P_3 will only be finite if it is finite at P_2. Hence, if P_2 is a point at which the intensity is zero owing to the lattice type ($F = 0$), double reflection will not add a finite intensity at P_3. This point is important in dealing with centred lattices and is illustrated in Fig. 15.

(c) The vectors $\mathbf{d}_1^*$, $\mathbf{d}_2^*$ will lie in the plane of the pattern near the centre but either, or both, may eventually tilt out slightly and therefore appear to take the pattern into a slightly different zone. The essential condition is that the points O, P_1, P_2, P_3 all lie approximately on the surface of the sphere of reflection.

The second of these points, illustrated in Fig. 15, shows, for example, that face-centred or body-centred lattice patterns are not affected (although double reflection could affect the relative intensities, see below). On the other hand, hexagonal or other structures can give additional spots at lattice points.

In the face-centred case, $(h_1k_1l_1)$ and $(h_2k_2l_2)$ must be all even or all odd, so that the vector sums must also satisfy the same rules. In a body-centred case, $(h_1+k_1+l_1)$, $(h_2+k_2+l_2)$ must be even and so must be the sums or differences likewise. In the hexagonal case the 'missing' reflections should be given by $(h+2k) = 3n$ and l odd. For d_3, $(h_1+h_2)+2(k_1+k_2) = 3n$, $l = 2m+1$, but a vector with these components can always be split as follows:

$$d_1: h_1+2k_1 = 3p+1; \quad l_1 = 2r$$

$$d_2: h_2+2k_2 = 3q-1; \quad l_1 = 2s+1$$

both of which are present. Hence the 'missing' reflections tend to be added to or filled in by double reflection. In other structures, such as cementite, the points may be added in one zone but not in another.

In general, double reflection will tend to enhance the intensities of weaker spots at the expense of stronger spots.

4.6.2 *Spurious extinctions*

H. J. Beattie[20] has noted that a pattern from a cementite $[20\bar{1}]$ containing rows of spots 020..., 204..., had missing reflections, which gave the (unexpected) appearance of a body-centred pattern. The same effect has been found with $M_aC_b = Fe_2MoC$ [21] and also in AlO_2H (or $Al_2O_3.H_2O$). An explanation of this effect is not known.

4.7 KIKUCHI LINES

In X-ray diffraction an increase of crystal size leads to sharp diffraction maxima becoming spotty. With electrons, as the thickness of the single crystal, or of the foil, increases so the two-dimensional spot pattern tends to fade out owing to increasing electron absorption, but simultaneously a pattern consisting of pairs of black and white parallel lines begins to appear. A considerable weakening of the straight-through beam is also observed.

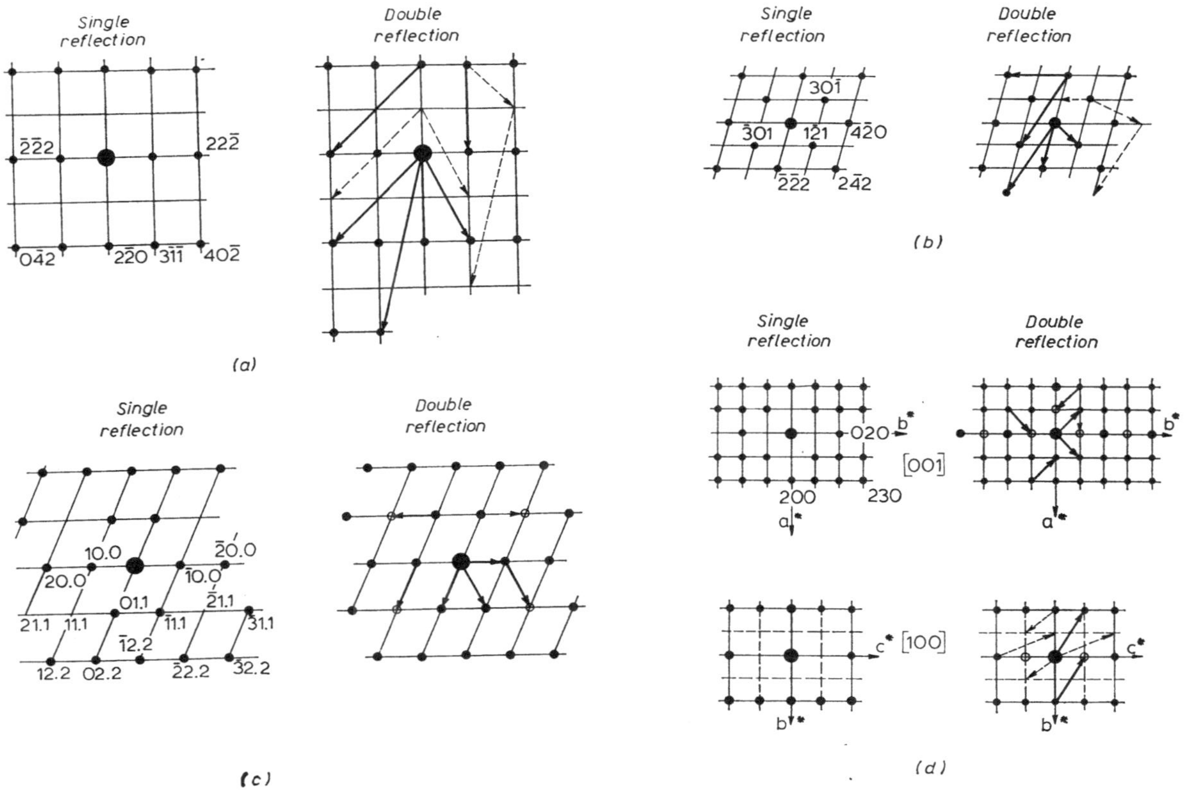

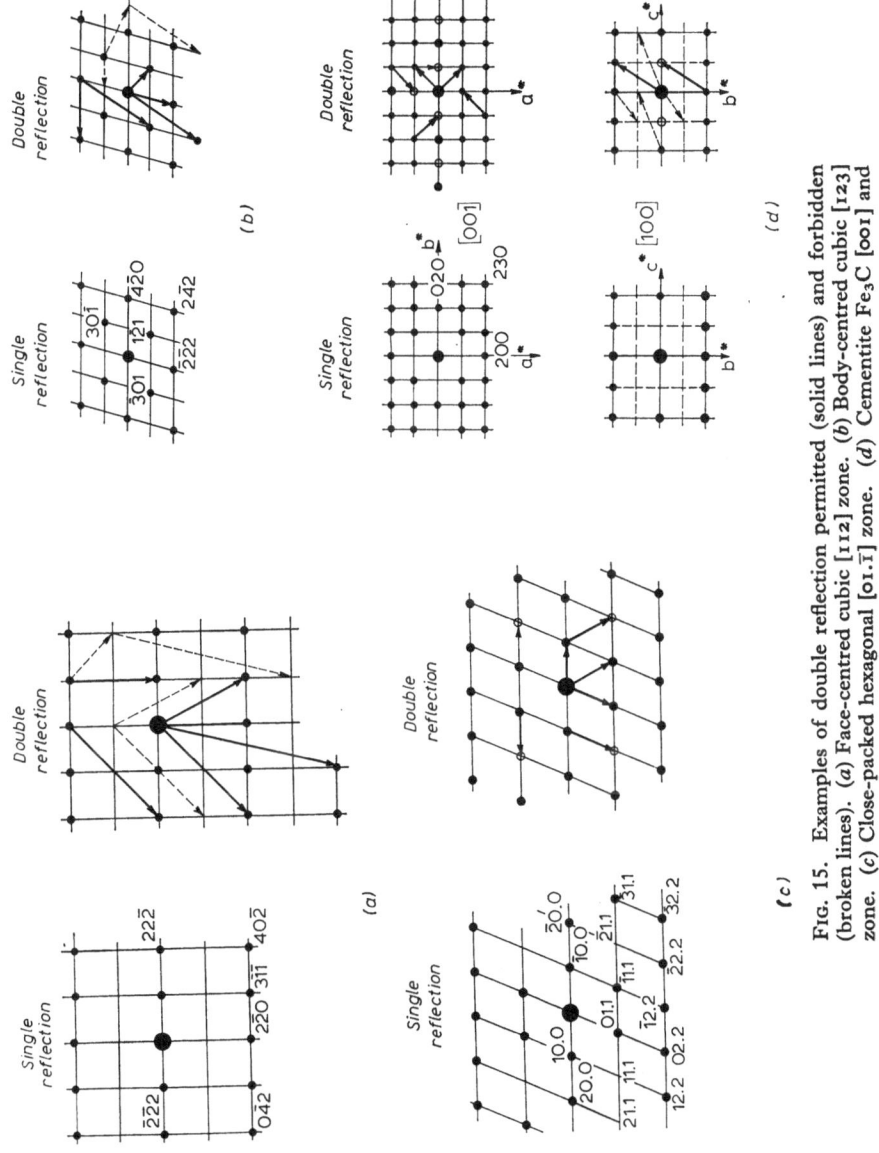

FIG. 15. Examples of double reflection permitted (solid lines) and forbidden (broken lines). (*a*) Face-centred cubic [112] zone. (*b*) Body-centred cubic [123] zone. (*c*) Close-packed hexagonal [01.$\bar{1}$] zone. (*d*) Cementite Fe_3C [001] and [100] zones.

These single-crystal Kikuchi line patterns can be used to determine the crystal orientation, the Laue symmetry, the lattice parameters and the magnitude of the deviation parameter s from the exact Bragg condition.

When the specimen is $\sim 10^{-5}$ cm thick, there is a strong interaction taking the form of inelastic collisions between the incident electron beam and the specimen. As a result the electron beam suffers a loss of energy and the emergent electrons make up the uniform background intensity of the photographic plate. Other inelastically scattered electrons suffer little or no energy loss and are scattered away from the electron beam by varying amounts. These electrons can, however, be diffracted by the crystal lattice planes providing they satisfy the Bragg condition. In Fig. 16, OA and OB represent the paths

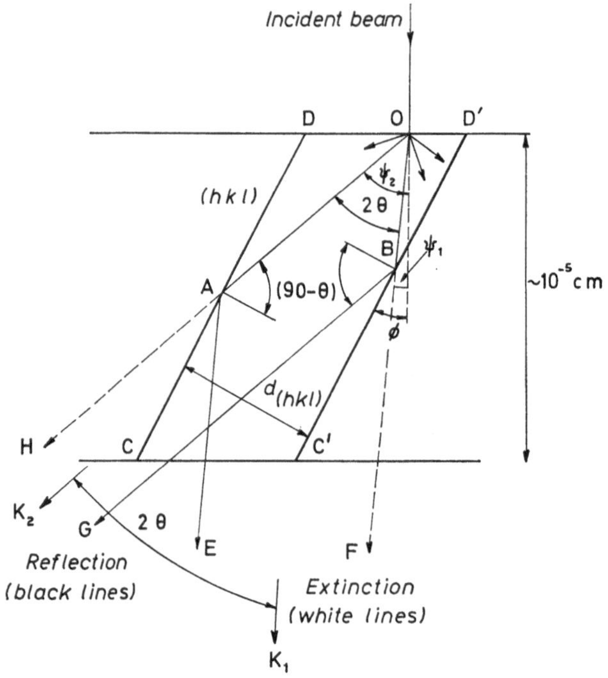

FIG. 16. The formation of Kikuchi lines.

of such electrons. These impinge on lattice planes CD and C′D′, whose separation is d_{hkl}, and are diffracted along AE and BG. Each of these diffracted beams forms a cone whose axis is the normal to CD and C′D′ and whose semi-vertical angle is $(90-\theta)$. These cones cut the plates in hyperbolae of *very small curvature* (since λ is small) and so the diffracted beams appear as lines. As a result of coherent scattering the beam OB follows the path BG instead of the path BF, and the beam OA follows the path AE instead of the path AH. Therefore the background will be weakened along BF and AH and reinforced along BG and AE since it is known that the intensity of coherently scattered radiation decreases with increasing angle of deviation from the initial direction, and hence there is a net decrease in the background along k_1 and an increase along k_2 giving rise to white and black lines respectively on the negative.

In order to use a Kikuchi line pattern to determine the orientation of the crystal etc., it is useful to refer to the reciprocal lattice. Fig. 17 shows the geometry of the construction. O is the crystal origin and OO′ the camera length L normal to the plate P′O′P. r^* is the reciprocal lattice vector of the (hkl) plane giving rise to the pair of Kikuchi lines. $r^* = ha^* + kb^* + lc^*$ and is measured along the normal to these planes, OP′,

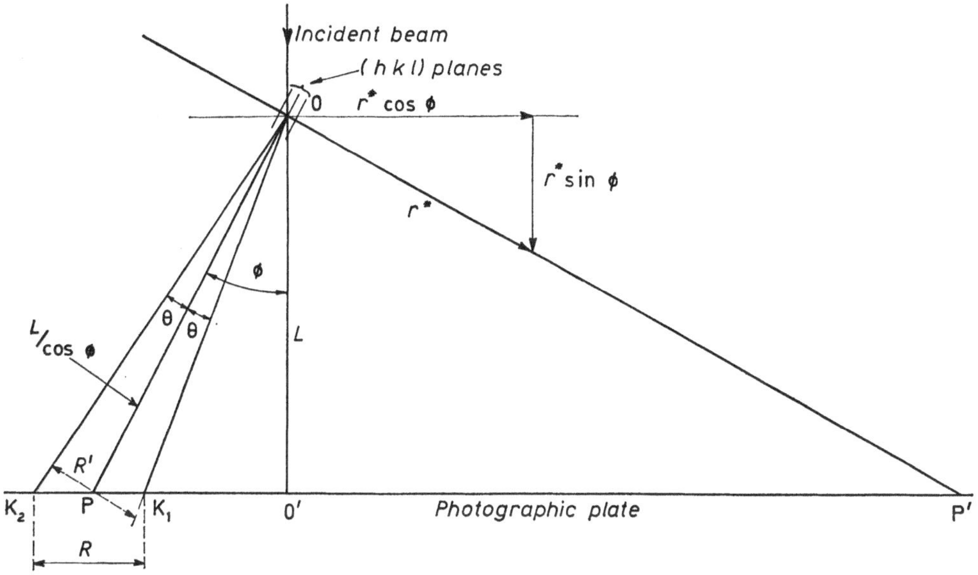

FIG. 17. Reciprocal lattice construction from the Kikuchi line pattern.

and is of length $1/d_{hkl}$. To define this point it is necessary to specify $r^* \cos\phi$ in the direction perpendicular to OO′ and $r^* \sin\phi$ along OO′.

$$R' = \frac{2L}{\cos\phi}.\tan\theta \simeq \frac{2L\theta}{\cos\phi}$$

$$R' = R\cos\phi$$

Therefore

$$r^* = \frac{1}{d} = \frac{2\sin\theta}{\lambda} \simeq \frac{2\theta}{\lambda}$$

$$\simeq \frac{2}{\lambda}.\frac{R'\cos\phi}{2L}$$

$$\simeq \frac{R\cos^2\phi}{\lambda L} \quad (39)$$

For small values of $\phi \simeq \theta$, equation (39) reduces to

$$Rd = \lambda L \quad (40)$$

Here, however, $R = D/2$ and therefore equation (40) reduces to

$$Dd = 2\lambda L \quad (14)$$

If sufficient pairs of lines are available, then different values of R will correspond to different d values, which will help to recognize the crystal if not already known and to establish its orientation as follows.

(*a*) Determine *R*, the shortest distance between any parallel pair of white and black lines, and calculate *d*.

(*b*) Assuming the identity of the diffracting medium to be known, index these spacings.

(*c*) Use the cross multiplication rule between any two pairs of sets of the Miller indices to obtain the zone axis, i.e. the approximate normal to the foil.

The position of a Kikuchi line with respect to its corresponding diffraction spot can be used to determine the sign of the deviation parameter *s* from the exact Bragg condition[33,34] and also its magnitude.[35] The sign of *s* can be used in determining the sense of lattice displacements.[36]

The interference effects which give rise to Kikuchi lines are an indication of the perfection of the crystal lattice. Electron diffraction patterns containing Kikuchi lines and bands have been obtained mostly from non-metallic crystals, e.g. NaCl, $CaCO_3$, ZnS, mica, etc. Kikuchi lines are not as common in patterns from metal single crystals. This indicates the probable presence of imperfections and the tendency to impose plastic deformation on the specimens during preparation. The lines have been observed in ferrite (α-iron), aluminium, germanium and other metals.

5 Geometrical effects

5.1 TILTING OF SPECIMEN AND RESULTING DIFFRACTION PATTERN

It is noted that diffraction spots decrease in size from the centre of the pattern where the spot caused by the direct beam is situated. This is usually the largest and brightest spot. The intensity of the other spots also decreases with increasing distance unless strong Laue zones are formed (see §3.4.3 and Plate 3). There are a number of reasons for this and some theory is available. It is, however, better to accept the facts and to regard each point on the reciprocal lattice as having its own fixed maximum diameter, assuming it to be a sphere for the present purpose. The actual spot recorded then depends on how the surface of the reflection sphere cuts these small spheres.

In Fig. 18 a flat surface is shown cutting such a sphere and giving an apparent spot radius r_1' smaller than the true radius r_1. This corresponds to tilting a sphere of infinite radius ($\lambda \to 0$) and gives a measure of how the diffraction pattern is affected by tilting the specimen. The analysis of Fig. 18 is quite simple.

A rotation of ρ about OX gives an apparent change in the position of a spot. Using the rotation of Fig. 18 for any spot such as P_1 (or P_2),

$$\cos\rho = \frac{S_1Q_1}{S_1P_1} = \frac{OS_1\tan\phi_1'}{OS_1\tan\phi_1} = \frac{\tan\phi_1'}{\tan\phi_1}$$

In general

$$\tan\phi' = \tan\phi.\cos\rho < \tan\phi \qquad (41)$$

i.e. $\phi' < \phi$.

The point moves in towards the axis of rotation as is obvious from the construction, but equation (41) provides a reminder that when $\phi = 0$, the axis of rotation coincides with OP, and when $\phi = 90°$, the displacement is zero. The maximum displacement is thus in between.

It is required to know how the angle between any two directions such as OP_1 and OP_2 (i.e. $\phi_2-\phi_1$) is affected by a small rotation ρ. The greatest displacement $\Delta(\phi_2-\phi_1)$ occurs when these two directions lie on either side of that for $\phi = 90°$. From equation (41)

$$\tan(\phi-\phi') = \frac{\tan\phi - \tan\phi'}{1+\tan\phi.\tan\phi'} = \frac{\tan\phi(1-\cos\rho)}{1+\tan^2\phi\cos\rho} \qquad (42)$$

The *maximum* value of this quantity occurs when

$$1-\tan^2\phi\cos\rho = 0, \text{ i.e. } \tan^2\phi = \sec\rho$$

and here

$$\tan(\phi-\phi') = \tfrac{1}{2}\tan\phi\,(1-\cos\rho)$$

For *small* rotations, the maximum shift occurs when $\phi \simeq 45°$ or $135°$ and, in radians, $(\phi-\phi')$ is small $\simeq \tan(\phi-\phi') \simeq \frac{1}{2}\tan\phi\,(1-\cos\rho) \simeq \frac{1}{2}(\sqrt{\sec\rho}-\sqrt{\cos\rho})$. Hence two directions at $45°$ and $135°$ to the axis of rotation will show the greatest separation, $\simeq (\sqrt{\sec\rho}-\sqrt{\cos\rho})$ radians; e.g. for a $10°$ rotation (ρ), the separation is 0·01532

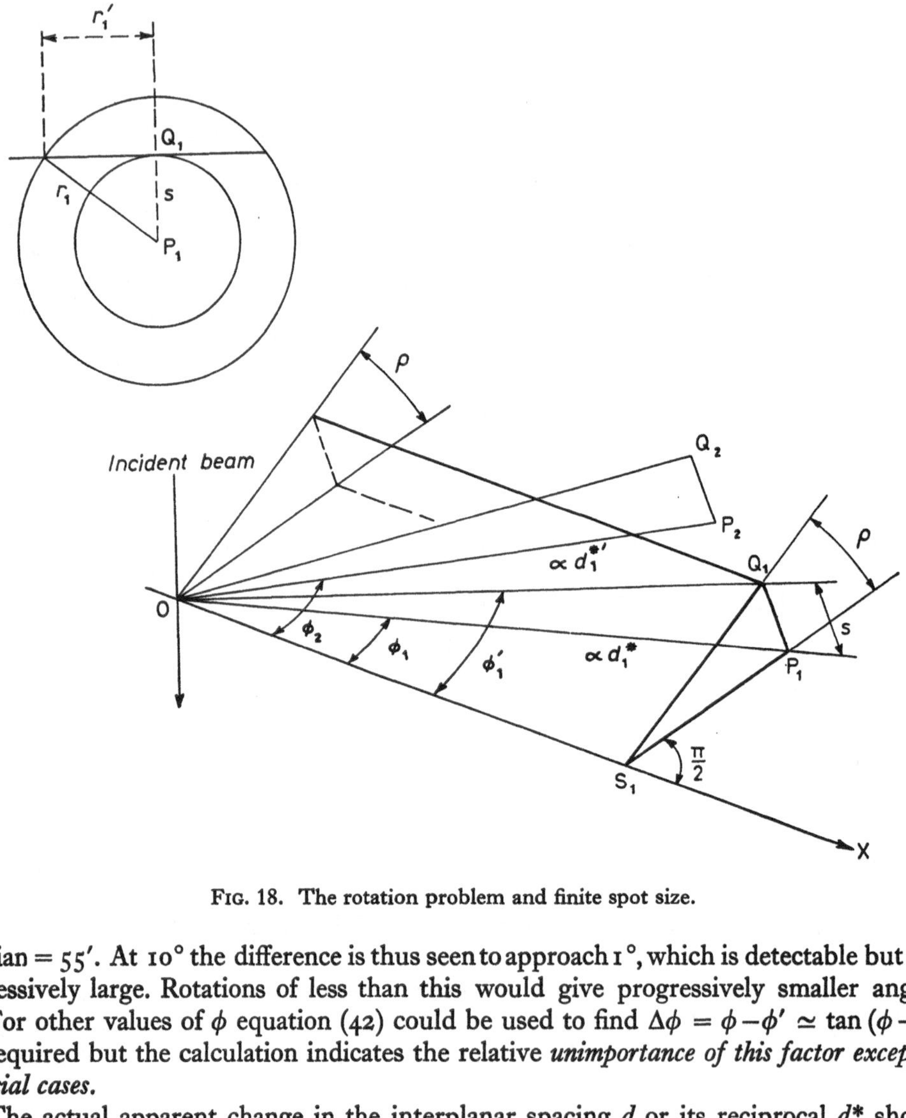

FIG. 18. The rotation problem and finite spot size.

radian = 55′. At 10° the difference is thus seen to approach 1°, which is detectable but not excessively large. Rotations of less than this would give progressively smaller angles.

For other values of ϕ equation (42) could be used to find $\Delta\phi = \phi-\phi' \simeq \tan(\phi-\phi')$ if required but the calculation indicates the relative *unimportance of this factor except in special cases.*

The actual apparent change in the interplanar spacing d or its reciprocal d^* should also be considered. Thus

$$-\frac{\Delta d}{d} = \frac{\Delta d^*}{d^*} = \frac{OP-OQ}{OP} = 1-\frac{OQ}{OP} = 1-\frac{\sec\phi}{\sec\phi'}$$

$$= 1-\frac{\cos\phi'}{\cos\phi} = 1-\left(\frac{\cos\phi.\cos\Delta\phi+\sin\phi.\sin\Delta\phi}{\cos\phi}\right)$$

(since $\phi' = \phi - \Delta\phi$)

$$= 1-(\cos \Delta\phi + \tan \phi . \sin \Delta\phi) \tag{43}$$

For the above rotation of $\rho = 10°$, $\Delta\phi \simeq 27' \simeq \frac{1}{2}°$ (considering only one of the two directions at $\sim 45°$ to the axis of rotation),

$$\frac{\Delta d^*}{d^*} = 0{\cdot}00782 < 1 \text{ per cent}$$

In a length of 10 cm = 100 mm the distance is $\simeq \frac{3}{4}$ mm and if spots are measured on both sides of its straight-through beam this could make a difference of $1\frac{1}{2}$ mm on the double separation. Hence, although the effect is *negligible for many purposes* it tends to become *important towards the outside of a diffraction pattern.* This effect *contributes* to the tendency for the apparent scale to alter towards the edge of a pattern. We note, however, that $\Delta\phi$ will decrease with ρ according to equation (42) and that the value of ρ chosen, viz. 10°, is quite arbitrary but certainly possible.

Whether or not the diffraction spot is recorded depends on its size and shape. It has an apparent radius (assuming sphericity) of

$$r' = (r^2 - s^2)^{\frac{1}{2}} \tag{44}$$

$$s = \text{PO} \sin \phi \sin \rho = d^* \sin \phi \sin \rho \tag{45}$$

$$(\text{using the scale } d^* = L \tan 2\theta)$$

Since the spot vanishes when $s = r$, $r/d^* \geqq s/d^*$ is a *measure* of the size of the diffracting unit and increases as the (equi-axed) particle size decreases. In fact, if $r/d^* \rightarrow \frac{1}{2}$, the particle size has diminished to 1 unit cell. Rotation through an angle of 2ρ for the largest value of $s = r$ thus provides a way of using the rotating stage for an approximate measure of particle size, e.g. with spots that just vanish at $\pm 10°$, $\sin 10° = 0{\cdot}174$ and with $\phi = 30°$ the spot vanishes when $r/d^* \geqq s/d^* = 0{\cdot}087$ and so the mean particle dimension is of the order of 6 unit cells. With particles of the order of 100–1000 Å and $a_0 \rightarrow 10$ Å, 10 to 100 unit cells correspond to a full rotation angle of

from $2\rho = 2 \sin^{-1} 0{\cdot}1$ to $2 \sin^{-1} 0{\cdot}01$

i.e. $= 2(5°44')$ to $2(0°34')$

As, however, the finite spot size is *not due to this cause alone*, these figures are only a guide to the differences that might arise between patterns of different particle size from otherwise identical materials.

5.2 FINITE WAVELENGTH EFFECT AND SPOT SIZE

In the above treatment of finite spot size the assumptions made are equivalent to infinitely small wavelength. The introduction of finite wavelength will not appreciably affect these conclusions, since, if two spots appear at all, the difference between a plane through OP′ and another through OQ, in Fig. 19, is closely similar to the effect of specimen rotation. Directions such as CQ or CP for different spots are radii of the sphere of reflection and it is the position of P relative to Q that may be different for one of a pair of spots, or the intrinsic spot diameter may differ.

In Fig. 19 the finite size allows the spot centre P′ to be somewhere else along the line CQ and diffraction occurs all over the circle which represents the intersection of the large sphere centre C and the small sphere centre P′. This circle of which SR is a diameter projects to a spot between S_1 and R_1, the mid-point of which is not P″ but P_1 slightly further away from O. For these positions the Bragg angle varies between $\theta' + \delta$ and $\theta' - \delta$. The actual angle of rotation of OP out of the sphere is, as above, given by ρ, which is slightly greater than $\hat{QOP}$. Now if the centre of the reciprocal spot lies on P and approaches point magnitude, the diffraction spot will appear at P_0. Many of the

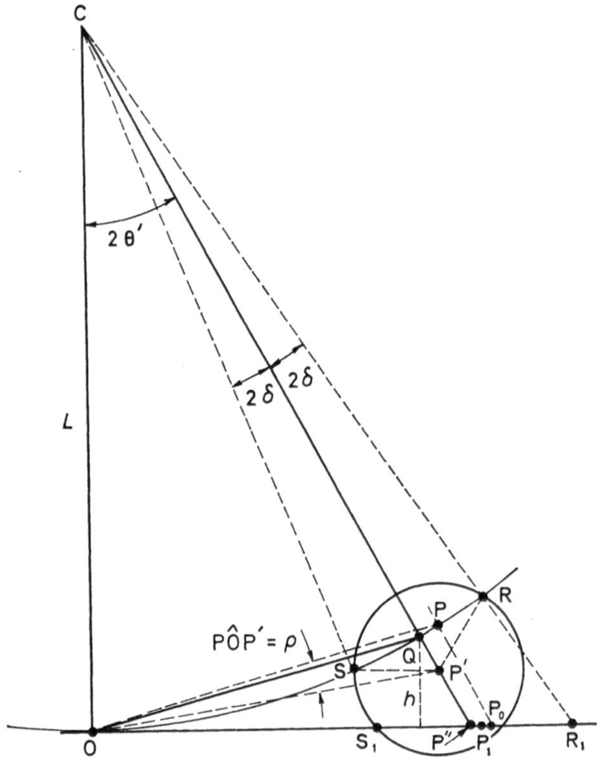

FIG. 19. Finite wavelength and finite spot size.

distances and magnitudes in Fig. 19 are represented by simple mathematical expressions. For many electron microscopes the patterns do not have values of θ exceeding $\sim 3°$. All the angles in Fig. 19 are of the order of 6° or less. Thus errors in length of the order of $\frac{1}{2}$ per cent or less occur if we take

$$2\theta \text{ radian} \simeq \sin 2\theta \simeq \tan 2\theta$$

providing $2\theta \leqq 6°$. For this value

$$2\theta = 0{\cdot}10472, \quad \sin 2\theta = 0{\cdot}10453, \quad \tan 2\theta = 0{\cdot}10510$$

The formulae from Fig. 19 for various lengths and angles are:

$$\left.\begin{array}{lll}
(a)\ \mathrm{OP} & = \mathrm{OP'} = 2L\sin\theta \simeq 2L\theta \\
(b)\ \mathrm{OQ} & = 2L\sin\theta' \simeq 2L\theta' \\
(c)\ \mathrm{OR} & = 2L\sin(\theta'+\delta) \simeq 2L(\theta'+\delta) \\
(d)\ \mathrm{OS} & = 2L\sin(\theta'-\delta) \simeq 2L(\theta'-\delta) \\
(e)\ \mathrm{OR_1} & = L\tan 2(\theta'+\delta) \simeq 2L(\theta'+\delta) \\
(f)\ \mathrm{OS_1} & = L\tan 2(\theta'-\delta) \simeq 2L(\theta'-\delta) \\
(g)\ \mathrm{OP_1} & = \tfrac{1}{2}(\mathrm{OR_1}+\mathrm{OS_1}) \simeq L\tan 2\theta' \simeq 2L\theta' \\
(h)\ \mathrm{OP''} & = L\tan 2\theta' \simeq 2L\theta' \\
(i)\ \mathrm{QP} & = 2L\sin(\theta-\theta') \simeq 2L(\theta-\theta') \simeq \mathrm{OP}-\mathrm{OQ} \\
(j)\ \mathrm{P''P_0} & = L(\tan 2\theta-\tan 2\theta') \simeq 2L(\theta-\theta')
\end{array}\right\} \tag{46}$$

The approximation to $L\tan 2\theta'$ for OP_1 is very close involving neglect of squares and higher powers of tangents of angles $<3°$. Hence, $OP_1 = OP''$ and this difference can be neglected for all practical purposes.

In Fig. 19 the application of standard formulae to the triangle QOP′ leads to the relation

$$\tan 2\theta' = \frac{2\sin\theta.\cos(\theta-\rho)}{1-2\sin(\theta-\rho)\sin\theta} \tag{47a}$$

or

$$\tan 2\theta' \simeq 2\sin\theta\cos(\theta-\rho) \tag{47b}$$

The error in neglecting powers of sines of small angles is here of the order of 1 in 10^{-4} or less, since $\theta = 3°$ and $\rho \to 3°$. From equation (47) it is found that the difference between θ and θ' is very small and cannot be calculated accurately with ordinary five-figure tables. On the other hand, the corresponding error in d^*, although very small, can be detected. Thus

$$\frac{\Delta d^*}{d^*} = \frac{\mathrm{P_0P''}}{\mathrm{OP_0}} = \frac{\tan 2\theta-\tan 2\theta'}{\tan 2\theta}$$

which on substitution of equation (47a) leads to the relation

$$\frac{\Delta d^*}{d^*} = \frac{\cos\theta-\cos(\theta+\rho)}{(1-2\sin\theta.\sin(\theta-\rho))\cos\theta} \tag{48a}$$

or

$$\simeq \cos 2\theta-\cos(2\theta+\rho)-\tan\theta(\sin 2\rho-\sin\rho) \tag{48b}$$

Thus the error for $\theta = 3°$ and $\rho = 3°$ is 4×10^{-3}, which in a length of 10 cm = 100 mm corresponds to $\sim \frac{1}{2}$ mm or 1 mm in the double distance. This length or greater can be found in a print enlargement. If it were due to particle size alone this difference would not probably be reached, since for values of ρ of the order of θ the limiting crystallite size is approached ($\sim$ one unit cell). An error which is just about detectable on a double measurement with a print enlargement arises when $\theta = 3°$ and $\rho = 2°$. In any case *other causes of finite spot size* make the above calculation necessary and an experimental value of ρ can be obtained by measuring the double angle 2ρ through which a spot must be rotated to appear and disappear. This value of ρ is thus primarily caused by the finite size of the electron beam and other factors, and it can obviously lead to detectable, but not important, differences when calculating interplanar spacings.

5.3 THE ZONE LAYER EFFECT AND DETECTION OF LATTICE BENDING OR CURVATURE

5.3.1 *The zone layer effect*

It has been seen in §§2.4 and 3.4.3 that it may be necessary to consider patterns for which equation (12) applies, viz.

$$hu+kv+lw = \pm C \neq 0$$

i.e. non-zero layers.

In §3.4.3 the connection between this relationship and the so-called Laue zones was considered. A practical example of the zone layer effect, in which the Laue zones (or zone layers) are not separated, is given in Plate 4 and is the diffraction pattern of M_6C previously referred to. All the spots except four can be interpreted according to a single zone axis with $C = 0, 2, 4, 6$. Only even values occur for the reasons given previously (§3.5.3).

Plate 4 and Fig. 20 suggest a further step which may be useful in interpretating patterns and in deciding whether an 'imperfect' pattern arises from a single fragment, or whether curvatures or slightly misoriented fragments are present. The distance I_{uvw} previously referred to (see also Table 1) is for the general case given by

$$I_{uvw} = \sqrt{[(au)^2+(bv)^2+(cw)^2+2abuv\cos\gamma+2bcvw\cos\alpha+2cawu\cos\beta]} \quad (49a)$$

for orthorhombic lattices by

$$I_{uvw} = \sqrt{[(au)^2+(bv)^2+(cw)^2]} \quad (49b)$$

and for cubic lattices

$$I_{uvw} = a\sqrt{[u^2+v^2+w^2]} \quad (49c)$$

These are the lengths of vectors in crystal space from the origin to lattice points $[uvw]$. The reciprocal layers have separations $1/I_{uvw}$.

As already seen in §3.4.3 these layers are closer together with higher lattice parameters or indices, so that spots on non-zero layers, or overlapping Laue zones, are more likely to occur. This situation is represented in Fig. 21 where a sphere of radius $1/\lambda$ cuts non-zero layers of [313] nearer to the centre than [111].

A *scale diagram* can be used to check this explanation in a given case but should normally not be necessary. The important point is to check whether the diffraction pattern can in fact be interpreted as arising exclusively from a single fragment.

5.3.2 *Specimen curvature or grain misorientations*

There may, however, be some parts of a pattern which cannot be explained in this way and these arise from fragments of crystal which are very slightly misoriented or from parts of a bent lattice. Fig. 22 shows the relationship between points which appear to belong to different zones, but which lie on the same sphere of radius $1/\lambda$ (the curvature is much exaggerated here). This diagram, as it stands, illustrates more generally the point already explained by Fig. 21, but it also enables the detection of lattice curvatures or misorientations for the following reasons. In Fig. 22, points at O and P_1 indicate zone $[u_1v_1w_1]$, at O and P_2 indicate zone $[u_2v_2w_2]$ and at P_1 and P_2 indicate zone $[u_3v_3w_3]$.

It is easy to deduce the angles between these vectors and the zone axes as indicated in the diagram. As before, the diffraction pattern gives the projections of P_1 and P_2, but this does not affect the conclusions. Any distances such as OP on this scale are given by

$$\mathrm{OP} = Md^* = K \sin\theta \simeq K\theta \simeq \tfrac{1}{2}K \tan 2\theta$$

where $K = 2M/\lambda$. The required test is to determine zone axes such as $[u_1 v_1 w_1]$ etc.

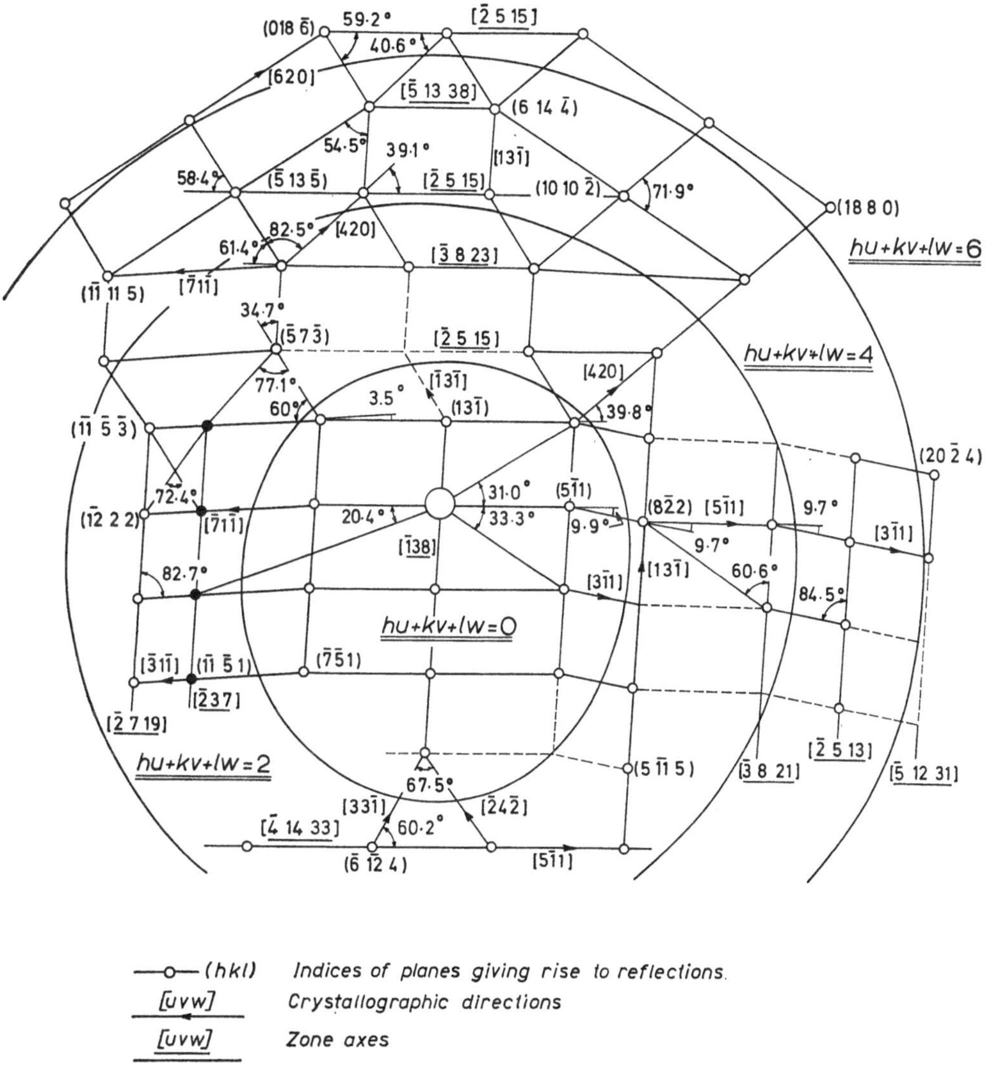

FIG. 20. Solution to Plate 4 showing zones, indexing, and the effect of curvature of the sphere of reflection.

The angle between any zone axis and the electron beam normal must be equal to the Bragg angle θ corresponding to a point in that zone which is in the reflecting position. Strictly only one set of points such as P_1, all with equal distances OP_1

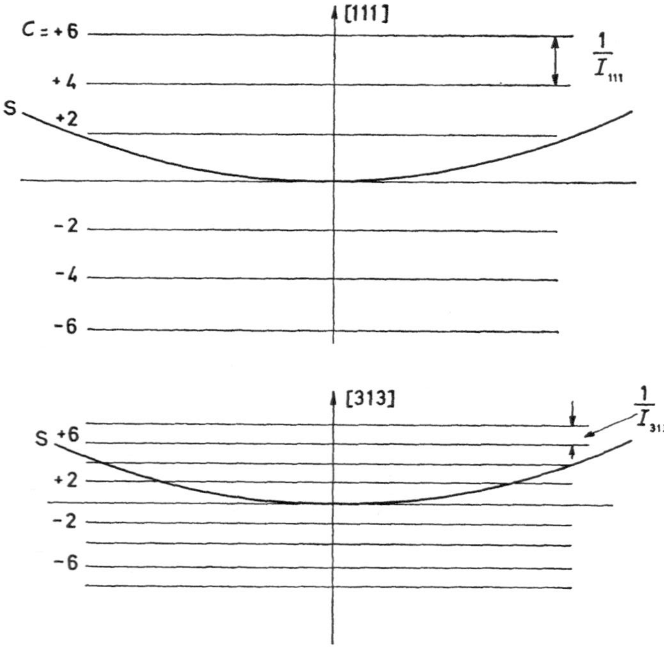

FIG. 21. Test of zone equation for $hu+kv+lw = \pm 1, \pm 2$, etc., illustrated for two cubic zones in which only even values of C occur. S is the sphere of radius $1/\lambda$ and has the same scale as $1/I_{uvw}$.

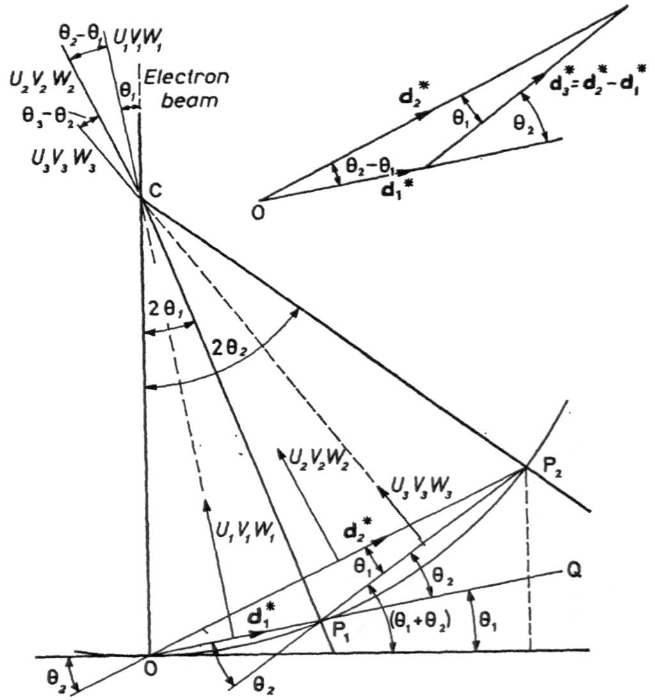

FIG. 22. Relation between zone axes.

(multiplicity), should fulfil this condition at any one time. In the diagram therefore the following angles should be found:

$$\begin{aligned} &\text{Electron beam with } [u_1v_1w_1] = \theta_1 \\ &\text{Electron beam with } [u_2v_2w_2] = \theta_2 \\ &[u_1v_1w_1] : [u_2v_2w_2] = \theta_2 - \theta_1 \\ &[u_3v_3w_3] : [u_2v_2w_2] = \theta_1 \\ &[u_3v_3w_3] : [u_1v_1w_1] = \theta_2 \end{aligned}$$

Actually any two out of the last three will provide a sufficient check. The first two cannot reasonably be used since the true electron beam direction relative to $[u_1v_1w_1]$ cannot be measured but should be fixed by θ_1. It is the differences between the zone axes themselves which should be studied. (It is of course necessary to make a suitable selection of primary directions, such as OP_1, OP_2, more or less in a straight line on the diffraction pattern.) It follows that any *actual lattice curvature* can be found by calculating zone axes for different regions of the pattern and subtracting the appropriate values of θ, since, if the angles are of the same order as θ, only reciprocal sphere curvature is involved, and angles in excess of θ must then be due to lattice curvature or misorientations. Diffraction spots on both sides of O should be examined since the excess angle will always add to θ on one side and be decreased by θ on the other, i.e. the larger inter-zonal angle on opposite sides indicates the direction of the curvature.

Part 2

TABLES AND DIAGRAMS FOR GENERAL USE

6 General

The use of standard and general data has frequently been referred to in Part 1. Parts 2 and 3 provide a collection of data, 2 concerning general cases and 3 specific substances.

Of the seven crystal systems, the orthorhombic, tetragonal, hexagonal and cubic are most frequently represented in metallurgical work. As $a \neq b \neq c$ in an orthorhombic cell, excessively large tabulations would be required to cover a practical range of angles between crystallographic planes. A range of standard stereograms for variations of two of the interaxial ratios could be constructed, but in this system it is better to deal with cases individually as they arise. In the tetragonal and hexagonal systems there is only one variable parameter between different unit cells. This is commonly expressed as c/a and can vary over a large range as shown in the following tables. For this reason it has been found useful to calculate the angles between certain crystallographic planes in each crystal system for a range of c/a values. The angles presented are sufficient to enable a stereographic projection of any crystal, with a unit cell whose c/a ratio falls within the limits specified, to be drawn. Combining this with a Wülff net enables many other angles to be calculated to within $\sim \frac{1}{2}°$. This information is based on data and procedures recommended in the *International Tables for X-ray Crystallography.*[5]

It so happens that the c/a ratio for the simple close-packed hexagonal structure of some metals is 1·633 for most compact spherical packing. It is convenient to have standard stereographic projections on different poles available for this ratio. Although in the tetragonal system there are many compounds with c/a near unity, no common value of c/a is found for as many substances as there is for 1·633 in the hexagonal system. The special case here is with $c/a = 1$, when the unit cell becomes exactly cubic.

In the cubic system the angles between the planes, and therefore the stereographic projection, are independent of the unit cell dimensions, i.e. they are the same for each cubic compound whether it has a face-centred, body-centred, or simple cubic cell.

7 Calculation of interplanar spacings and angles

Although there are tabulations of lattice spacings[37] and parameters[38,39] which assist the identification of unknown materials, it is equally necessary, especially with spot patterns, to have tables of interplanar and interzonal angles. It is often found useful to re-arrange or extend tables of spacings. Table 1 summarizes the formulae used to obtain any d-spacing or angle. Tables 2 to 7 show applications of these formulae to give the angles between certain planes in the tetragonal and hexagonal systems, and cover a wide range of interaxial ratios. Tables 3 and 7 are included in Part 2 because they are specific examples of the general tables. Table 9 covers all the angles between crystallographic planes in the cubic crystal system for h, k and $l \leqq 5$. Massalski and King[40,41] have published tables for quick calculation of d-spacings in the tetragonal and hexagonal systems for given c/a ratios. Cubic d-spacings can be easily calculated to slide-rule accuracy.

Even so, a computer programme provides the most rapid method of obtaining a table of d-spacings for a specific phase, particularly one of low crystal symmetry. Programmes for calculating spacings and angles have been written at the authors' laboratory for use with an Elliot 803 computer. The spacing programme gives values of d and $(1/d)^2$ within selected limits and can reject any d-spacing that does not reflect because of the crystal symmetry. However, it is advisable to have these printed for face- and body-centred lattices since ordering and, in some cases, double reflection are able to fill these in. Programmes for use with other computers have been listed.[42]

Additional material from *Interpretation of electron diffraction patterns*,
ISBN 978-1-4899-6228-7, is available at http://extras.springer.com

8 Tetragonal system

8.1 ANGLES AND OTHER DATA

A cubic unit cell elongated or reduced along one dimension becomes a tetragonal unit cell, or a tetragonal crystal with $c/a = 1$ is a cubic crystal. The (small-scale) projection (Fig. 23) is for (001) and shows two great circles which determine the positions of poles 101, 111, 011. A tetragonal distortion from the cube, by, for example, increasing c/a from unity, therefore destroys the cubic symmetry including the threefold axis along [111], but the symmetry of this diagram itself is unaltered. The angles Ψ and Φ increase as c/a increases, and when c/a is less than 1 they decrease. Either angle alone is sufficient to determine the complete stereographic projection (but both are useful) for any tetragonal crystal.

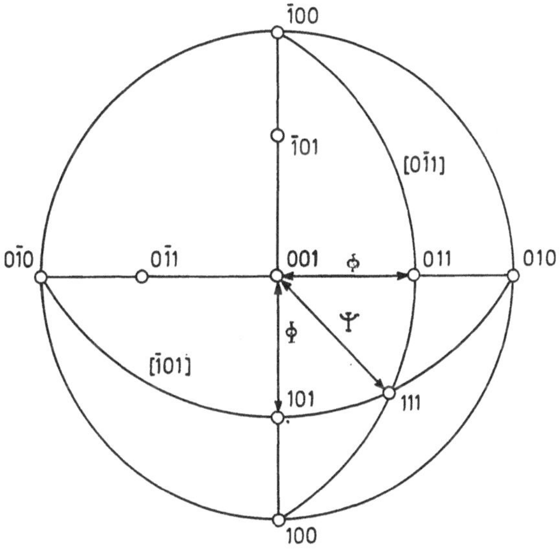

Fig. 23. Stereographic projection for tetragonal crystals showing principal poles and angles.

For Fig. 23, $a = b \neq c$, $\alpha = \beta = \gamma = 90$, where a and c are the edges of the tetragonal unit cell. Interplanar angles are calculated from the general formula

$$\cos\phi = \frac{(h_1h_2+k_1k_2)\frac{1}{a^2}+l_1l_2\frac{1}{c^2}}{\sqrt{\left\{\left[(h_1^2+k_1^2)\frac{1}{a^2}+l_1^2\frac{1}{c^2}\right]\left[(h_2^2+k_2^2)\frac{1}{a^2}+l_2^2\frac{1}{c^2}\right]\right\}}} \tag{50a}$$

where $(h_1k_1l_1)$ and $(h_2k_2l_2)$ are the two crystal planes.

Ψ is the angle between (001) and {111} and Φ is the angle between (001) and {101}, and therefore Ψ and Φ are calculated from

$$\cos \Psi = \frac{1}{\sqrt{[2(c/a)^2+1]}} \quad \text{i.e. } \tan \Psi = \sqrt{2}(c/a) \qquad (50b)$$

and

$$\cos \Phi = \frac{1}{\sqrt{[(c/a)^2+1]}} \quad \text{i.e. } \tan \Phi = c/a \qquad (50c)$$

and so

$$\tan \Psi = \sqrt{2} \tan \Phi \qquad (50d)$$

Note: It is not considered necessary to provide a full stereographic projection since the only axial ratios to which particular significance is attached are covered by the cubic projections in Chapter 10. These are at or near 1·0 or near 1·414 = $\sqrt{2}$. The latter occurs for a body-centred tetragonal lattice near to a face-centred cube.

TABLE 2: Angles for establishing stereographic projections of tetragonal crystals

c/a	Φ	Ψ	c/a	Φ	Ψ
0·200	11°19′	15°47½′	1·500	56°19′	64°45½′
0·300	16°42′	22°59′	1·600	58°0′	66°9½′
0·400	21°48′	29°30′	1·700	59°32′	67°25′
0·500	26°34′	35°16′	1·800	60°57′	68°33′
0·600	30°58′	40°19′	1·900	62°14½′	69°35′
0·700	34°59½′	44°43′	2·000	63°26′	70°32′
0·800	38°40′	48°32′	2·500	68°12′	74°12′
0·900	41°59′	51°51′	3·000	71°34′	76°44′
1·000	45°	54°44′	3·500	74°4′	78°34′
1·100	47°44′	57°16′	4·000	76°4′	79°58′
1·200	50°12′	59°29½	4·500	77°28′	81°4′
1·300	52°26′	61°27′	5·000	78°42′	81°57′
1·400	54°28′	63°12′			

The figures in the frame are for a cubic crystal.

TABLE 3. Angles between planes for selected tetragonal materials (in degrees)†

		Axial ratio c/a for			
		β-Sn	SnO_2	Pb_3O_4	TeO_2
$h_1k_1l_1$	$h_2k_2l_2$	0·5457	0·6728	0·7447	1·5833
001	105	6·22	7·67	8·47	17·59
	103	10·32	12·64	13·94	27·82
	102	15·26	18·59	20·43	38·37
	203	19·99	24·16	26·40	46·55
	101	28·62	33·93	36·68	57·72
	302	39·30	45·26	48·17	67·17
	201	47·50	53·38	56·13	72·47
	301	58·58	63·65	65·89	78·11
	501	69·87	73·45	74·97	82·80
	100	90·00	90·00	90·00	90·00
	511	70·23	73·75	75·25	82·94
	510	90·00	90·00	90·00	90·00
	313	29·91	35·35	38·13	59·07
	312	40·79	46·77	49·66	68·23
	311	59·91	64·83	67·00	78·70
	310	90·00	90·00	90·00	90·00
	213	22·13	26·64	29·03	49·73
	212	31·39	36·95	39·79	60·54
	211	50·67	56·39	59·01	74·23
	210	90·00	90·00	90·00	90·00
	321	63·06	67·60	69·57	80·06
	320	90·00	90·00	90·00	90·00
	115	8·78	10·77	11·90	24·12
	113	14·43	17·60	19·34	36·73
	112	21·10	25·44	27·78	48·23
	111	37·66	43·58	46·49	65·93
	221	57·06	62·28	64·60	77·41
	331	66·64	70·69	72·44	81·53
	110	90·00	90·00	90·00	90·00

† From R. E. Frounfelker and W. M. Hirthe, *Trans. A.I.M.E.*, 1962, **224**, 196.

TABLE 4. Angles between planes in tetragonal crystals (in degrees)†

$(h_1k_1l_1) = (001)$ with $(h_2k_2l_2)$

Axial ratio c/a

$h_1k_1l_1$	$h_2k_2l_2$	0·500	0·550	0·600	0·650	0·700	0·750	0·800	0·850	0·900	0·950	1·000	1·050	1·100	1·150	1·200
001	105	5·711	6·277	6·843	7·407	7·970	8·531	9·090	9·648	10·204	10·758	11·310	11·860	12·407	12·953	13·496
	103	9·462	10·389	11·310	12·225	13·134	14·036	14·931	15·819	16·699	17·571	18·435	19·290	20·136	20·973	21·801
	102	14·036	15·376	16·699	18·004	19·290	20·556	21·801	23·025	24·228	25·408	26·565	27·699	28·811	29·899	30·964
	203	18·435	20·136	21·801	23·429	25·017	26·565	28·072	29·539	30·964	32·347	33·690	34·992	36·254	37·476	38·660
	101	26·565	28·811	30·964	33·024	34·992	36·870	38·660	40·365	41·987	43·531	45·000	46·397	47·726	48·991	50·194
	302	36·870	39·523	41·987	44·275	46·397	48·366	50·194	51·892	53·471	54·941	56·310	57·588	58·782	59·899	60·945
	201	45·000	47·726	50·194	52·431	54·462	56·310	57·995	59·534	60·945	62·241	63·435	64·537	65·556	66·501	67·380
	301	56·310	58·782	60·945	62·850	64·537	66·038	67·380	68·587	69·677	70·665	71·565	72·387	73·142	73·836	74·476
	501	68·199	70·017	71·565	72·897	74·055	75·069	75·964	76·659	77·471	78·111	78·690	79·216	79·695	80·134	80·538
	100	90·000	90·000	90·000	90·000	90·000	90·000	90·000	90·000	90·000	90·000	90·000	90·000	90·000	90·000	90·000
	511	68·583	70·375	71·899	73·211	74·349	75·346	76·226	77·008	77·707	78·336	78·904	79·420	79·891	80·322	80·718
	510	90·000	90·000	90·000	90·000	90·000	90·000	90·000	90·000	90·000	90·000	90·000	90·000	90·000	90·000	90·000
	313	27·791	30·103	32·312	34·417	36·422	38·329	40·140	41·860	43·492	45·040	46·508	47·902	49·224	50·479	51·671
	312	38·329	41·011	43·492	45·784	47·902	49·860	51·671	53·348	54·903	56·347	57·688	58·938	60·103	61·191	62·209
	311	57·688	60·103	62·209	64·057	65·689	67·138	68·432	69·593	70·640	71·589	72·452	73·239	73·961	74·625	75·237
	310	90·000	90·000	90·000	90·000	90·000	90·000	90·000	90·000	90·000	90·000	90·000	90·000	90·000	90·000	90·000
	213	20·439	22·291	24·095	25·849	27·553	29·206	30·807	32·356	33·855	35·302	36·699	38·048	39·348	40·602	41·810
	212	29·206	31·588	33·855	36·007	38·048	39·981	41·810	43·541	45·178	46·726	48·190	49·574	50·885	52·126	53·301
	211	48·190	50·885	53·301	55·471	57·426	59·193	60·794	62·250	63·577	64·791	65·905	66·930	67·875	68·750	69·561
	210	90·000	90·000	90·000	90·000	90·000	90·000	90·000	90·000	90·000	90·000	90·000	90·000	90·000	90·000	90·000
	321	60·983	63·239	65·191	66·892	68·386	69·706	70·879	71·929	72·872	73·725	74·499	75·204	75·849	76·441	76·986
	320	90·000	90·000	90·000	90·000	90·000	90·000	90·000	90·000	90·000	90·000	90·000	90·000	90·000	90·000	90·000
	115	8·049	8·842	9·632	10·417	11·199	11·977	12·750	13·518	14·282	15·040	15·793	16·541	17·282	18·018	18·748
	113	13·263	14·535	15·793	17·036	18·262	19·471	20·663	21·836	22·990	24·124	25·239	26·334	27·409	28·463	29·496
	112	19·471	21·251	22·990	24·684	26·334	27·938	29·496	31·008	32·473	33·891	35·264	36·592	37·876	39·117	40·316
	111	35·264	37·876	40·316	42·590	44·711	46·686	48·527	50·243	51·844	53·339	54·736	56·042	57·266	58·414	59·491
	221	54·736	57·266	59·491	61·457	63·203	64·761	66·157	67·415	68·553	69·587	70·529	71·391	72·182	72·911	73·584
	331	64·761	66·802	68·553	70·068	71·391	72·554	73·584	74·501	75·324	76·066	76·737	77·348	77·906	78·417	78·888
	110	90·000	90·000	90·000	90·000	90·000	90·000	90·000	90·000	90·000	90·000	90·000	90·000	90·000	90·000	90·000

† From R. E. Frounfelker, and W. M. Hirthe, *Trans. A.I.M.E.*, 1962, **224**, 196.

$h_1k_1l_1$	$h_2k_2l_2$	Axial ratio c/a 1.250	1.300	1.350	1.400	1.450	1.500	1.550	1.600	1.650	1.700	1.750	1.800	1.850	1.900	1.950	2.000
001	105	14.036	14.574	15.110	15.642	16.172	16.699	17.223	17.745	18.263	18.778	19.290	19.799	20.304	20.807	21.306	21.801
	103	22.620	23.429	24.228	25.017	25.796	26.565	27.324	28.072	28.811	29.539	30.256	30.964	31.661	32.347	33.024	33.690
	102	32.005	33.024	34.019	34.992	35.942	36.870	37.776	38.660	39.523	40.365	41.186	41.987	42.769	43.531	44.275	45.000
	203	39.806	40.914	41.987	43.025	44.029	45.000	45.939	46.848	47.726	48.576	49.399	50.194	50.964	51.710	52.431	53.130
	101	51.340	52.431	53.471	54.462	55.408	56.310	57.171	57.995	58.782	59.534	60.255	60.945	61.607	62.241	62.850	63.435
	302	61.928	62.850	63.719	64.537	65.308	66.038	66.727	67.380	67.999	68.587	69.146	69.677	70.183	70.665	71.125	71.565
	201	68.199	68.962	69.677	70.346	70.974	71.565	72.121	72.646	73.142	73.610	74.055	74.476	74.876	75.256	75.619	75.964
	301	75.069	75.619	76.130	76.607	77.053	77.471	77.863	78.232	78.579	78.906	79.216	79.509	79.786	80.049	80.300	80.538
	501	80.910	81.254	81.573	81.870	82.147	82.405	82.648	82.875	83.089	83.290	83.480	83.660	83.830	83.991	84.144	84.289
	100	90.000	90.000	90.000	90.000	90.000	90.000	90.000	90.000	90.000	90.000	90.000	90.000	90.000	90.000	90.000	90.000
	511	81.083	81.421	81.734	82.026	82.297	82.551	82.789	83.012	83.222	83.419	83.606	83.782	83.949	84.107	84.257	84.400
	510	90.000	90.000	90.000	90.000	90.000	90.000	90.000	90.000	90.000	90 000	90.000	90.000	90.000	90.000	90.000	90.000
	313	52.803	53.880	54.903	55.877	56.805	57.688	58.531	59.335	60.103	60.836	61.538	62.209	62.851	63.467	64.057	64.623
	312	63.162	64.057	64.898	65.689	66.434	67.138	67.803	68.432	69.028	69.593	70.130	70.640	71.126	71.589	72.030	72.452
	311	75.803	76.328	76.817	77.272	77.697	78.095	78.469	78.820	79.151	79.462	79.757	80.036	80.300	80.551	80.789	81.015
	310	90.000	90.000	90.000	90.000	90.000	90.000	90.000	90.000	90.000	90.000	90.000	90.000	90.000	90.000	90.000	90.000
	213	42.975	44.097	45.178	46.219	47.223	48.190	49.121	50.019	50.885	51.719	52.524	53.301	54.050	54.773	55.471	56.145
	212	54.415	55.471	56.474	57.426	58.332	59.193	60.013	60.794	61.539	62.250	62.928	63.577	64.197	64.791	65.360	65.905
	211	70.314	71.016	71.672	72.285	72.859	73.398	73.906	74.384	74.835	75.261	75.665	76.047	76.410	76.755	77.083	77.396
	210	90.000	90.000	90.000	90.000	90.000	90.000	90.000	90.000	90.000	90.000	90.000	90.000	90.000	90.000	90.000	90.000
	321	77.490	77.957	78.390	78.794	79.171	79.524	79.855	80.166	80.458	80.734	80.994	81.241	81.474	81.695	81.905	82.105
	320	90.000	90.000	90.000	90.000	90.000	90.000	90.000	90.000	90.000	90.000	90.000	90.000	90.000	90.000	90.000	90.000
	115	19.471	20.188	20.899	21.603	22.300	22.990	23.673	24.349	25.018	25.680	26.334	26.981	27.621	28.254	28.879	29.496
	113	30.509	31.501	32.473	33.423	34.354	35.264	36.155	37.025	37.876	38.708	39.521	40.316	41.092	41.850	42.590	43.314
	112	41.473	42.590	43.669	44.711	45.716	46.686	47.623	48.527	49.400	50.243	51.058	51.844	52.604	53.339	54.049	54.736
	111	60.504	61.457	62.355	63.203	64.003	64.761	65.478	66.157	66.802	67.415	67.998	68.553	69.082	69.587	70.068	70.529
	221	74.207	74.786	75.324	75.827	76.397	76.737	77.151	77.540	77.906	78.252	78.578	78.888	79.181	79.459	79.723	79.975
	331	79.322	79.723	80.096	80.443	80.767	81.070	81.353	81.620	81.870	82.160	82.329	82.540	82.739	82.928	83.108	83.279
	110	90.000	90.000	90.000	90.000	90.000	90.000	90.000	90.000	90.000	90.000	90.000	90.000	90.000	90.000	90.000	90.000

9 Hexagonal and rhombohedral systems

9.1 ANGLES AND OTHER DATA

Fig. 24 and Tables 5, 6 and 7 provide for calculations and stereographic projections for hexagonal crystals over a wide range of axial ratios.† Rhombohedral lattices referring to these hexagonal axes include the three cubic lattices with the axial ratios given in Table 5. In Fig. 24 (0001) = (00.1) hexagonal or (111) rhombohedral (cubic) is the plane of projection. The four-index system is used in the diagrams to emphasize the difference between the H and R axes but only three indices are desirable for transformations and calculations of angles etc. (see §9.2). The following notes may be of help when using Fig. 24 and Tables 5, 6 and 7:

The rhombohedral faces have poles which fall in the position shown at equal angles ρ_2 from the threefold axis.

Half-way between any pair of these planes, e.g. (100) and (010), is located the corresponding (110). This provides the angle $\lambda/2$, which in Fig. 24 is measured along the great circle.

The {110} poles are at an angle ρ_1 from the centre.

It is clear that a knowledge of any one of the three angles ρ_1, ρ_2 or λ is sufficient to determine the other two and to establish the stereographic projection completely for either a rhombohedral or a hexagonal lattice.

The angles are functions of the hexagonal axial ratio c/a and are deduced from the general formulae given below.

The angle between the rhombohedral cell edges is given by α ($\alpha' = 180-\alpha$), which is seen to be the angle between the great circles for the zones containing the poles of the cell faces. (These great circles intersect at the same angle as their zone axes, which contain the cell faces, in pairs, and so are themselves the cell edges.)

Table 7 is a supplementary table for commonly occurring phases with close-packed hexagonal structures.

The indices of planes in the rhombohedral lattice ($h_R k_R l_R$) are expressed in terms of the hexagonal indices ($hk.l$) by $h_R = \frac{1}{3}(2h+k+l)$, $k_R = \frac{1}{3}(-h+k+l)$, $l_R = \frac{1}{3}(-h-2k+l)$. If a hexagonal lattice shows only reflections which satisfy these transformations in integers, the rhombohedral lattice is present. Any of the formulae represents a sufficient condition, since the other two rhombohedral indices differ by integers. Thus, $2h+k+l$ = a multiple of 3 implies all three conditions.

Reciprocal lattice vectors which already satisfy such conditions can only add to or subtract from other vectors which also satisfy the same conditions. Hence double reflection cannot add points to a rhombohedral diffraction pattern and so confuse the recognition of such a lattice.

† The standard projections in Figs. 25–28 refer only to the axial ratio 1·633 for spherical close-packing.

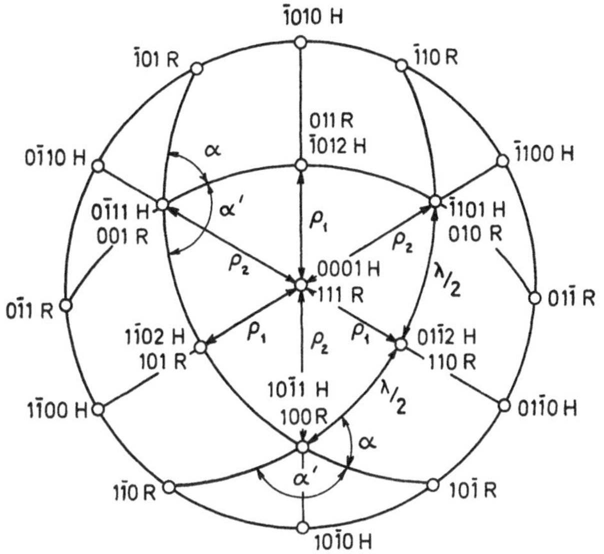

Fig. 24. Stereographic projection for hexagonal crystals showing principal poles and angles.

The orientation in the diagram as shown is clearly not sufficient to distinguish a phase by itself or to discriminate between rhombohedral and plain hexagonal. Patterns in other orientations are needed.

One use of the table of angles is in connection with phases which result from rhombohedral distortions of cubic phases or are the close-packed hexagonal analogues.

The interplanar angles of the hexagonal unit cell are calculated from:

$$\cos\phi = \frac{h_1h_2+k_1k_2+\frac{1}{2}(h_1k_2+k_1h_2)+\frac{3}{4}(a^2/c^2)l_1l_2}{\sqrt{([h_1^2+k_1^2+h_1k_1+\frac{3}{4}(a^2/c^2)l_1^2][h_2^2+k_2^2+h_2k_2+\frac{3}{4}(a^2/c^2)l_2^2])}} \tag{51a}$$

ρ_1 is the angle between (0001) and $\{10\bar{1}1\}$, ρ_2 is the angle between (0001) and $\{01\bar{1}2\}$ and λ is the angle between $\{10\bar{1}1\}$ and $\{\bar{1}101\}$. These angles are given by the following relationships:

$$\cos\rho_1 = \frac{1}{\sqrt{[\frac{4}{3}(c^2/a^2)+1]}} \quad \text{i.e. } \tan\rho_1 = \frac{2c}{\sqrt{3}(a)} \tag{51b}$$

$$\cos\rho_2 = \frac{2}{\sqrt{[\frac{4}{3}(c^2/a^2)+4]}} \quad \text{i.e. } \tan\rho_2 = \frac{c}{\sqrt{3}(a)} \tag{51c}$$

$$\text{i.e. } \tan\rho_1 = 2\tan\rho_2 \tag{51d}$$

$$\cos\lambda = \frac{\frac{3}{4}(a^2/c^2)-\frac{1}{2}}{\frac{3}{4}(a^2/c^2)+1} \tag{51e}$$

$$\frac{a_R}{a} = \frac{1}{\sqrt{3}\cos\rho_2} \tag{52}$$

$$\cos\frac{\alpha'}{2} = \frac{\sqrt{3}}{2}\cos\rho_2 \tag{53}$$

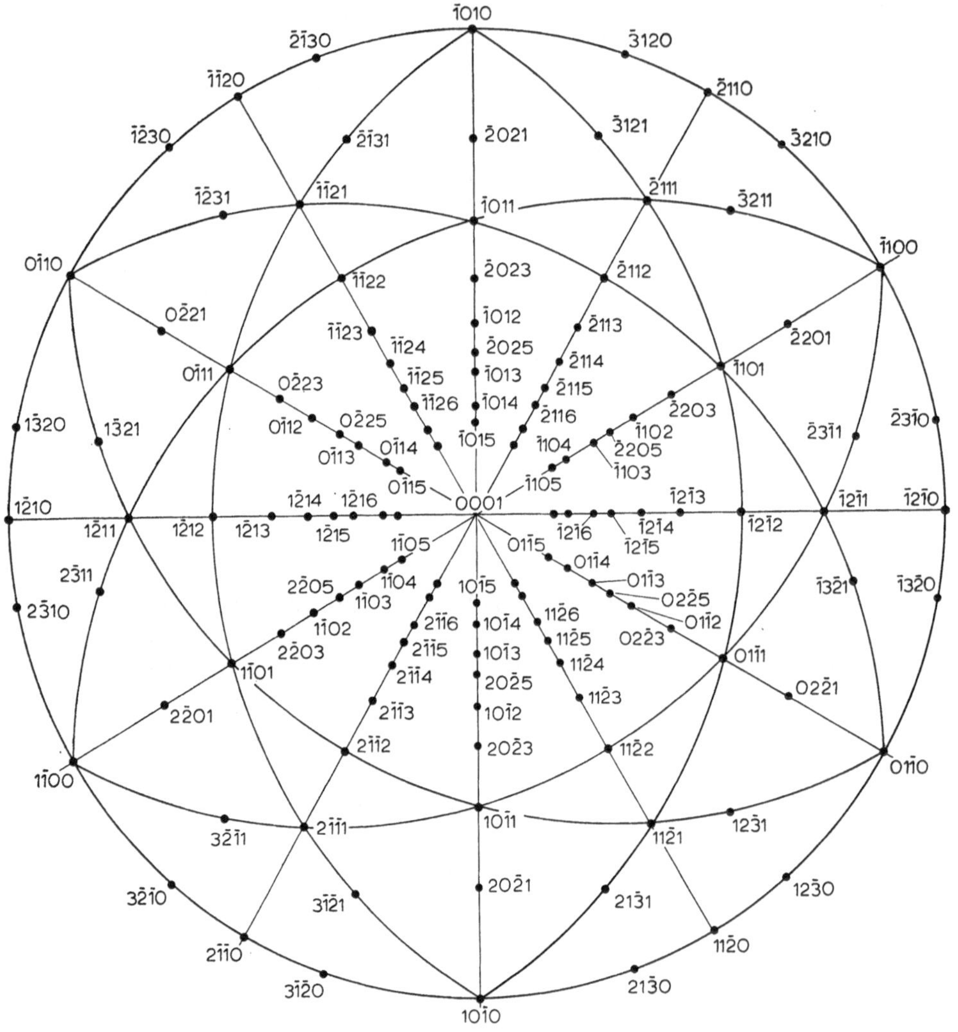

FIG. 25. Stereographic projection for hexagonal crystals on (00.1) $\equiv$ (0001), with $c/a = 1{\cdot}633$.

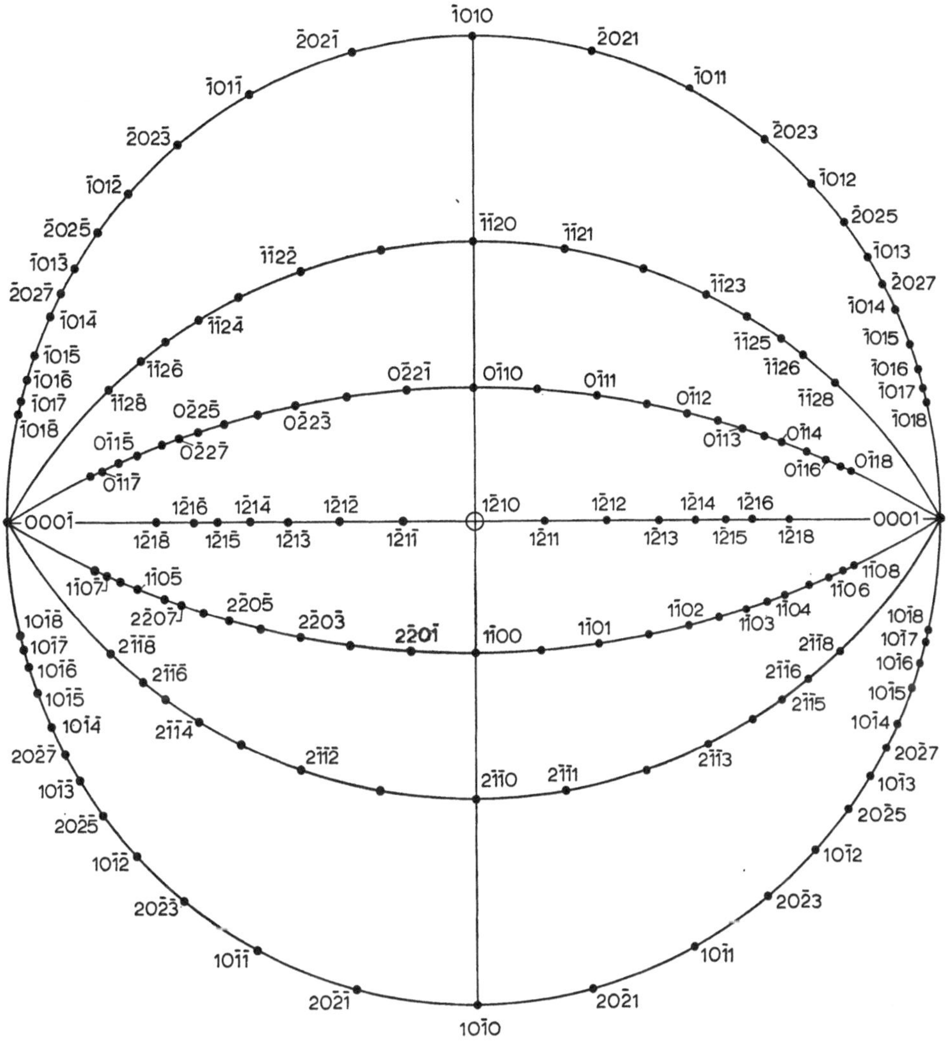

Fig. 26. Stereographic projection for hexagonal crystals on ($1\bar{2}10$), with $c/a = 1{\cdot}633$.

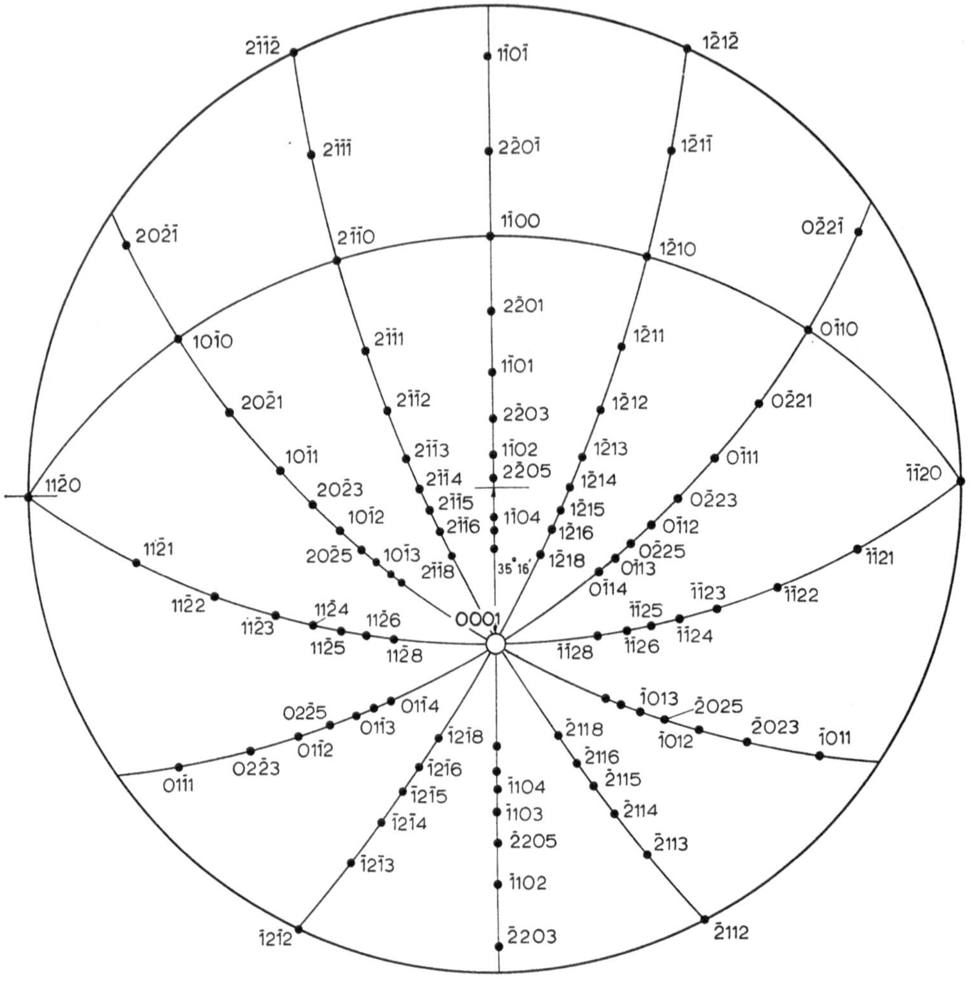

Fig. 27. Intermediate projection for hexagonal crystals with $c/a = 1{\cdot}633$ obtained by rotating Fig. 25 about [11.0] by 35° 16′. (This rotation about a twofold axis is identical with that needed to rotate Fig. 30 into Fig. 31, but here the rotation does not bring a plane of simple indices into the plane of projection.)

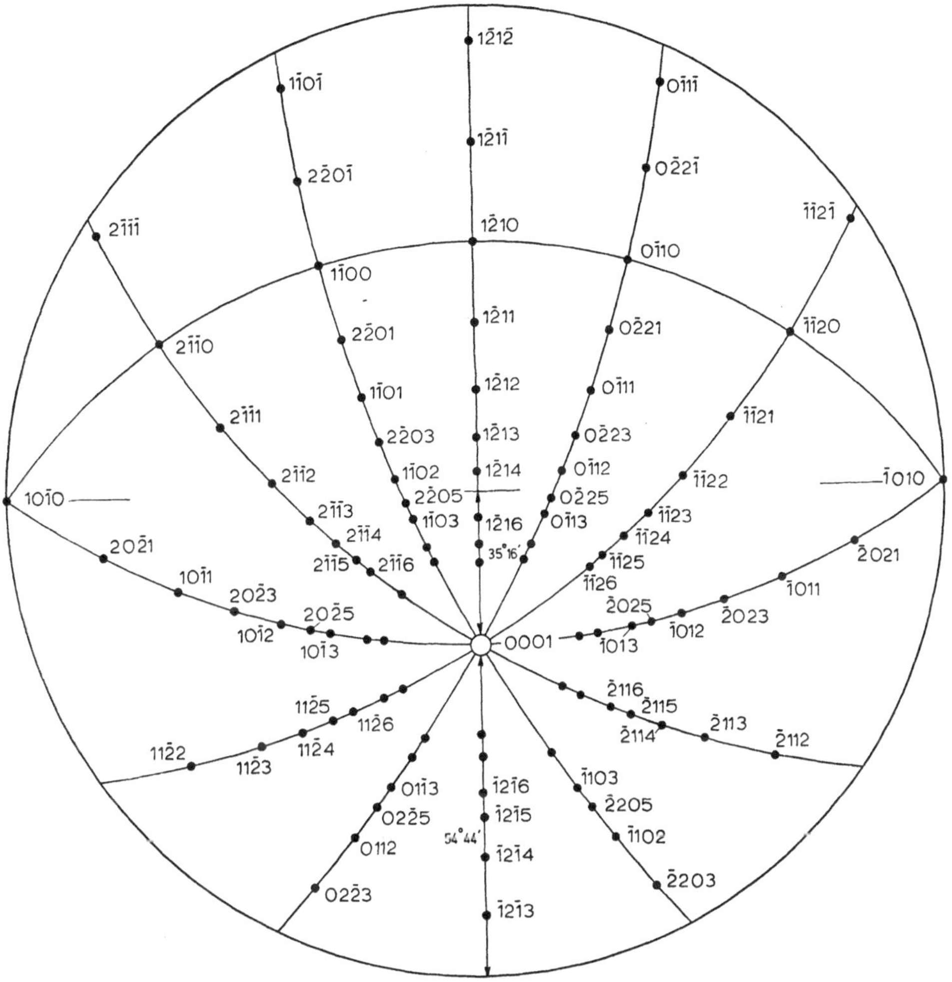

FIG. 28. Intermediate projection for hexagonal crystals with $c/a = 1{\cdot}633$ obtained by rotating Fig. 25 about [21·0] by 35° 16′. (This rotation takes place about an alternative twofold axis and again the plane of projection does not have simple indices.)

TABLE 5. Angles for establishing stereographic projections of hexagonal and rhombohedral crystals†

Axial ratio c/a	ρ_1	ρ_2	λ	α	a_R/a	Comments
0·20	13°0′	6°35′	22°28½′	118°42′	0·5812	
0·25	16°6′	8°13′	27°48′	117°59½′	0·5833	
0·30	19°6½′	9°49½′	32°56′	117°9′	0·5859	
0·35	22°0½′	11°25½′	37°52′	116°10½′	0·5890	
0.40	24°47½′	13°0′	42°35′	115°5½′	0·5925	
0·45	27°27½′	14°34′	47°4′	113°54′	0·5965	
0.50	30°0′	16°6′	51°19′	112°37′	0·6009	
0·55	32°25′	17°37′	55°19½′	111°15½′	0·6058	
0·60	34°43′	19°6½′	59°6′	109°50′	0·6110	
0·6123725	35°16′	19°31′	60°0′	109°28′	0·6124	Body-centred cubic
0·65	36°53½′	20°34′	62°39′	108°21′	0·6167	
0·70	38°57′	22°0½′	65°58′	106°49′	0·6227	
0·75	40°53½′	23°25′	69°4½′	105°15½′	0·6292	
0·80	42°44′	24°47½′	71°59′	103°40′	0·6360	
0·85	44°28′	26°8½′	74°41½′	102°3½′	0·6431	
0·90	46°6′	27°27½′	77°13½′	100°26′	0·6506	
0·95	47°39′	28°45′	79°35′	98°48½′	0·6585	
1·00	48°6½′	30°0′	81°47′	97°11′	0·6667	
1·05	50°29′	31°13½′	83°54½′	95°33½′	0·6752	
1·10	51°47′	32°25′	85°45½′	93°57′	0·6839	
1·15	53°1′	33°35′	87°33′	92°21′	0·6930	
1·20	54°11′	34°43′	89°13′	90°46½′	0·7024	
1·224745	54°44′	35°16′	90°0′	90°0′	0·7071	Simple cubic
1·25	55°17′	35°49′	90°46½′	89°13′	0·7120	
1·30	56°20′	36°53½′	92°14′	87°41′	0·7219	
1·35	57°19′	37°56′	93°35½′	86°10′	0·7320	
1·40	58°15½′	38°57′	94°52′	84°41′	0·7424	
1·45	59°9′	39°56′	96°4′	83°13′	0·7530	
1·50	60°0′	40°53½′	97°11′	81°47′	0·7638	
1·55	60°48½′	41°49½′	98°14′	80°23′	0·7748	
1·60	61°34½′	42°44′	99°13′	79°0½′	0·7860	
1·6330065	62°4′	43°19′	99°50′	78°7′	0·7935	Close-packed hexagonal
1·65	61°18½′	43°36½′	100°8½′	77°40′	0·7974	
1·70	63°0′	44°28′	101°0½′	76°21′	0·8090	
1·75	63°40′	45°18′	101°49½′	75°4′	0·8207	
1·80	64°18½′	46°6′	102°35½′	73°48½′	0·8327	
1·85	64°55′	46°53′	103°19′	72°35′	0·8448	
1·90	65°30′	47°39′	104°0′	71°23′	0·8570	
1·95	66°3′	48°23′	104°39′	70°13′	0·8694	
2·00	66°35′	49°6½′	105°15½′	69°4½′	0·8819	
2·05	67°6′	49°48′	105°50′	67°58′	0·8946	
2·10	67°35½′	50°29′	106°23′	66°52½′	0·9074	
2·15	68°4′	51°9′	106°56½′	65°49′	0·9203	
2·20	68°31′	51°47′	107°23′	64°47′	0·9333	
2·25	68°57′	52°24½′	107°51′	63°46½′	0·9465	
2·30	69°22′	53°1′	108°17′	62°48′	0·9597	
2·35	69°46′	53°36½′	108°42′	61°50′	0·9731	
2·40	70°9½′	54°11′	109°6′	60°54′	0·9866	

† From Vol. II of *International Tables for X-ray Crystallography*.[5]

Table 5 (cont.)

Axial ratio c/a	ρ_1	ρ_2	λ	α	a_R/a	Comments
2·4495098	70°32′	54°44′	109°28′	60°0′	1·0000	Close-packed cubic
2·50	70°53½′	55°17′	109°50′	59°6′	1·0138	
2·60	71°35′	56°20′	110°30′	57°23½′	1·0414	
2·70	72°13′	57°19′	111°6′	55°45½′	1·0693	
2.80	72°49′	58°15½′	111°39½′	54°12½′	1·0975	
2·90	73°22½′	59°9′	112°9½′	52°43½′	1·1260	
3·00	73°54′	60°0′	112°37′	51°19′	1·1547	
3·20	74°51½′	61°34½′	113°26′	48°41½′	1·2129	
3·40	75°42½′	63°0′	114°7′	46°18′	1·2719	
3·60	76°28½′	64°18½′	114°42′	44°6½′	1·3317	
3·80	77°10′	65°30′	115°12½′	42°6′	1·3920	
4·00	77°47′	66°35′	115°39′	40°15½′	1·4530	
4·50	79°6½′	68°57′	116°31′	36°15′	1·6073	
5·00	80°10½′	70°53½′	117°9′	32°56′	1·7638	
6·00	81°47′	73°54′	117°59½′	27°48′	2·0817	
8·00	83°49′	77°47′	118°51½′	21°7′	2·7285	
10·00	85°3′	80°10½′	119°16′	17°0′	3·3830	
15·00	86°42′	83°25′	119°40′	11°24′	5·0332	
20·00	87°31′	85°3′	119°49′	8°34′	6·6916	

TABLE 6. Angles between planes in hexagonal crystals (in degrees)†

$h_1k_1i_1l_1$	$h_2k_2i_2l_2$	Axial ratio c/a																
		0·50	0·51	0·52	0·53	0·54	0·55	0·56	0·57	0·58	0·59	0·60	0·61	0·62	0·63	0·64	0·65	0·66
0001	$10\bar{1}8$	4·128	4·210	4·292	4·375	4·457	4·539	4·621	4·703	4·785	4·867	4·950	5·032	5·144	5·196	5·278	5·360	5·442
	$10\bar{1}7$	4·715	4·809	4·903	4·997	5·090	5·184	5·278	5·371	5·465	5·559	5·652	5·746	5·840	5·933	6·027	6·120	6·213
	$10\bar{1}6$	5·496	5·606	5·715	5·824	5·933	6·042	6·151	6·260	6·369	6·478	6·587	6·696	6·804	6·913	7·022	7·130	7·239
	$10\bar{1}5$	6·587	6·717	6·848	6·978	7·109	7·239	7·369	7·499	7·629	7·759	7·889	8·019	8·148	8·278	8·408	8·537	8·666
	$10\bar{1}4$	8·213	8·375	8·537	8·699	8·860	9·022	9·183	9·344	9·505	9·666	9·826	9·987	10·147	10·307	10·467	10·627	10·787
	$20\bar{2}7$	9·367	9·551	9·735	9·918	10·101	10·285	10·467	10·650	10·833	11·015	11·197	11·379	11·560	11·742	11·923	12·103	12·284
	$10\bar{1}3$	10·893	11·106	11·318	11·530	11·742	11·953	12·164	12·374	12·584	12·794	13·004	13·213	13·422	13·630	13·839	14·046	14·254
	$20\bar{2}5$	13·004	13·255	13·505	13·755	14·005	14·254	14·502	14·750	14·997	15·244	15·490	15·735	15·980	16·224	16·468	16·711	16·953
	$10\bar{1}2$	16·102	16·407	16·711	17·014	17·316	17·617	17·917	18·216	18·514	18·811	19·107	19·401	19·695	19·988	20·279	20·570	20·859
	$20\bar{2}3$	21·052	21·435	21·816	22·195	22·572	22·947	23·320	23·691	24·060	24·427	24·791	25·154	25·514	25·872	26·228	26·582	26·934
	$10\bar{1}1$	30·000	30·494	30·982	31·466	31·945	32·419	32·888	33·352	33·811	34·266	34·715	35·160	35·599	36·034	36·465	36·890	37·311
	$20\bar{2}1$	49·107	49·667	50·215	50·751	51·275	51·787	52·288	52·777	53·256	53·724	54·183	54·631	55·069	55·498	55·918	56·330	56·732
	$10\bar{1}0$	90·000	90·000	90·000	90·000	90·000	90·000	90·000	90·000	90·000	90·000	90·000	90·000	90·000	90·000	90·000	90·000	90·000
	$21\bar{3}2$	37·371	37·920	38·461	38·993	39·518	40·035	40·544	41·046	41·540	42·026	42·506	42·978	43·443	43·901	44·352	44·796	45·233
	$21\bar{3}1$	56·789	57·307	57·811	58·301	58·777	59·241	59·693	60·133	60·562	60·979	61·386	61·782	62·168	62·545	62·913	63·271	63·621
	$21\bar{3}0$	90·000	90·000	90·000	90·000	90·000	90·000	90·000	90·000	90·000	90·000	90·000	90·000	90·000	90·000	90·000	90·000	90·000
	$11\bar{2}8$	7·125	7·266	7·407	7·548	7·688	7·829	7·970	8·110	8·250	8·391	8·531	8·671	8·811	8·951	9·090	9·230	9·369
	$11\bar{2}6$	9·462	9·648	9·834	10·019	10·204	10·389	10·574	10·758	10·942	11·126	11·310	11·493	11·677	11·860	12·043	12·225	12·407
	$11\bar{2}4$	14·036	14·306	14·574	14·842	15·110	15·376	15·642	15·908	16·172	16·436	16·699	16·962	17·223	17·484	17·745	18·004	18·263
	$11\bar{2}2$	26·565	27·022	27·474	27·924	28·369	28·811	29·249	29·683	30·114	30·541	30·964	31·383	31·799	32·211	32·619	33·024	33·425
	$11\bar{2}0$	90·000	90·000	90·000	90·000	90·000	90·000	90·000	90·000	90·000	90·000	90·000	90·000	90·000	90·000	90·000	90·000	90·000
$10\bar{1}0$	$21\bar{3}0$	19·107	19·107	19·107	19·107	19·107	19·107	19·107	19·107	19·107	19·107	19·107	19·107	19·107	19·107	19·107	19·107	19·107
	$11\bar{2}0$	30·000	30·000	30·000	30·000	30·000	30·000	30·000	30·000	30·000	30·000	30·000	30·000	30·000	30·000	30·000	30·000	30·000
	$01\bar{1}0$	60·000	60·000	60·000	60·000	60·000	60·000	60·000	60·000	60·000	60·000	60·000	60·000	60·000	60·000	60·000	60·000	60·000

$h_1k_1i_1l_1$	$h_2k_2i_2l_2$	Axial ratio c/a																
		0·67	0·68	0·69	0·70	0·71	0·72	0·73	0·74	0·75	0·76	0·77	0·78	0·79	0·80	0·81	0·82	0·83
0001	$10\bar{1}8$	5·524	5·606	5·688	5·769	5·851	5·933	6·015	6·097	6·178	6·260	6·342	6·424	6·505	6·587	6·668	6·750	6·831
	$10\bar{1}7$	6·307	6·400	6·494	6·587	6·680	6·773	6·866	6·960	7·053	7·146	7·239	7·332	7·425	7·518	7·611	7·703	7·796
	$10\bar{1}6$	7·347	7·456	7·564	7·672	7·781	7·889	7·997	8·105	8·213	8·321	8·429	8·537	8·645	8·753	8·860	8·968	9·075
	$10\bar{1}5$	8·796	8·925	9·054	9·183	9·312	9·441	9·569	9·698	9·826	9·955	10·083	10·211	10·339	10·467	10·595	10·723	10·851
	$10\bar{1}4$	10·947	11·106	11·265	11·424	11·583	11·742	11·900	12·058	12·216	12·374	12·532	12·689	12·847	13·004	13·161	13·318	13·474
	$20\bar{2}7$	12·464	12·644	12·824	13·004	13·183	13·362	13·541	13·720	13·898	14·076	14·254	14·431	14·608	14·785	14·962	15·138	15·314
	$10\bar{1}3$	14·461	14·667	14·873	15·079	15·285	15·490	15·694	15·898	16·102	16·305	16·508	16·711	16·913	17·115	17·316	17·517	17·717
	$20\bar{2}5$	17·195	17·436	17·677	17·917	18·156	18·395	18·633	18·870	19·107	19·343	19·578	19·812	20·046	20·279	20·512	20·744	20·975
	$10\bar{1}2$	21·148	21·435	21·721	22·006	22·290	22·572	22·854	23·134	23·413	23·691	23·968	24·244	24·518	24·791	25·063	25·334	25·604
	$20\bar{2}3$	27·283	27·630	27·976	28·318	28·659	28·998	29·334	29·668	30·000	30·330	30·657	30·982	31·306	31·626	31·945	32·262	32·576
	$10\bar{1}1$	37·727	38·139	38·546	38·948	39·346	39·740	40·129	40·513	40·893	41·269	41·641	42·008	42·372	42·731	43·085	43·436	43·783
	$20\bar{2}1$	57·126	57·512	57·890	58·260	58·622	58·977	59·325	59·666	60·000	60·328	60·649	60·963	61·272	61·575	61·872	62·163	62·449
	$10\bar{1}0$	90·000	90·000	90·000	90·000	90·000	90·000	90·000	90·000	90·000	90·000	90·000	90·000	90·000	90·000	90·000	90·000	90·000

	$21\bar{3}2$	45·664	46·088	46·506	46·917	47·323	47·722	48·115	48·502	48·883	49·259	49·629	49·993	50·352	50·706	51·054	51·398	51·736
	$21\bar{3}1$	63·962	64·296	64·621	64·939	65·249	65·553	65·849	66·139	66·422	66·699	66·970	67·235	67·494	67·748	67·996	68·239	68·477
	$21\bar{3}0$	90·000	90·000	90·000	90·000	90·000	90·000	90·000	90·000	90·000	90·000	90·000	90·000	90·000	90·000	90·000	90·000	90·000
	$11\bar{2}8$	9·509	9·648	9·787	9·926	10·065	10·204	10·343	10·481	10·620	10·758	10·896	11·034	11·172	11·310	11·448	11·585	11·723
	$11\bar{2}6$	12·589	12·771	12·953	13·134	13·315	13·496	13·676	13·856	14·036	14·216	14·395	14·574	14·753	14·931	15·110	15·287	15·465
	$11\bar{2}4$	18·521	18·778	19·034	19·290	19·545	19·799	20·052	20·304	20·556	20·807	21·057	21·306	21·554	21·801	22·048	22·294	22·538
	$11\bar{2}2$	33·822	34·216	34·606	34·992	35·375	35·754	36·129	36·501	36·870	37·235	37·596	37·954	38·309	38·660	39·008	39·352	39·693
	$11\bar{2}0$	90·000	90·000	90·000	90·000	90·000	90·000	90·000	90·000	90·000	90·000	90·000	90·000	90·000	90·000	90·000	90·000	90·000
$10\bar{1}0$	$21\bar{3}0$	19·107	19·107	19·107	19·107	19·107	19·107	19·107	19·107	19·107	19·107	19·107	19·107	19·107	19·107	19·107	19·107	19·107
	$11\bar{2}0$	30·000	30·000	30·000	30·000	30·000	30·000	30·000	30·000	30·000	30·000	30·000	30·000	30·000	30·000	30·000	30·000	30·000
	$01\bar{1}0$	60·000	60·000	60·000	60·000	60·000	60·000	60·000	60·000	60·000	60·000	60·000	60·000	60·000	60·000	60·000	60·000	60·000

		Axial ratio c/a																
$h_1k_1i_1l_1$	$h_2k_2i_2l_2$	0·84	0·85	0·86	0·87	0·88	0·89	0·90	0·91	0·92	0·93	0·94	0·95	0·96	0·97	0·98	0·99	1·00
0001	$10\bar{1}8$	6·913	6·994	7·076	7·157	7·239	7·320	7·401	7·483	7·564	7·645	7·727	7·808	7·889	7·970	8·051	8·132	8·213
	$10\bar{1}7$	7·889	7·982	8·074	8·167	8·260	8·352	8·445	8·537	8·629	8·722	8·814	8·906	8·999	9·091	9·183	9·275	9·367
	$10\bar{1}6$	9·183	9·290	9·398	9·505	9·612	9·719	9·826	9·933	10·040	10·147	10·254	10·361	10·467	10·574	10·681	10·787	10·893
	$10\bar{1}5$	10·978	11·106	11·233	11·361	11·488	11·615	11·742	11·868	11·995	12·122	12·248	12·374	12·500	12·626	12·752	12·878	13·004
	$10\bar{1}4$	13·630	13·787	13·942	14·098	14·254	14·409	14·564	14·719	14·873	15·028	15·182	15·336	15·490	15·643	15·796	15·949	16·102
	$20\bar{2}7$	15·490	15·665	15·840	16·015	16·189	16·364	16·537	16·711	16·884	17·057	17·230	17·402	17·574	17·746	17·917	18·088	18·258
	$10\bar{1}3$	17·917	18·116	18·315	18·514	18·712	18·909	19·107	19·303	19·499	19·695	19·890	20·085	20·280	20·473	20·667	20·859	21·052
	$20\bar{2}5$	21·205	21·435	21·664	21·892	22·120	22·346	22·572	22·798	23·022	23·246	23·469	23·691	23·913	24·134	24·354	24·573	24·791
	$10\bar{1}2$	25·872	26·139	26·405	26·670	26·934	27·196	27·457	27·717	27·976	28·233	28·489	28·744	28·998	29·250	29·501	29·751	30·000
	$20\bar{2}3$	32·888	33·198	33·506	33·811	34·115	34·416	34·713	35·012	35·307	35·599	35·890	36·178	36·465	36·749	37·031	37·311	37·589
	$10\bar{1}1$	44·126	44·465	44·800	45·131	45·459	45·782	46·102	46·418	46·731	47·040	47·346	47·648	47·946	48·241	48·533	48·821	49·107
	$20\bar{2}1$	62·729	63·005	63·275	63·540	63·800	64·056	64·307	64·553	64·795	65·033	65·267	65·496	65·722	65·944	66·162	66·376	66·587
	$10\bar{1}0$	90·000	90·000	90·000	90·000	90·000	90·000	90·000	90·000	90·000	90·000	90·000	90·000	90·000	90·000	90·000	90·000	90·000
	$21\bar{3}2$	52·069	52·397	52·721	53·039	53·354	53·663	53·968	54·269	54·565	54·857	55·145	55·429	55·709	55·985	56·257	56·525	56·789
	$21\bar{3}1$	68·710	68·939	69·163	69·382	69·597	69·807	70·014	70·216	70·415	70·610	70·801	70·988	71·172	71·353	71·530	71·704	71·875
	$21\bar{3}0$	90·000	90·000	90·000	90·000	90·000	90·000	90·000	90·000	90·000	90·000	90·000	90·000	90·000	90·000	90·000	90·000	90·000
	$11\bar{2}8$	11·860	11·997	12·134	12·271	12·407	12·544	12·680	12·817	12·953	13·089	13·225	13·360	13·496	13·631	13·766	13·901	14·036
	$11\bar{2}6$	15·642	15·819	15·996	16·172	16·348	16·524	16·699	16·874	17·049	17·223	17·398	17·571	17·745	17·918	18·091	18·263	18·435
	$11\bar{2}4$	22·782	23·026	23·268	23·509	23·750	23·989	24·228	24·466	24·702	24·938	25·174	25·408	25·641	25·873	26·105	26·335	26·565
	$11\bar{2}2$	40·030	40·365	40·696	41·023	41·348	41·669	41·987	42·302	42·614	42·923	43·229	43·531	43·831	44·128	44·421	44·712	45·000
	$11\bar{2}0$	90·000	90·000	90·000	90·000	90·000	90·000	90·000	90·000	90·000	90·000	90·000	90·000	90·000	90·000	90·000	90·000	90·000
$10\bar{1}0$	$21\bar{3}0$	19·107	19·107	19·107	19·107	19·107	19·107	19·107	19·107	19·107	19·107	19·107	19·107	19·107	19·107	19·107	19·107	19·107
	$11\bar{2}0$	30·000	30·000	30·000	30·000	30·000	30·000	30·000	30·000	30·000	30·000	30·000	30·000	30·000	30·000	30·000	30·000	30·000
	$01\bar{1}0$	60·000	60·000	60·000	60·000	60·000	60·000	60·000	60·000	60·000	60·000	60·000	60·000	60·000	60·000	60·000	60·000	60·000

† From R. E. Frounfelker, M. A. Seitz and W. M. Hirthe, *Nuclear Science and Engineering*, 1962, 14 (2), 192.

TABLE 6 (cont.)

$h_1k_1i_1l_1$	$h_2k_2i_2l_2$	Axial ratio c/a 1·01	1·02	1·03	1·04	1·05	1·06	1·07	1·08	1·09	1·10	1·11	1·12	1·13	1·14	1·15	1·16	1·17
0001	$10\bar{1}8$	8·294	8·375	8·456	8·537	8·618	8·699	8·779	8·860	8·941	9·022	9·102	9·183	9·263	9·344	9·425	9·505	9·585
	$10\bar{1}7$	9·459	9·551	9·643	9·735	9·826	9·918	10·010	10·101	10·193	10·285	10·376	10·467	10·559	10·650	10·741	10·833	10·924
	$10\bar{1}6$	11·000	11·106	11·212	11·318	11·424	11·530	11·636	11·742	11·847	11·953	12·058	12·164	12·269	12·374	12·479	12·585	12·689
	$10\bar{1}5$	13·129	13·255	13·380	13·505	13·630	13·755	13·880	14·005	14·129	14·254	14·378	14·502	14·626	14·750	14·873	14·997	15·120
	$10\bar{1}4$	16·255	16·407	16·559	16·711	16·863	17·014	17·165	17·316	17·467	17·617	17·767	17·917	18·066	18·216	18·365	18·514	18·662
	$20\bar{2}7$	18·429	18·599	18·768	18·938	19·107	19·275	19·443	19·611	19·779	19·946	20·113	20·279	20·446	20·611	20·777	20·942	21·107
	$10\bar{1}3$	21·244	21·435	21·626	21·816	22·006	22·195	22·384	22·572	22·760	22·947	23·134	23·320	23·506	23·691	23·876	24·060	24·244
	$20\bar{2}5$	25·009	25·226	25·442	25·658	25·872	26·086	26·299	26·511	26·723	26·934	27·144	27·353	27·561	27·769	27·976	28·182	28·387
	$10\bar{1}2$	30·247	30·494	30·739	30·982	31·225	31·466	31·706	31·945	32·183	32·419	32·654	32·888	33·121	33·352	33·582	33·811	34·039
	$20\bar{2}3$	37·865	38·139	38·411	38·681	38·948	39·214	39·478	39·740	39·999	40·257	40·513	40·767	41·019	41·269	41·518	41·764	42·008
	$10\bar{1}1$	49·389	49·667	49·943	50·215	50·485	50·751	51·014	51·275	51·532	51·787	52·039	52·288	52·534	52·777	53·018	53·256	53·491
	$20\bar{2}1$	66·794	66·998	67·198	67·395	67·589	67·780	67·968	68·152	68·334	68·513	68·689	68·863	69·033	69·201	69·367	69·530	69·691
	$10\bar{1}0$	90·000	90·000	90·000	90·000	90·000	90·000	90·000	90·000	90·000	90·000	90·000	90·000	90·000	90·000	90·000	90·000	90·000
	$21\bar{3}2$	57·050	57·307	57·561	57·811	58·057	58·301	58·541	58·777	59·011	59·241	59·469	59·693	59·915	60·133	60·349	60·562	60·772
	$21\bar{3}1$	72·043	72·208	72·370	72·529	72·686	72·839	72·991	73·139	73·285	73·429	73·570	73·709	73·845	73·980	74·112	74·242	74·370
	$21\bar{3}0$	90·000	90·000	90·000	90·000	90·000	90·000	90·000	90·000	90·000	90·000	90·000	90·000	90·000	90·000	90·000	90·000	90·000
	$11\bar{2}8$	14·171	14·306	14·440	14·574	14·708	14·842	14·976	15·110	15·243	15·376	15·509	15·642	15·775	15·908	16·040	16·172	16·304
	$11\bar{2}6$	18·607	18·778	18·949	19·120	19·290	19·460	19·630	19·799	19·968	20·136	20·304	20·472	20·640	20·807	20·974	21·140	21·306
	$11\bar{2}4$	26·794	27·022	27·248	27·474	27·699	27·924	28·147	28·369	28·590	28·811	29·030	29·249	29·466	29·683	29·899	30·114	30·328
	$11\bar{2}2$	45·285	45·567	45·847	46·123	46·397	46·668	46·937	47·203	47·466	47·726	47·984	48·240	48·493	48·743	48·991	49·236	49·479
	$11\bar{2}0$	90·000	90·000	90·000	90·000	90·000	90·000	90·000	90·000	90·000	90·000	90·000	90·000	90·000	90·000	90·000	90·000	90·000
$10\bar{1}0$	$21\bar{3}0$	19·107	19·107	19·107	19·107	19·107	19·107	19·107	19·107	19·107	19·107	19·107	19·107	19·107	19·107	19·107	19·107	19·107
	$11\bar{2}0$	30·000	30·000	30·000	30·000	30·000	30·000	30·000	30·000	30·000	30·000	30·000	30·000	30·000	30·000	30·000	30·000	30·000
	$01\bar{1}0$	60·000	60·000	60·000	60·000	60·000	60·000	60·000	60·000	60·000	60·000	60·000	60·000	60·000	60·000	60·000	60·000	60·000

$h_1k_1i_1l_1$	$h_2k_2i_2l_2$	Axial ratio c/a 1·18	1·19	1·20	1·21	1·22	1·23	1·24	1·25	1·26	1·27	1·28	1·29	1·30	1·31	1·32	1·33	1·34
0001	$10\bar{1}8$	9·666	9·746	9·826	9·907	9·987	10·067	10·147	10·227	10·307	10·388	10·467	10·547	10·627	10·707	10·787	10·867	10·947
	$10\bar{1}7$	11·015	11·106	11·197	11·288	11·379	11·469	11·560	11·651	11·742	11·832	11·923	12·013	12·103	12·194	12·284	12·374	12·464
	$10\bar{1}6$	12·794	12·899	13·004	13·109	13·213	13·318	13·422	13·526	13·630	13·735	13·839	13·942	14·046	14·150	14·254	14·357	14·461
	$10\bar{1}5$	15·244	15·367	15·490	15·612	15·735	15·858	15·980	16·102	16·224	16·346	16·468	16·589	16·711	16·832	16·953	17·074	17·195
	$10\bar{1}4$	18·811	18·959	19·107	19·254	19·401	19·548	19·695	19·842	19·988	20·134	20·279	20·425	20·570	20·715	20·859	21·004	21·148
	$20\bar{2}7$	21·271	21·435	21·599	21·762	21·925	22·087	22·249	22·411	22·572	22·733	22·894	23·054	23·214	23·373	23·533	23·691	23·850
	$10\bar{1}3$	24·427	24·609	24·791	24·973	25·154	25·334	25·514	25·693	25·872	26·050	26·228	26·405	26·582	26·758	26·934	27·109	27·283
	$20\bar{2}5$	28·591	28·795	28·998	29·200	29·401	29·601	29·801	30·000	30·198	30·395	30·592	30·788	30·982	31·177	31·370	31·562	31·754
	$10\bar{1}2$	34·265	34·491	34·715	34·938	35·160	35·380	35·599	35·818	36·034	36·250	36·465	36·678	36·890	37·101	37·311	37·520	37·727
	$20\bar{2}3$	42·251	42·492	42·731	42·968	43·203	43·436	43·668	43·898	44·126	44·352	44·577	44·800	45·021	45·241	45·459	45·675	45·889
	$10\bar{1}1$	53·724	53·955	54·183	54·408	54·631	54·851	55·069	55·285	55·498	55·710	55·918	56·125	56·330	56·532	56·732	56·930	57·126
	$20\bar{2}1$	69·849	70·005	70·158	70·310	70·459	70·606	70·751	70·893	71·034	71·173	71·310	71·445	71·578	71·709	71·838	71·966	72·092
	$10\bar{1}0$	90·000	90·000	90·000	90·000	90·000	90·000	90·000	90·000	90·000	90·000	90·000	90·000	90·000	90·000	90·000	90·000	90·000

$h_1k_1l_1i_1$	$h_2k_2i_2l_2$																	
	$21\bar{3}2$	60.979	51.184	61.386	61.585	61.782	61.976	62.168	62.358	62.545	62.730	62.913	63.093	63.271	63.447	63.621	63.793	63.962
	$21\bar{3}1$	74.496	74.620	74.743	74.863	74.981	75.098	75.213	75.326	75.437	75.547	75.656	75.762	75.867	75.971	76.073	76.174	76.273
	$21\bar{3}0$	90.000	90.000	90.000	90.000	90.000	90.000	90.000	90.000	90.000	90.000	90.000	90.000	90.000	90.000	90.000	90.000	90.000
	$11\bar{2}8$	16.436	16.568	16.699	16.831	16.962	17.093	17.223	17.354	17.484	17.615	17.745	17.875	18.004	18.134	18.263	18.392	18.521
	$11\bar{2}6$	21.471	21.637	21.801	21.966	22.130	22.294	22.457	22.620	22.782	22.945	23.106	23.268	23.429	23.589	23.750	23.909	24.069
	$11\bar{2}4$	30.541	30.753	30.964	31.174	31.383	31.592	31.799	32.005	32.211	32.416	32.619	32.822	33.024	33.225	33.425	33.624	33.822
	$11\bar{2}2$	49.720	49.958	50.194	50.428	50.660	50.889	51.116	51.340	51.563	51.783	52.001	52.217	52.431	52.643	52.853	53.061	53.267
	$11\bar{2}0$	90.000	90.000	90.000	90.000	90.000	90.000	90.000	90.000	90.000	90.000	90.000	90.000	90.000	90.000	90.000	90.000	90.000
$10\bar{1}0$	$21\bar{3}0$	19.107	19.107	19.107	19.107	19.107	19.107	19.107	19.107	19.107	19.107	19.107	19.107	19.107	19.107	19.107	19.107	19.107
	$11\bar{2}0$	30.000	30.000	30.000	30.000	30.000	30.000	30.000	30.000	30.000	30.000	30.000	30.000	30.000	30.000	30.000	30.000	30.000
	$01\bar{1}0$	60.000	60.000	60.000	60.000	60.000	60.000	60.000	60.000	60.000	60.000	60.000	60.000	60.000	60.000	60.000	60.000	60.000

		Axial ratio c/a																
$h_1k_1l_1i_1$	$h_2k_2i_2l_2$	1.35	1.36	1.37	1.38	1.39	1.40	1.41	1.42	1.43	1.44	1.45	1.46	1.47	1.48	1.49	1.50	1.51
0001	$10\bar{1}8$	11.026	11.106	11.186	11.265	11.345	11.424	11.504	11.583	11.662	11.742	11.821	11.900	11.979	12.058	12.137	12.216	12.295
	$10\bar{1}7$	12.554	12.644	12.734	12.824	12.914	13.004	13.094	13.183	13.273	13.362	13.452	13.541	13.630	13.720	13.809	13.898	13.987
	$10\bar{1}6$	14.564	14.667	14.770	14.873	14.976	15.079	15.182	15.285	15.387	15.490	15.592	15.694	15.796	15.898	16.000	16.102	16.204
	$10\bar{1}5$	17.316	17.436	17.557	17.677	17.797	17.917	18.037	18.156	18.276	18.395	18.514	18.633	18.751	18.870	18.988	19.107	19.225
	$10\bar{1}4$	21.291	21.435	21.578	21.721	21.864	22.006	22.148	22.290	22.431	22.572	22.713	22.854	22.994	23.134	23.274	23.413	23.552
	$20\bar{2}7$	24.007	24.165	24.322	24.479	24.635	24.791	24.947	25.102	25.257	25.411	25.565	25.719	25.872	26.025	26.177	26.330	26.481
	$10\bar{1}3$	27.457	27.630	27.803	27.976	28.147	28.318	28.489	28.659	28.829	28.998	29.166	29.334	29.501	29.668	29.834	30.000	30.165
	$20\bar{2}5$	31.945	32.135	32.325	32.513	32.701	32.888	33.074	33.260	33.444	33.628	33.811	33.994	34.175	34.356	34.536	34.715	34.893
	$10\bar{1}2$	37.934	38.139	38.343	38.546	38.748	38.948	39.148	39.346	39.543	39.740	39.935	40.129	40.321	40.513	40.704	40.893	41.082
	$20\bar{2}3$	46.102	46.313	46.523	46.731	46.937	47.142	47.346	47.547	47.747	47.946	48.143	48.339	48.533	48.726	48.917	49.107	49.295
	$10\bar{1}1$	57.320	57.512	57.702	57.890	58.075	58.260	58.442	58.622	58.800	58.977	59.152	59.325	59.496	59.666	59.834	60.000	60.165
	$20\bar{2}1$	72.216	72.339	72.460	72.579	72.697	72.813	72.928	73.042	73.153	73.264	73.373	73.481	73.587	73.692	73.796	73.898	73.999
	$10\bar{1}0$	90.000	90.000	90.000	90.000	90.000	90.000	90.000	90.000	90.000	90.000	90.000	90.000	90.000	90.000	90.000	90.000	90.000
	$21\bar{3}2$	64.130	64.296	64.459	64.621	64.781	64.939	65.095	65.249	65.402	65.553	65.702	65.849	65.995	66.139	66.281	66.422	66.561
	$21\bar{3}1$	76.371	76.467	76.563	76.656	76.749	76.840	76.931	77.019	77.107	77.194	77.279	77.364	77.447	77.529	77.610	77.690	77.769
	$21\bar{3}0$	90.000	90.000	90.000	90.000	90.000	90.000	90.000	90.000	90.000	90.000	90.000	90.000	90.000	90.000	90.000	90.000	90.000
	$11\bar{2}8$	18.650	18.778	18.906	19.034	19.162	19.290	19.418	19.545	19.672	19.799	19.926	20.052	20.178	20.304	20.430	20.556	20.682
	$11\bar{2}6$	24.228	24.386	24.545	24.702	24.860	25.017	25.174	25.330	25.486	25.641	25.796	25.951	26.105	26.259	26.412	26.565	26.718
	$11\bar{2}4$	34.019	34.216	34.411	34.606	34.799	34.992	35.184	35.375	35.565	35.754	35.942	36.129	36.316	36.501	36.686	36.870	37.053
	$11\bar{2}2$	53.471	53.673	53.873	54.072	54.268	54.462	54.655	54.846	55.035	55.222	55.408	55.592	55.774	55.954	56.133	56.310	56.485
	$11\bar{2}0$	90.000	90.000	90.000	90.000	90.000	90.000	90.000	90.000	90.000	90.000	90.000	90.000	90.000	90.000	90.000	90.000	90.000
$10\bar{1}0$	$21\bar{3}0$	19.107	19.107	19.107	19.107	19.107	19.107	19.107	19.107	19.107	19.107	19.107	19.107	19.107	19.107	19.107	19.107	19.107
	$11\bar{2}0$	30.000	30.000	30.000	30.000	30.000	30.000	30.000	30.000	30.000	30.000	30.000	30.000	30.000	30.900	30.000	30.000	30.000
	$01\bar{1}0$	60.000	60.000	60.000	60.000	60.000	60.000	60.000	60.000	60.000	60.000	60.000	60.000	60.000	60.000	60.000	60.000	60.000

Table 6 (cont.)

$h_1k_1i_1l_1$	$h_2k_2i_2l_2$	Axial ratio c/a 1·52	1·53	1·54	1·55	1·56	1·57	1·58	1·59	1·60	1·61	1·62	1·63	1·64	1·65	1·66	1·67	1·68
0001	$10\bar{1}8$	12·374	12·453	12·532	12·611	12·689	12·768	12·847	12·925	13·004	13·082	13·161	13·239	13·318	13·396	13·474	13·552	13·630
	$10\bar{1}7$	14·076	14·165	14·254	14·342	14·431	14·520	14·608	14·697	14·785	14·873	14·962	15·050	15·138	15·226	15·314	15·402	15·490
	$10\bar{1}6$	16·305	16·407	16·508	16·610	16·711	16·812	16·913	17·014	17·115	17·215	17·316	17·416	17·517	17·617	17·717	17·817	17·917
	$10\bar{1}5$	19·343	19·460	19·578	19·695	19·812	19·929	20·046	20·163	20·279	20·396	20·512	20·628	20·744	20·859	20·975	21·090	21·205
	$10\bar{1}4$	23·691	23·830	23·968	24·106	24·244	24·381	24·518	24·655	24·791	24·927	25·063	25·199	25·334	25·469	25·604	25·738	25·872
	$20\bar{2}7$	26·632	26·783	26·934	27·084	27·233	27·383	27·531	27·680	27·829	27·976	28·123	28·270	28·416	28·562	28·708	28·853	28·998
	$10\bar{1}3$	30·330	30·494	30·657	30·820	30·982	31·144	31·306	31·466	31·626	31·786	31·945	32·104	32·262	32·419	32·576	32·732	32·888
	$20\bar{2}5$	35·071	35·248	35·424	35·599	35·774	35·948	36·121	36·293	36·465	36·635	36·806	36·975	37·143	37·311	37·478	37·644	37·810
	$10\bar{1}2$	41·269	41·456	41·641	41·825	42·008	42·190	42·372	42·552	42·731	42·909	43·085	43·261	43·436	43·610	43·783	43·955	44·126
	$20\bar{2}3$	49·482	49·667	49·851	50·034	50·215	50·395	50·574	50·751	50·927	51·101	51·275	51·447	51·617	51·787	51·955	52·122	52·287
	$10\bar{1}1$	60·328	60·489	60·649	60·807	60·963	61·119	61·272	61·424	61·575	61·724	61·872	62·018	62·163	62·307	62·449	62·590	62·729
	$20\bar{2}1$	74·099	74·198	74·295	74·392	74·487	74·581	74·674	74·766	74·857	74·946	75·035	75·123	75·210	75·295	75·380	75·464	75·546
	$10\bar{1}0$	90·000	90·000	90·000	90·000	90·000	90·000	90·000	90·000	90·000	90·000	90·000	90·000	90·000	90·000	90·000	90·000	90·000
	$2\bar{1}\bar{3}2$	66·699	66·835	66·970	67·103	67·235	67·365	67·494	67·622	67·748	67·872	67·996	68·118	68·239	68·359	68·477	68·594	68·710
	$2\bar{1}\bar{3}1$	77·847	77·924	78·000	78·076	78·150	78·223	78·296	78·367	78·438	78·508	78·577	78·645	78·713	78·779	78·845	78·910	78·975
	$2\bar{1}\bar{3}0$	90·000	90·000	90·000	90·000	90·000	90·000	90·000	90·000	90·000	90·000	90·000	90·000	90·000	90·000	90·000	90·000	90·000
	$11\bar{2}8$	20·807	20·932	21·057	21·181	21·306	21·430	21·554	21·678	21·801	21·925	22·048	22·171	22·294	22·416	22·538	22·661	22·782
	$11\bar{2}6$	26·870	27·022	27·173	27·324	27·474	27·625	27·774	27·924	28·072	28·221	28·369	28·517	28·664	28·811	28·957	29·103	29·249
	$11\bar{2}4$	37·235	37·416	37·596	37·776	37·954	38·132	38·309	38·485	38·660	38·834	39·007	39·180	39·352	39·523	39·693	39·862	40·030
	$11\bar{2}2$	56·659	56·832	57·002	57·172	57·339	57·505	57·670	57·833	57·995	58·155	58·314	58·471	58·627	58·782	58·935	59·087	59·237
	$11\bar{2}0$	90·000	90·000	90·000	90·000	90·000	90·000	90·000	90·000	90·000	90·000	90·000	90·000	90·000	90·000	90·000	90·000	90·000
$10\bar{1}0$	$2\bar{1}\bar{3}0$	19·107	19·107	19·107	19·107	19·107	19·107	19·107	19·107	19·107	19·107	19·107	19·107	19·107	19·107	19·107	19·107	19·107
	$11\bar{2}0$	30·000	30·000	30·000	30·000	30·000	30·000	30·000	30·000	30·000	30·000	30·000	30·000	30·000	30·000	30·000	30·000	30·000
	$01\bar{1}0$	60·000	60·000	60·000	60·000	60·000	60·000	60·000	60·000	60·000	60·000	60·000	60·000	60·000	60·000	60·000	60·000	60·000

$h_1k_1i_1l_1$	$h_2k_2i_2l_2$	Axial ratio c/a 1·69	1·70	1·71	1·72	1·73	1·74	1·75	1·76	1·77	1·78	1·79	1·80	1·81	1·82	1·83	1·84	1·85
0001	$10\bar{1}8$	13·709	13·787	13·865	13·942	14·020	14·098	14·176	14·254	14·331	14·409	14·486	14·564	14·641	14·719	14·796	14·873	14·951
	$10\bar{1}7$	15·577	15·665	15·753	15·840	15·927	16·015	16·102	16·189	16·276	16·364	16·450	16·537	16·624	16·711	16·798	16·884	16·971
	$10\bar{1}6$	18·017	18·116	18·216	18·315	18·415	18·514	18·613	18·712	18·811	18·909	19·008	19·107	19·205	19·303	19·401	19·499	19·597
	$10\bar{1}5$	21·320	21·435	21·549	21·664	21·778	21·892	22·006	22·120	22·233	22·346	22·459	22·572	22·685	22·798	22·910	23·022	23·134
	$10\bar{1}4$	26·006	26·139	26·273	26·405	26·538	26·670	26·802	26·934	27·065	27·196	27·327	27·457	27·587	27·717	27·846	27·976	28·104
	$20\bar{2}7$	29·142	29·286	29·430	29·573	29·716	29·858	30·000	30·142	30·283	30·423	30·564	30·704	30·843	30·982	31·121	31·259	31·397
	$10\bar{1}3$	33·043	33·198	33·352	33·506	33·659	33·811	33·963	34·115	34·266	34·416	34·566	34·715	34·864	35·012	35·160	35·307	35·453
	$20\bar{2}5$	37·975	38·139	38·302	38·465	38·627	38·788	38·948	39·108	39·267	39·425	39·583	39·740	39·896	40·051	40·206	40·360	40·513
	$10\bar{1}2$	44·296	44·465	44·633	44·800	44·966	45·131	45·295	45·459	45·621	45·782	45·943	46·102	46·261	46·418	46·575	46·731	46·886
	$20\bar{2}3$	52·452	52·615	52·777	52·938	53·098	53·256	53·413	53·569	53·724	53·878	54·031	54·182	54·333	54·482	54·631	54·778	54·924
	$10\bar{1}1$	62·868	63·004	63·140	63·275	63·408	63·540	63·670	63·800	63·928	64·056	64·182	64·307	64·430	64·553	64·675	64·795	64·915
	$20\bar{2}1$	75·629	75·710	75·790	75·869	75·948	76·025	76·102	76·178	76·253	76·327	76·401	76·474	76·546	76·617	76·688	76·757	76·826
	$10\bar{1}0$	90·000	90·000	90·000	90·000	90·000	90·000	90·000	90·000	90·000	90·000	90·000	90·000	90·000	90·000	90·000	90·000	90·000

	$21\bar{3}2$	68·825	68·939	69·051	69·163	69·273	69·382	69·490	69·597	69·703	69·807	69·911	70·014	70·116	70·216	70·316	70·415	70·513
	$21\bar{3}1$	79·038	79·101	79·164	79·225	79·286	79·346	79·406	79·464	79·523	79·580	79·637	79·693	79·749	79·804	79·859	79·913	79·966
	$21\bar{3}0$	90·000	90·000	90·000	90·000	90·000	90·000	90·000	90·000	90·000	90·000	90·000	90·000	90·000	90·000	90·000	90·000	90·000
	$11\bar{2}8$	22·904	23·026	23·147	23·268	23·388	23·509	23·629	23·750	23·869	23·989	24·109	24·228	24·347	24·466	24·584	24·702	24·821
	$11\bar{2}6$	29·394	29·539	29·683	29·827	29·971	30·114	30·256	30·399	30·541	30·682	30·823	30·964	31·104	31·244	31·383	31·522	31·661
	$11\bar{2}4$	40·198	40·365	40·530	40·696	40·859	41·023	41·186	41·348	41·509	41·669	41·829	41·987	42·145	42·302	42·459	42·614	42·769
	$11\bar{2}2$	59·387	59·534	59·681	59·826	59·971	60·113	60·255	60·396	60·535	60·673	60·810	60·945	61·080	61·213	61·346	61·477	61·607
	$11\bar{2}0$	90·000	90·000	90·000	90·000	90·000	90·000	90·000	90·000	90·000	90·000	90·000	90·000	90·000	90·000	90·000	90·000	90·000
$10\bar{1}0$	$21\bar{3}0$	19·107	19·107	19·107	19·107	19·107	19·107	19·107	19·107	19·107	19·107	19·107	19·107	19·107	19·107	19·107	19·107	19·107
	$11\bar{2}0$	30·000	30·000	30·000	30·000	30·000	30·000	30·000	30·000	30·000	30·000	30·000	30·000	30·900	30·000	30·000	30·000	30·000
	$01\bar{1}0$	60·000	60·000	60·000	60·000	60·000	60·000	60·000	60·000	60·000	60·000	60·000	60·000	60·000	60·000	60·000	60·000	60·000

		Axial ratio c/a														
$h_1k_1i_1l_1$	$h_2k_2i_2l_2$	1·86	1·87	1·88	1·89	1·90	1·91	1·92	1·93	1·94	1·95	1·96	1·97	1·98	1·99	2·00
0001	$10\bar{1}8$	15·028	15·105	15·182	15·259	15·336	15·413	15·490	15·566	15·643	15·720	15·796	15·873	15·949	16·026	16·102
	$10\bar{1}7$	17·057	17·143	17·230	17·316	17·402	17·488	17·574	17·660	17·746	17·831	17·917	18·002	18·088	18·173	18·258
	$10\bar{1}6$	19·695	19·793	19·890	19·988	20·085	20·182	20·279	20·376	20·473	20·570	20·667	20·763	20·859	20·956	21·052
	$10\bar{1}5$	23·246	23·358	23·469	23·580	23·691	23·802	23·913	24·023	24·134	24·244	24·354	24·463	24·573	24·682	24·791
	$10\bar{1}4$	28·233	28·361	28·489	28·617	28·744	28·871	28·998	29·124	29·250	29·376	29·501	29·626	29·751	29·876	30·000
	$20\bar{2}7$	31·535	31·672	31·809	31·945	32·081	32·217	32·352	32·486	32·621	32·755	32·888	33·021	33·154	33·286	33·418
	$10\bar{1}3$	35·599	35·745	35·890	36·034	36·178	36·322	36·465	36·607	36·749	36·890	37·031	37·171	37·311	37·450	37·589
	$20\bar{2}5$	40·666	40·818	40·969	41·119	41·269	41·418	41·567	41·715	41·862	42·008	42·154	42·299	42·444	42·587	42·731
	$10\bar{1}2$	47·040	47·193	47·345	47·497	47·648	47·797	47·946	48·094	48·241	48·388	48·533	48·678	48·821	48·964	49·107
	$20\bar{2}3$	55·069	55·213	55·356	55·498	55·639	55·779	55·918	56·056	56·193	56·330	56·465	56·599	56·732	56·864	56·996
	$10\bar{1}1$	65·033	65·150	65·267	65·382	65·496	65·610	65·722	65·833	65·944	66·053	66·162	66·269	66·376	66·482	66·587
	$20\bar{2}1$	76·895	76·963	77·030	77·096	77·161	77·226	77·291	77·355	77·418	77·480	77·542	77·603	77·664	77·724	77·784
	$10\bar{1}0$	90·000	90·000	90·000	90·000	90·000	90·000	90·000	90·000	90·000	90·000	90·000	90·000	90·000	90·000	90·000
	$21\bar{3}2$	70·610	70·706	70·801	70·895	70·988	71·081	71·172	71·263	71·353	71·442	71·530	71·618	71·704	71·790	71·875
	$21\bar{3}1$	80·019	80·071	80·123	80·174	80·225	80·275	80·325	80·374	80·423	80·471	80·519	80·566	80·613	80·659	80·705
	$21\bar{3}0$	90·000	90·000	90·000	90·000	90·000	90·000	90·000	90·000	90·000	90·000	90·000	90·000	90·000	90·000	90·000
	$11\bar{2}8$	24·938	25·056	25·174	25·291	25·408	25·524	25·641	25·757	25·873	25·989	26·105	26·220	26·335	26·450	26·565
	$11\bar{2}6$	31·799	31·937	32·074	32·211	32·347	32·484	32·619	32·755	32·889	33·024	33·158	33·292	33·425	33·558	33·690
	$11\bar{2}4$	42·923	43·076	43·229	43·380	43·531	43·681	43·831	43·980	44·128	44·275	44·421	44·567	44·712	44·856	45·000
	$11\bar{2}2$	61·736	61·864	61·991	62·117	62·241	62·365	62·488	62·610	62·731	62·850	62·969	63·087	63·204	63·320	63·435
	$11\bar{2}0$	90·000	90·000	90·000	90·000	90·000	90·000	90·000	90·000	90·000	90·000	90·000	90·000	90·000	90·000	90·000
$10\bar{1}0$	$21\bar{3}0$	19·107	19·107	19·107	19·107	19·107	19·107	19·107	19·107	19·107	19·107	19·107	19·107	19·107	19·107	19·107
	$11\bar{2}0$	30·000	30·000	30·000	30·000	30·000	30·000	30·000	30·000	30·000	30·000	30·000	30·000	30·000	30·000	30·000
	$01\bar{1}0$	60·000	60·000	60·000	60·000	60·000	60·000	60·000	60·000	60·000	60·000	60·000	60·000	60·000	60·000	60·000

Table 7. Angles between planes for selected hexagonal elements (in degrees)†

		Axial ratio c/a for									
		Dy	Hf	Be	Gd	Ti	Y	Zr	Mg	Zn	Cd
$h_1k_1i_1l_1$	$h_2k_2i_2l_2$	1·5790	1·5822	1·5847	1·5870	1·5873	1·5880	1·5893	1·6235	1·8563	1·8859
0001	$10\bar{1}8$	12·839	12·864	12·884	12·902	12·90	12·910	12·92	13·19	15·00	15·23
	$10\bar{1}7$	14·599	14·628	14·650	14·670	14·67	14·679	14·69	14·99	17·03	17·28
	$10\bar{1}6$	16·903	16·935	16·960	16·984	16·99	16·994	17·01	17·35	19·66	19·95
	$10\bar{1}5$	20·035	20·072	20·101	20·128	20·13	20·140	20·15	20·55	23·21	23·53
	$10\bar{1}4$	24·504	24·548	24·582	24·614	24·62	24·627	24·65	25·11	28·19	28·56
	$20\bar{2}7$	27·517	27·564	27·601	27·635	27·64	27·650	27·67	28·17	31·48	31·89
	$10\bar{1}3$	31·289	31·341	31·381	31·418	31·42	31·434	31·45	32·00	35·55	35·98
	$20\bar{2}5$	36·104	36·159	36·202	36·242	36·25	36·259	36·29	36·87	40·61	41·06
	$10\bar{1}2$	42·353	42·411	42·456	42·498	42·50	42·516	42·54	43·15	46·98	47·43
	$20\bar{2}3$	50·556	50·613	50·657	50·698	50·70	50·716	50·74	51·31	55·02	55·44
	$10\bar{1}1$	61·257	61·306	61·344	61·379	61·38	61·394	61·41	61·92	64·99	65·33
	$20\bar{2}1$	74·665	74·694	74·717	74·738	74·74	74·748	74·76	75·07	76·87	77·07
	$10\bar{1}0$	90·000	90·000	90·000	90·000	90·00	90·000	90·00	90·00	90·00	90·00
	$21\bar{3}2$	67·481	67·522	67·554	67·583	67·59	67·596	67·61	68·04	70·57	70·86
	$21\bar{3}1$	78·288	78·311	78·329	78·346	78·35	78·353	78·36	78·60	80·00	80·15
	$21\bar{3}0$	90·000	90·000	90·000	90·000	90·00	90·000	90·00	90·00	90·00	90·00
	$11\bar{2}8$	21·542	21·581	21·612	21·641	21·64	21·653	21·71	22·09	24·89	25·24
	$11\bar{2}6$	27·759	27·807	27·845	27·879	27·88	27·894	27·91	28·42	31·75	32·16
	$11\bar{2}4$	38·291	38·348	38·392	38·432	38·44	38·450	38·47	39·07	42·87	43·32
	$11\bar{2}2$	57·653	57·706	57·747	57·784	57·79	57·800	57·82	58·37	61·69	62·07
	$11\bar{2}0$	90·000	90·000	90·000	90·000	90·00	90·000	90·00	90·00	90·00	90·00
$10\bar{1}0$	$21\bar{3}0$	19·107	19·107	19·107	19·107	19.11	19·107	19·11	19·11	19·11	19·11
	$11\bar{2}0$	30·000	30·000	30·000	30·000	30·00	30·000	30·00	30·00	30·00	30·00
	$01\bar{1}0$	60·000	60·000	60·000	60·000	60·00	60·000	60·00	60·00	60·00	60·00

† From A. Taylor and S. Lieber, *Trans. A.I.M.E., J. Metals, N.Y.*, 1954, **6**(2), 190, and R. E. Frounfelker. M. A. Seitz and W. M. Hirthe, *Nuclear Science and Engineering*, 1962, **14**(2) 192.

9.2 NOTES TO AID THE INDEXING OF HEXAGONAL CRYSTAL PATTERNS

9.2.1 *Three-index system*

For the interpretation of hexagonal patterns it is advisable to follow the convention adopted by the compilers of the *International Tables for X-ray Crystallography*, Vol. II.[5]

Planes are denoted by $(hk.l)$ where the dot signifies that the third index $i = -(h+k)$ has been omitted.

Zone axes are strictly represented by $[uv0w]$. If in place of the zero there is a dot, e.g. $[uv.w]$, this indicates that there is no third index $-(u+v)$.

The zone equation (12)

$$hu+kv+lw = 0\ (1, 2, 3, \ldots)$$

then still applies as for other systems. Similarly, one can use the cross-multiplication rule to find the common zone to a pair of planes or vice versa.

The reasons for the above difference between zones and planes in regard to the three co-planar axes are:

(*a*) In the case of planes, the two intersections of the plane $hx+ky+lz = 1$ with the x or y axis also define completely its intersection with the basal plane. There is no need to specify its intersection along the third co-planar axis at 120°. If the third axis replaces x or y, this gives a corresponding equation referring to the new axes, with h or k replaced by $-(h+k)$. If this substitution is made in the expressions for d-spacings, they are unchanged since they contain (h^2+hk+k^2) which is unaltered by replacing h or k by $-(h+k)$.

(*b*) In the case of zones, however, a lattice point is reached by a vector from the origin whose components are $u.\mathbf{a}_1$, $v.\mathbf{a}_2$, and $w.\mathbf{c}$. The same point cannot be reached if a third component parallel to $\mathbf{a}_3$ is then added. Alternatively, the vector distance contains (u^2-uv+v^2) which is changed if $-(u+v)$ is substituted for u or v.

With the three-index convention for zone axes, all points on the lattice in real space are defined by vectors $[uv.w]$ or $[uv0w]$ and all the points in reciprocal space by reciprocal vectors $(hk.l)$ *without the third index*, i. Also, in reciprocal space [with the axes necessarily defined (at 60°) as in Fig. 33], the reciprocal lattice planes $[uv.w]$ give intersections a_1^*/u and a_2^*/v ($|a_1^*| = |a_2^*|$) along the x^* and y^* axes. (One should not attempt to use the simple third co-planar index unless there are three equal axes at 120°.)

A direction $[uv.w]$ perpendicular to a plane $(hk.l)$ can be found from

$$\frac{u}{(2h+k)} = \frac{v}{(h+2k)} = \frac{2w}{3l}\left(\frac{c}{a}\right)^2 \qquad (54a)$$

9.2.2 *Four-index system*

In spite of the obvious advantages of using only three indices, there is one disadvantage in regard to zones—directions which are *geometrically equivalent* (but not always structurally equivalent) may be represented by *different* sets of indices. The four-index system for planes $(hkil)$ is useful in this respect and presents no problems *as long as it is not used to define reciprocal vectors*. A four-index system can, however, be used for zones in crystal space, but it must clearly involve different components parallel to the crystal axes. Such a four-index system was used by Weber who employed the symbols $[UVJW]$ where the third, J, was not redundant and had always to be included. The transformation formulae do, however, give fractional values if one takes $W = w$. Hence another system without fractions is denoted by $[pqrs]$, i.e. $p = 3U$ etc. [A. Taylor[3] (pp. 112–4) used $[pqr]$ as a three-index system and $[u, v, \overline{u+v}, r]$ for the four-index system so that his symbols are reversed. We prefer to use the international convention here.] The vector from the origin to a lattice point has length I_{uvw} and this can be represented by $\mathbf{I}$. Therefore

$$3\mathbf{I} = 3(u\mathbf{a}_1+v\mathbf{a}_2+w\mathbf{c})$$

and

$$3\mathbf{I} = p\mathbf{a}_1+q\mathbf{a}_2+r\mathbf{a}_3+s\mathbf{c}$$

The threes are included to avoid fractions in the second set of indices. As the three unit vectors in the plane add up to zero, i.e.

$$\mathbf{a}_1+\mathbf{a}_2+\mathbf{a}_3 = 0$$

one may substitute for $\mathbf{a}_3$

$$3\mathbf{I} = (p-r)\mathbf{a}_1+(q-r)\mathbf{a}_2+s\mathbf{c}$$

and on equating coefficients

$$3u = p-r, \quad 3v = q-r, \quad 3w = s$$

Alternative choices of pairs of axes will therefore give alternative indices for $3u$, $3v$, viz.

$$(r-q) \text{ and } (p-q) \text{ or } (q-p) \text{ and } (r-p)$$

In each case, $9(u^2-uv+v^2)$ is identically equal to the symmetrical expression $(p+q+r)^2-3(pq+qr+rp)$. As a point in a plane only requires two co-ordinates, one of the three, p, q, or r, is not independent. The simplest symmetrical (and linear) relationship between them is $p+q+r = 0$, corresponding to $h+k+i = 0$. This substitution gives

$$9(u^2-uv+v^2) = 3(p^2+pq+q^2)$$

which should then be, and is in fact, invariant if p or q is replaced by $r = -(p+q)$. The indices are therefore given by

$$u = \frac{2p+q}{3}, \quad v = \frac{2q+p}{3}, \quad w = \frac{s}{3} \tag{55a}$$

and conversely

$$p = 2u-v, \quad q = 2v-u, \quad r = \overline{p+q} = \overline{u+v}, \quad s = 3w \tag{55b}$$

This set of equations enables the change from the three- to the four-index system to be made. Equation (12) can thus be replaced by

$$hp+kq+ir+ls = 0$$

which on substituting for i and r gives [consistently with (12)]

$$h(2p+q)+k(2q+p)+ls = 0$$

or

$$p(2h+k)+q(2k+h)+ls = 0$$

The zone multiplication law gives the direction relationships for finding p, q, r, s, from two planes as follows:

$$\left.\begin{aligned} p &= l_2(2k_1+h_1)-l_1(2k_2+h_2) \\ q &= l_1(2h_2+k_2)-l_2(2h_1+k_1) \\ r &= -(p+q) = l_2(h_1-k_1)-l_1(h_2-k_2) \\ s &= 3(h_1k_2-k_1h_2) \end{aligned}\right\} \tag{56}$$

Equation (52a) becomes

$$\frac{p}{h} = \frac{q}{k} = \frac{r}{i} = \frac{2s}{3l}\left(\frac{c}{a}\right)^2 \tag{54b}$$

The following table records indices for the first few integers and should be sufficient for most practical purposes. Apart from the axes themselves, it will be seen that there are twice as many directions of one type in the positive (120°) sector (and also in the corresponding opposite sector with x and y negative). The alternative sectors enclosed by $\pm x$ and $\mp y$ axes only cover one third of the area in the basal plane (60° sectors).

Three- and four-index systems for directions

Two positive indices		One negative index				
$[uv.w]$	Number†	$[uv.w]$	Number†	$[pqrs]$	u^2-uv+v^2	p^2+pq+q^2
$[10.w]$	2	($[10.w]$	2)	$[2\bar{1}\bar{1}, 3w]$	1	3
$[11.w]$	1			$[11\bar{2}, 3w]$		
		$[1\bar{1}.w]$	1	$[3\bar{3}0, 3w] =$	3	9
				$3[1\bar{1}0w]$		
$[12.w]$	2			$[03\bar{3}, 3w]$		
		$[1\bar{2}.w]$	2	$[4\bar{5}1, 3w]$		
$[13.w]$	2			$[\bar{1}5\bar{4}, 3w]$	7	21
$[23.w]$	2			$[14\bar{5}, 3w]$		
		$[1\bar{3}.w]$	2	$[5\bar{7}2, 3w]$		
$[14.w]$	2			$[\bar{2}7\bar{5}, 3w]$	13	39
$[34.w]$	2			$[25\bar{7}, 3w]$		
		$[2\bar{3}.w]$	2	$[7\bar{8}1, 3w]$		
$[25.w]$	2			$[\bar{1}8\bar{7}, 3w]$	19	57
$[35.w]$	2			$[17\bar{8}, 3w]$		
		$[1\bar{4}.w]$	2	$[6\bar{9}3, 3w] =$		
				$3[2\bar{3}1w]$		
$[15.w]$	2			$[\bar{3}9\bar{6}, 3w]$	21	63
$[45.w]$	2			$[36\bar{9}, 3w]$		

† (a) For through reversals of sign, the number is doubled but these are equal and opposite directions.
(b) For $w \neq 0$, the numbers are again doubled corresponding to $+w$ and $-w$.

If frequent work is anticipated with the hexagonal system, it is desirable to plot these directions (omitting w) on a trilinear co-ordinate chart. It is also useful to prepare the corresponding reciprocal net as in Fig. 33.

9.2.3 *Practical points relating to reciprocal patterns*

Once the d-spacings have been calculated by using equation (14), and if the crystal structure and identity of the diffracting medium are known, it will be possible to index the reflections. It should be remembered that, although the three-index system may be in use, planes which are denoted by different permutations of h, k, and i with a given l are identical in all respects including their d-spacing. Thus

$$(31.2) = (3\bar{4}.2) = (\bar{4}1.2) \neq (43.2)$$

The formula for the angles between two lattice planes $(h_1k_1.l_1)$ and $(h_2k_2.l_2)$ contains the terms

$$h_1h_2 + k_1k_2 + \tfrac{1}{2}(h_1k_2 + k_1h_2)$$

Thus, given $(h_1k_1.l_1)$ and $(h_2k_2.l_2)$, and since it is possible to permute h, k, and i, more than one angle can be found between any such pair of indices. This is the effect of multiplicity so common with cubic compounds. For this reason the standard stereograms of this crystal system (Figs. 25 to 28) have been constructed showing the four-index system for the planes, although only three indices are used in the reciprocal lattice.

10 Cubic system

10.1 ANGLES AND OTHER DATA

In §3.4 it was pointed out that the interplanar spacings of cubic crystals depend on $\sqrt{N}$, where $N = h^2+k^2+l^2$. All possible values of N up to 64 are given in Table 8. Diffraction with any of these indices can occur for a primitive lattice. The occurrences for the body-centred cubic (b.c.c.) and face-centred cubic (f.c.c.) lattices are denoted by ×. The values of $\sqrt{N}$ given here should be sufficiently accurate for most calculations of interplanar spacings. Solution of diffraction patterns necessitates the determination and comparison of angles as well as spacings. Table 9 lists the angles between planes in cubic crystals calculated from the formula

$$\cos\phi = \frac{h_1h_2+k_1k_2+l_1l_2}{\sqrt{(h_1^2+k_1^2+l_1^2)}\,\sqrt{(h_2^2+k_2^2+l_2^2)}}$$

where $(h_1k_1l_1)$ and $(h_2k_2l_2)$ are the indices of the two planes and ϕ is the angle between them. The formula applies equally well to the angle ρ between two zone axes with the same indices, since the direction of these axes will be normal to the two planes.

The table gives the angle between any pair of planes providing the indices involved are not greater in value than 5 except where the three indices contain a common integer factor. In such a case the indices are divided by the common factor and the resulting set of indices used. This table should therefore suffice for the angles found necessary to construct a stereographic projection on most (hkl) planes of the cubic crystal system. Examples of such projections are Figs. 29, 30 and 31.

In order to limit the size of Table 9 only general indices are given. However, for a given pair of $\{hkl\}$ planes the denominator of the above expression for $\cos\phi$ is independent of the permutation of hkl and of sign. Thus different angles for a given pair of $\{hkl\}$ values are characterized only by differences in the numerator of the above expression. The magnitude of the numerator is therefore given in brackets for each value in the table. This aids the solution of electron diffraction patterns. The measured d-spacings indicate the general values of $\{h_1k_1l_1\}$, $\{h_2k_2l_2\}$, but the measured angle between the vectors limits the choice to a certain value of Σh_1h_2 in the numerator. The specific permutations of hkl and sign in each case must therefore fit the numerator value for the angle found. This may help to decide which particular planes in each set can be present or to confirm indexing already suggested by the general arrangement of the diffraction spots.

This table is based on one provided by R. J. Peavler and J. L. Lenusky and published by the American Institute of Mining, Metallurgical and Petroleum Engineers as I.M.D. Special Report No. 8. The angles have now been completely recalculated on an Elliot 803 digital computer and the numerator values added. Any discrepancies in the angles have been individually rechecked. The present table is therefore believed to be complete and accurate.

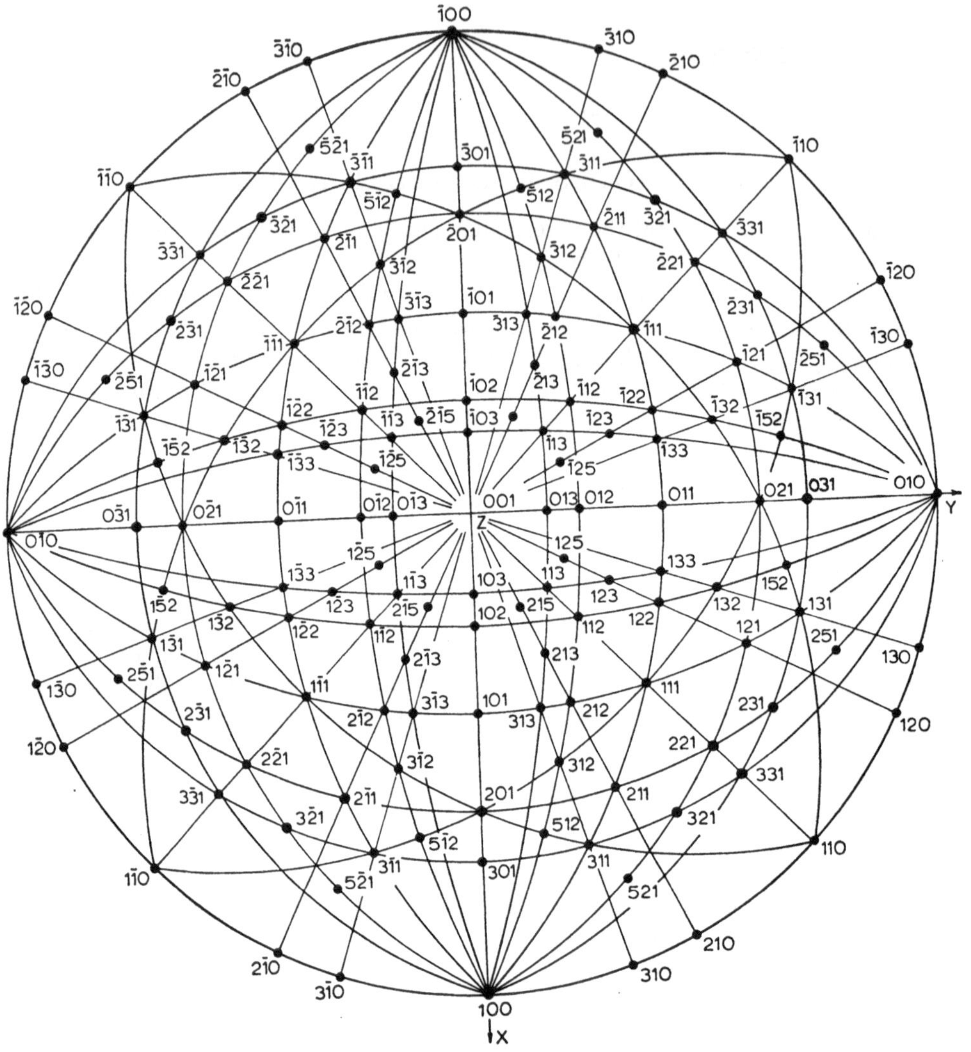

Fig. 29. Stereographic projection for cubic crystals on (001).

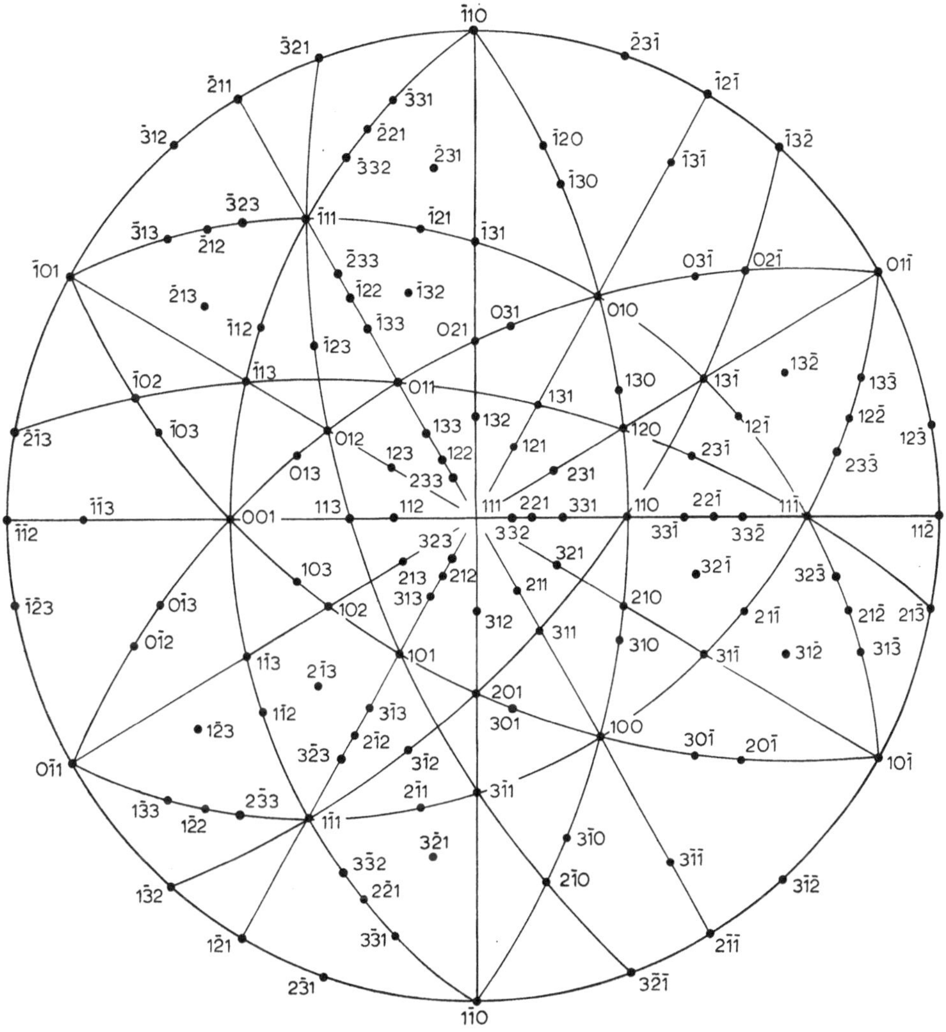

FIG. 30. Stereographic projection for cubic crystals on (111).

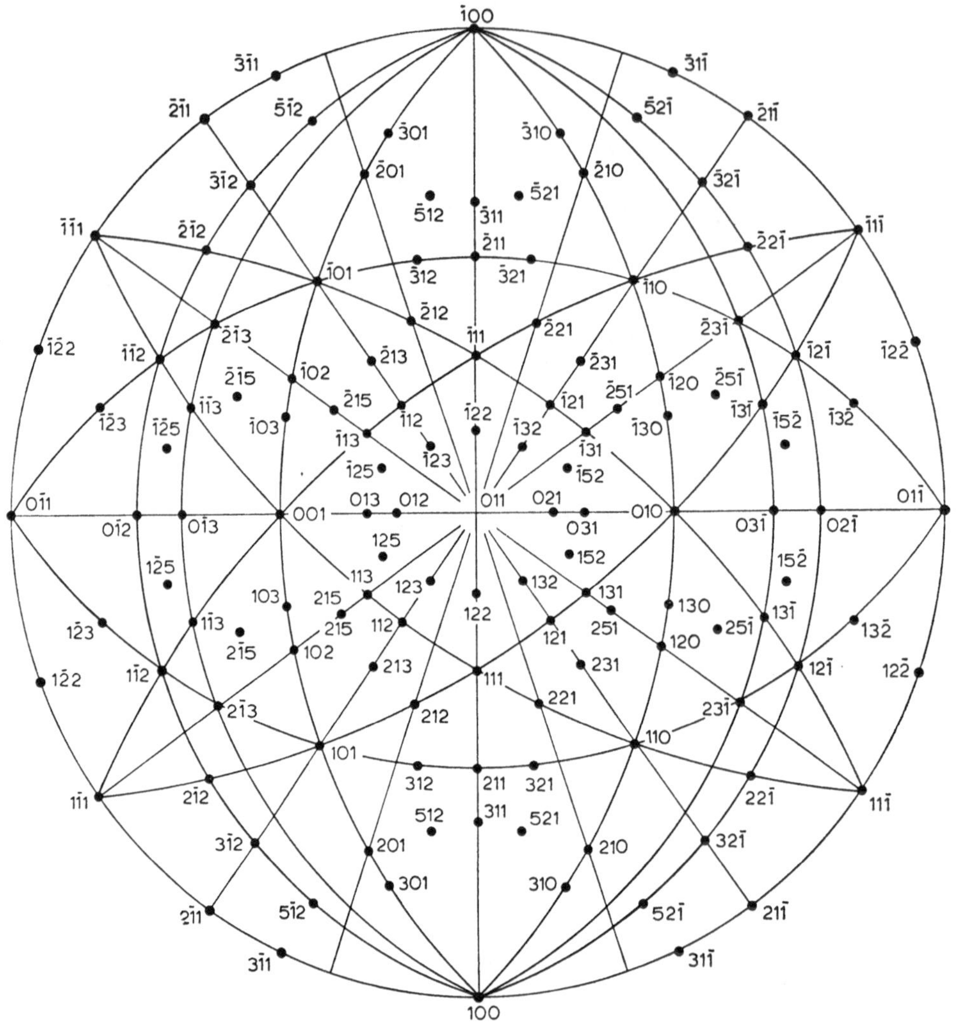

FIG. 31. Stereographic projection for cubic crystals on (011).

TABLE 8. Occurrence of indices for cubic crystals

Line no. $N = (h^2+k^2+l^2)$	*hkl* indices	$\sqrt{N} = \sqrt{(h^2+k^2+l^2)}$	b.c.c.	f.c.c.	Line no. $N = (h^2+k^2+l^2)$	*hkl* indices	$\sqrt{N} = \sqrt{(h^2+k^2+l^2)}$	b.c.c.	f.c.c.
1	100	1·00			33	522, 441	5·745		
2	110	1·414	×		34	530, 433	5·831	×	
3	111	1·732		×	35	531	5·916		×
4	200	2·00	×	×	36	600, 442	6·00	×	×
5	210	2·236			37	610	6·083		
6	211	2·450	×		38	611, 532	6·164	×	
7	—	—			39	—	—		
8	220	2·828	×	×	40	620	6·325	×	×
9	300, 221	3·00			41	621, 540, 443	6·403		
10	310	3·162	×		42	541	6·481	×	
11	311	3·317		×	43	533	6·557		×
12	222	3·464	×	×	44	622	6·633	×	×
13	320	3·606			45	630, 542	6·708		
14	321	3·742	×		46	631	6·782	×	
15	—	—			47	—	—		
16	400	4·00	×	×	48	444	6·928	×	×
17	410, 322	4·123			49	700, 632	7·00		
18	411, 330	4·243	×		50	710, 550, 543	7·071	×	
19	331	4·359		×	51	711, 551	7·141		×
20	420	4·472	×	×	52	640	7·211	×	×
21	421	4·583			53	720, 641	7·280		
22	332	4·690	×		54	721, 633, 552	7·349	×	
23	—	—			55	—	—		
24	422	4·899	×	×	56	642	7·483	×	×
25	500, 430	5·00			57	722, 544	7·550		
26	510, 431	5·099	×		58	730	7·616	×	
27	511, 333	5·196		×	59	731, 553	7·681		×
28	—	—			60	—	—		
29	520, 432	5·385			61	650, 643	7·810		
30	521	5·477	×		62	732, 651	7·874	×	
31	—	—			63	—	—		
32	440	5·657	×	×	64	800	8·00	×	×

Table 9. Angles between planes or zones in cubic crystals (in degrees)

h_1	k_1	l_1	h_2	k_2	l_2					
1	0	0	1	0	0	90·00 (0)				
			1	1	0	45·00 (1)	90·00 (0)			
			1	1	1	54·74 (1)				
			2	1	0	26·57 (2)	63·43 (1)	90·00 (0)		
			2	1	1	35·26 (2)	65·91 (1)			
			2	2	1	48·19 (2)	70·53 (1)			
			3	1	0	18·43 (3)	71·57 (1)	90·00 (0)		
			3	1	1	25·24 (3)	72·45 (1)			
			3	2	0	33·69 (3)	56·31 (2)	90·00 (0)		
			3	2	1	36·70 (3)	57·69 (2)	74·50 (1)		
			3	2	2	43·31 (3)	60·98 (2)			
			3	3	1	46·51 (3)	76·74 (1)			
			3	3	2	50·24 (3)	64·76 (2)			
			4	1	0	14·04 (4)	75·96 (1)	90·00 (0)		
			4	1	1	19·47 (4)	76·37 (1)			
			4	2	1	29·21 (4)	64·12 (2)	77·40 (1)		
			4	3	0	36·87 (4)	53·13 (3)	90·00 (0)		
			4	3	1	38·33 (4)	53·96 (3)	78·69 (1)		
			4	3	2	42·03 (4)	56·15 (3)	68·20 (2)		
			4	3	3	46·69 (4)	59·04 (3)			
			4	4	1	45·87 (4)	79·98 (1)			
			4	4	3	51·34 (4)	62·06 (3)			
			5	1	0	11·31 (5)	78·69 (1)	90·00 (0)		
			5	1	1	15·79 (5)	78·90 (1)			
			5	2	0	21·80 (5)	68·20 (2)	90·00 (0)		
			5	2	1	24·09 (5)	68·58 (2)	79·48 (1)		
			5	2	2	29·50 (5)	69·63 (2)			
			5	3	0	30·96 (5)	59·04 (3)	90·00 (0)		
			5	3	1	32·31 (5)	59·53 (3)	80·27 (1)		
			5	3	2	35·80 (5)	60·88 (3)	71·07 (2)		
			5	3	3	40·32 (5)	62·77 (3)			
			5	4	0	38·66 (5)	51·34 (4)	90·00 (0)		
			5	4	1	39·51 (5)	51·89 (4)	81·12 (1)		
			5	4	2	41·81 (5)	53·40 (4)	72·65 (2)		
			5	4	3	45·00 (5)	55·55 (4)	64·90 (3)		
			5	4	4	48·53 (5)	58·01 (4)			
			5	5	1	45·56 (5)	81·95 (1)			
			5	5	2	47·12 (5)	74·21 (2)			
			5	5	3	49·39 (5)	67·01 (3)			
			5	5	4	52·01 (5)	60·50 (4)			

h_1	k_1	l_1	h_2	k_2	l_2					
1	1	0	1	1	0	60·00 (1)	90·00 (0)			
			1	1	1	35·26 (2)	90·00 (0)			
			2	1	0	18·43 (3)	50·77 (2)	71·57 (1)		
			2	1	1	30·00 (3)	54·74 (2)	73·22 (1)	90·00 (0)	
			2	2	1	19·47 (4)	45·00 (3)	76·37 (1)	90·00 (0)	
			3	1	0	26·57 (4)	47·87 (3)	63·43 (2)	77·08 (1)	
			3	1	1	31·48 (4)	64·76 (2)	90·00 (0)		
			3	2	0	11·31 (5)	53·96 (3)	66·91 (2)	78·69 (1)	
			3	2	1	19·11 (5)	40·89 (4)	55·46 (3)	67·79 (2)	79·11 (1)

3	2	2	30·96 (5)	46·69 (4)	80·13 (1)	90·00 (0)			
3	3	1	13·26 (6)	49·54 (4)	71·07 (2)	90·00 (0)			
3	3	2	25·24 (6)	41·08 (5)	81·33 (1)	90·00 (0)			
4	1	0	30·96 (5)	46·69 (4)	59·04 (3)	80·13 (1)			
4	1	1	33·56 (5)	60·00 (3)	70·53 (2)	90·00 (0)			
4	2	1	22·21 (6)	39·51 (5)	62·42 (3)	72·02 (2)	81·12 (1)		
4	3	0	8·13 (7)	55·55 (4)	64·90 (3)	81·87 (1)			
4	3	1	13·90 (7)	46·10 (5)	56·31 (4)	65·42 (3)	73·90 (2)	82·03 (1)	
4	3	2	23·20 (7)	38·02 (6)	48·96 (5)	74·77 (2)	82·45 (1)		
4	3	3	31·91 (7)	43·31 (6)	83·03 (1)	90·00 (0)			
4	4	1	10·02 (8)	52·01 (5)	68·33 (3)	90·00 (0)			
4	4	3	27·94 (8)	39·37 (7)	83·66 (1)	90·00 (0)			
5	1	0	33·69 (6)	46·10 (5)	56·31 (4)	82·03 (1)			
5	1	1	35·26 (6)	57·02 (4)	74·21 (2)	90·00 (0)			
5	2	0	23·20 (7)	48·96 (5)	66·80 (3)	74·77 (2)			
5	2	1	25·35 (7)	39·23 (6)	58·91 (4)	67·21 (3)	82·58 (1)		
5	2	2	30·50 (7)	60·50 (4)	68·33 (3)	90·00 (0)			
5	3	0	14·04 (8)	52·67 (5)	68·67 (5)	75·96 (2)			
5	3	1	17·02 (8)	44·18 (6)	61·44 (4)	76·17 (2)			
5	3	2	23·41 (8)	36·59 (7)	55·00 (5)	69·87 (3)	76·74 (2)	83·41 (1)	
5	3	3	30·38 (8)	49·68 (6)	77·55 (2)	90·00 (0)			
5	4	0	6·34 (9)	56·48 (5)	63·79 (4)	83·66 (1)			
5	4	1	10·89 (9)	49·11 (6)	56·94 (5)	64·12 (4)	70·89 (3)	83·74 (1)	
5	4	2	18·43 (9)	42·45 (7)	50·77 (6)	71·57 (3)	77·83 (2)	83·95 (1)	
5	4	3	25·84 (9)	36·87 (8)	45·57 (7)	78·46 (2)	84·26 (1)		
5	4	4	32·55 (9)	41·47 (8)	84·63 (1)	90·00 (0)			
5	5	1	8·05 (10)	53·55 (6)	66·67 (4)	90·00 (0)			
5	5	2	15·79 (10)	47·66 (7)	73·22 (3)	90·00 (0)			
5	5	3	22·99 (10)	42·57 (8)	79·39 (2)	90·00 (0)			
5	5	4	29·50 (10)	38·43 (9)	85·01 (1)	90·00 (0)			

h_1	k_1	l_1	h_2	k_2	l_2				
1	1	1	1	1	1	70·53 (1)			
			2	1	0	39·23 (3)	75·04 (1)		
			2	1	1	19·47 (4)	61·87 (2)	90·00 (0)	
			2	2	1	15·79 (5)	54·74 (3)	78·90 (1)	
			3	1	0	43·09 (4)	68·58 (2)		
			3	1	1	29·50 (5)	58·52 (3)	79·98 (1)	
			3	2	0	36·81 (5)	80·79 (1)		
			3	2	1	22·21 (6)	51·89 (4)	72·02 (2)	90·00 (0)
			3	2	2	11·42 (7)	65·16 (3)	81·95 (1)	
			3	3	1	22·00 (7)	48·53 (5)	82·39 (1)	
			3	3	2	10·02 (8)	60·50 (4)	75·75 (2)	
			4	1	0	45·56 (5)	65·16 (3)		
			4	1	1	35·26 (6)	57·02 (4)	74·21 (2)	
			4	2	1	28·13 (7)	50·95 (5)	67·79 (3)	82·76 (1)
			4	3	0	36·07 (7)	83·37 (1)		
			4	3	1	25·07 (8)	47·21 (6)	76·91 (2)	90·00 (0)
			4	3	2	15·23 (9)	57·58 (5)	71·24 (3)	83·85 (1)
			4	3	3	8·05 (10)	66·67 (4)	78·58 (2)	
			4	4	1	25·24 (9)	45·29 (7)	84·23 (1)	
			4	4	3	7·33 (11)	63·20 (5)	74·31 (3)	
			5	1	0	47·21 (6)	63·07 (4)		

Table 9 (cont.)

h_1	k_1	l_1	h_2	k_2	l_2								
1 (cont.)	1	1	5	1	1	38·94 (7)	56·25 (5)	70·53 (3)					
			5	2	0	41·37 (7)	71·24 (3)						
			5	2	1	32·51 (8)	50·77 (6)	65·06 (4)	77·83 (2)				
			5	2	2	25·24 (9)	59·83 (5)	84·23 (1)					
			5	3	0	37·62 (8)	78·58 (2)						
			5	3	1	28·56 (9)	46·91 (7)	72·98 (3)	84·40 (1)				
			5	3	2	20·51 (10)	55·81 (6)	68·00 (4)	90·00 (0)				
			5	3	3	14·42 (11)	63·88 (5)	84·95 (1)					
			5	4	0	35·76 (9)	84·83 (1)						
			5	4	1	27·02 (10)	44·55 (8)	79·74 (2)	90·00 (0)				
			5	4	2	18·79 (11)	52·95 (7)	75·04 (3)	85·06 (1)				
			5	4	3	11·54 (12)	60·67 (6)	70·94 (4)	80·60 (2)				
			5	4	4	6·21 (13)	67·52 (5)	76·74 (3)					
			5	5	1	27·21 (11)	43·31 (9)	85·36 (1)					
			5	5	2	19·47 (12)	51·06 (8)	80·96 (2)					
			5	5	3	12·27 (13)	58·25 (7)	76·97 (3)					
			5	5	4	5·77 (14)	64·76 (6)	73·48 (4)					

h_1	k_1	l_1	h_2	k_2	l_2								
2	1	0	2	1	0	36·87 (4)	53·13 (3)	66·42 (2)	78·46 (1)	90·00 (0)			
			2	1	1	24·09 (5)	43·09 (4)	56·79 (3)	79·48 (1)	90·00 (0)			
			2	2	1	26·57 (6)	41·81 (5)	53·40 (4)	63·43 (3)	72·65 (2)	90·00 (0)		
			3	1	0	8·13 (7)	31·95 (6)	45·00 (5)	64·90 (3)	73·57 (2)	81·87 (1)		
			3	1	1	19·29 (7)	47·61 (5)	66·14 (3)	82·25 (1)				
			3	2	0	7·13 (8)	29·74 (7)	41·91 (6)	60·26 (4)	68·15 (3)	75·64 (2)	82·87 (1)	
			3	2	1	17·02 (8)	33·21 (7)	53·30 (5)	61·44 (4)	68·99 (3)	83·14 (1)	90·00 (0)	
			3	2	2	29·81 (8)	40·60 (7)	49·40 (6)	64·29 (4)	77·47 (2)	83·77 (1)		
			3	3	1	22·57 (9)	44·10 (7)	59·14 (5)	72·07 (3)	84·11 (1)			
			3	3	2	30·89 (9)	40·29 (8)	48·13 (7)	67·58 (4)	73·38 (3)	84·53 (1)		
			4	1	0	12·53 (9)	29·81 (8)	40·60 (7)	49·40 (6)	64·29 (4)	77·47 (2)	83·77 (1)	
			4	1	1	18·43 (9)	42·45 (7)	50·77 (6)	71·57 (3)	77·83 (2)	83·95 (1)		
			4	2	1	12·60 (10)	28·56 (9)	38·67 (8)	46·91 (7)	54·16 (6)	60·79 (5)	67·02 (4)	72·98 (3)
						78·74 (2)	90·00 (0)						
			4	3	0	10·30 (11)	26·57 (10)	44·31 (8)	57·54 (6)	63·43 (5)	69·04 (4)	74·44 (3)	79·70 (2)
			4	3	1	15·26 (11)	28·71 (10)	37·87 (9)	52·13 (7)	58·25 (6)	63·99 (5)	79·90 (2)	84·97 (1)
			4	3	2	24·01 (11)	33·85 (10)	48·37 (8)	54·46 (7)	60·11 (6)	65·47 (5)	70·60 (4)	80·44 (2)
						85·24 (1)	90·00 (0)						
			4	3	3	32·47 (11)	39·92 (10)	46·35 (9)	67·45 (5)	76·70 (3)	81·18 (2)		
			4	4	1	20·90 (12)	45·52 (9)	56·98 (7)	62·15 (6)	71·86 (4)	81·04 (2)		
			4	4	3	33·06 (12)	39·80 (11)	45·70 (10)	69·56 (5)	73·78 (4)	81·97 (2)		
			5	1	0	15·26 (11)	28·71 (10)	37·87 (9)	52·13 (7)	63·99 (5)	74·74 (3)	79·90 (2)	84·97 (1)
			5	1	1	18·79 (11)	39·23 (9)	52·95 (7)	75·04 (3)	85·06 (1)			
			5	2	0	4·76 (12)	33·85 (10)	41·63 (9)	48·37 (8)	65·47 (5)	70·60 (4)	80·44 (2)	85·24 (1)
			5	2	1	11·54 (12)	26·08 (11)	42·71 (9)	49·22 (8)	55·14 (7)	65·91 (5)	70·94 (4)	75·82 (3)
						85·32 (1)	90·00 (0)						
			5	2	2	20·90 (12)	45·52 (9)	51·48 (8)	62·15 (6)	81·04 (2)	85·54 (1)		
			5	3	0	4·40 (13)	32·47 (11)	39·92 (10)	57·53 (7)	62·60 (6)	67·45 (5)	76·70 (3)	85·60 (1)
			5	3	1	10·67 (13)	33·74 (11)	47·13 (9)	58·05 (7)	67·79 (5)	76·89 (3)	85·66 (1)	
			5	3	2	19·42 (13)	29·47 (12)	37·06 (11)	49·24 (9)	54·52 (8)	59·48 (7)	73·13 (4)	85·84 (1)
			5	3	3	27·55 (13)	41·39 (11)	52·14 (9)	61·48 (7)	78·19 (3)	86·09 (1)		
			5	4	0	12·09 (14)	24·78 (13)	45·70 (10)	56·03 (8)	65·22 (6)	69·56 (5)	73·78 (4)	77·91 (3)
			5	4	1	14·96 (14)	26·22 (13)	40·62 (11)	51·61 (9)	61·12 (7)	65·54 (6)	78·05 (3)	82·07 (2)

h_1	k_1	l_1	h_2	k_2	l_2								
			5	4	2	21·04 (14)	29·93 (13)	36·87 (12)	48·19 (10)	53·13 (9)	57·77 (8)	66·42 (6)	78·46 (3)
						86·18 (1)	90·00 (0)						
			5	4	3	27·69 (14)	34·70 (13)	45·92 (11)	50·77 (10)	63·72 (7)	67·70 (6)	71·57 (5)	79·06 (3)
						82·73 (2)	86·37 (1)						
			5	4	4	33·97 (14)	39·64 (13)	44·70 (12)	69·18 (6)	76·29 (4)	79·76 (3)		
			5	5	1	20·06 (15)	46·46 (11)	55·69 (9)	64·00 (7)	71·75 (5)	79·17 (3)		
			5	5	2	24·09 (15)	43·09 (12)	56·79 (9)	60·87 (8)	72·28 (5)	86·51 (1)		
			5	5	3	29·15 (15)	40·81 (13)	50·18 (11)	65·95 (7)	73·08 (5)	86·66 (1)		
			5	5	4	34·34 (15)	39·59 (14)	44·31 (13)	70·71 (6)	74·02 (5)	80·49 (3)		
h_1	k_1	l_1	h_2	k_2	l_2								
2	1	1	2	1	1	33·56 (5)	48·19 (4)	60·00 (3)	70·53 (2)	80·41 (1)			
			2	2	1	17·72 (7)	35·26 (6)	47·12 (5)	65·91 (3)	74·21 (2)	82·18 (1)		
			3	1	0	25·35 (7)	49·80 (5)	58·91 (4)	75·04 (2)	82·58 (1)			
			3	1	1	10·02 (8)	42·39 (6)	60·50 (4)	75·75 (2)	90·00 (0)			
			3	2	0	25·07 (8)	37·57 (7)	55·52 (5)	63·07 (4)	83·50 (1)			
			3	2	1	10·89 (9)	29·21 (8)	40·20 (7)	49·11 (6)	56·94 (5)	70·89 (3)	77·40 (2)	83·74 (1)
						90·00 (0)							
			3	2	2	8·05 (10)	26·98 (9)	53·55 (6)	60·33 (5)	72·72 (3)	78·58 (2)	84·32 (1)	
			3	3	1	20·51 (10)	41·47 (8)	68·00 (4)	79·20 (2)				
			3	3	2	16·78 (11)	29·50 (10)	52·46 (7)	64·20 (5)	69·63 (4)	79·98 (2)	85·01 (1)	
			4	1	0	26·98 (9)	46·12 (7)	53·55 (6)	60·33 (5)	72·72 (3)	78·58 (2)		
			4	1	1	15·79 (10)	39·66 (8)	47·66 (7)	54·74 (6)	61·24 (5)	73·22 (3)	84·48 (1)	
			4	2	1	11·49 (11)	36·70 (9)	44·55 (8)	51·42 (7)	63·55 (5)	69·12 (4)	84·89 (1)	90·00 (0)
			4	3	0	26·08 (11)	35·26 (10)	55·14 (7)	65·91 (5)	80·60 (2)	85·32 (1)		
			4	3	1	16·10 (12)	28·27 (11)	36·81 (10)	43·90 (9)	61·29 (6)	66·40 (5)	71·32 (4)	76·10 (3)
						85·41 (1)							
			4	3	2	9·76 (13)	24·53 (12)	33·50 (11)	46·98 (9)	52·66 (8)	57·95 (7)	67·73 (5)	72·35 (4)
						76·85 (3)	90·00 (0)						
			4	3	3	11·42 (14)	24·47 (13)	55·94 (8)	60·65 (7)	69·51 (5)	81·95 (2)	85·99 (1)	
			4	4	1	22·50 (13)	38·58 (11)	44·71 (10)	64·76 (6)	69·19 (5)	77·69 (3)	81·83 (2)	
			4	4	3	16·99 (15)	26·80 (14)	54·98 (9)	63·49 (7)	67·51 (6)	82·67 (2)	86·34 (1)	
			5	1	0	28·27 (11)	43·90 (9)	55·91 (7)	61·29 (6)	71·32 (4)	76·10 (3)		
			5	1	1	19·47 (12)	38·22 (10)	51·06 (8)	61·87 (6)	71·68 (4)	80·96 (2)		
			5	2	0	24·53 (12)	46·98 (9)	52·66 (8)	57·95 (7)	76·85 (3)	85·65 (1)		
			5	2	1	14·31 (13)	34·93 (11)	41·81 (10)	47·87 (9)	53·40 (8)	58·55 (7)	68·12 (5)	81·43 (2)
						85·73 (1)	90·00 (0)						
			5	2	2	5·77 (14)	38·58 (11)	44·71 (10)	60·17 (7)	64·76 (6)	77·69 (3)	85·92 (1)	
			5	3	0	24·47 (13)	39·63 (11)	55·94 (8)	60·65 (7)	81·95 (2)	85·99 (1)		
			5	3	1	14·96 (14)	34·10 (12)	46·36 (10)	56·49 (8)	65·54 (6)	73·98 (4)	82·07 (2)	90·00 (0)
			5	3	2	6·59 (15)	30·58 (13)	37·37 (12)	43·24 (11)	53·41 (9)	66·59 (6)	70·66 (5)	74·64 (4)
						78·54 (3)	82·39 (2)	86·20 (1)					
			5	3	3	5·05 (16)	29·35 (14)	51·50 (10)	60·13 (8)	75·58 (4)	82·85 (2)		
			5	4	0	26·80 (14)	34·02 (13)	54·98 (9)	67·51 (6)	78·97 (3)	86·34 (1)		
			5	4	1	19·11 (15)	28·13 (14)	35·02 (13)	40·89 (12)	46·14 (11)	63·83 (7)	71·64 (5)	75·41 (4)
						79·11 (3)	82·76 (2)	86·39 (1)					
			5	4	2	13·16 (16)	24·09 (15)	37·71 (13)	43·09 (12)	47·98 (11)	60·87 (8)	72·28 (5)	75·91 (4)
						79·48 (3)	86·51 (1)						
			5	4	3	11·04 (17)	22·52 (16)	30·00 (15)	50·57 (11)	54·74 (10)	58·69 (9)	66·16 (7)	69·73 (6)
						73·22 (5)	80·03 (3)	90·00 (0)					
			5	4	4	13·26 (18)	23·18 (17)	57·27 (10)	60·88 (9)	67·76 (7)	83·79 (2)	86·90 (1)	
			5	5	1	23·84 (16)	36·84 (14)	46·69 (12)	62·79 (8)	69·94 (6)	76·78 (4)	83·43 (2)	
			5	5	2	19·19 (17)	38·94 (14)	43·76 (13)	67·11 (7)	70·53 (6)	77·16 (4)	80·41 (3)	
			5	5	3	16·92 (18)	31·75 (16)	50·37 (12)	64·84 (8)	71·40 (6)	77·73 (4)	83·90 (2)	
			5	5	4	17·30 (19)	25·24 (18)	56·44 (11)	63·11 (9)	66·30 (8)	84·23 (2)	87·12 (1)	

Table 9 (cont.)

h_1	k_1	l_1	h_2	k_2	l_2								
2	2	1	2	2	1	27·27 (8)	38·94 (7)	63·61 (4)	83·62 (1)	90·00 (0)			
			3	1	0	32·51 (8)	42·45 (7)	58·19 (5)	65·06 (4)	83·95 (1)			
			3	1	1	25·24 (9)	45·29 (7)	59·83 (5)	72·45 (3)	84·23 (1)			
			3	2	0	22·41 (10)	42·30 (8)	49·67 (7)	68·30 (4)	79·34 (2)	84·70 (1)		
			3	2	1	11·49 (11)	27·02 (10)	36·70 (9)	57·69 (6)	63·55 (5)	74·50 (3)	79·74 (2)	84·89 (1)
			3	2	2	14·04 (12)	27·21 (11)	49·70 (8)	66·16 (5)	71·13 (4)	75·96 (3)	90·00 (0)	
			3	3	1	6·21 (13)	32·73 (11)	57·64 (7)	67·52 (5)	85·61 (1)			
			3	3	2	5·77 (14)	22·50 (13)	44·71 (10)	60·17 (7)	69·19 (5)	81·83 (2)	85·92 (1)	
			4	1	0	36·06 (10)	43·31 (9)	55·53 (7)	60·98 (6)	80·69 (2)			
			4	1	1	30·20 (11)	45·00 (9)	51·06 (8)	56·63 (7)	66·87 (5)	71·68 (4)	90·00 (0)	
			4	2	1	18·98 (13)	29·21 (12)	36·86 (11)	43·33 (10)	54·41 (8)	64·12 (6)	68·67 (5)	73·08 (4)
						77·40 (3)	81·64 (2)						
			4	3	0	21·04 (14)	42·83 (11)	48·19 (10)	70·53 (5)	82·34 (2)			
			4	3	1	11·31 (15)	31·81 (13)	38·33 (12)	53·96 (9)	58·47 (8)	62·77 (7)	74·84 (4)	78·69 (3)
						86·25 (1)	90·00 (0)						
			4	3	2	7·96 (16)	21·80 (15)	29·94 (14)	42·03 (12)	56·15 (9)	64·32 (7)	68·20 (6)	75·66 (4)
						82·89 (2)	86·45 (1)	90·00 (0)					
			4	3	3	13·63 (17)	23·84 (16)	51·04 (11)	62·79 (8)	73·39 (5)	76·78 (4)	86·72 (1)	
			4	4	1	9·45 (17)	29·50 (15)	35·67 (14)	54·53 (10)	69·63 (6)	83·34 (2)	86·67 (1)	
			4	4	3	8·47 (19)	20·44 (18)	47·41 (13)	58·63 (10)	71·80 (6)	81·02 (3)	84·02 (2)	
			5	1	0	38·33 (12)	44·02 (11)	53·96 (9)	58·47 (8)	62·77 (7)	78·69 (3)		
			5	1	1	33·49 (13)	45·12 (11)	54·74 (9)	63·32 (7)	71·29 (5)	86·32 (1)		
			5	2	0	29·94 (14)	42·03 (12)	56·15 (9)	60·32 (8)	68·20 (6)	86·45 (1)		
			5	2	1	24·09 (15)	31·57 (14)	37·71 (13)	47·98 (11)	52·51 (10)	64·79 (7)	68·58 (6)	72·28 (5)
						79·48 (3)	86·51 (1)						
			5	2	2	21·81 (16)	41·03 (13)	45·87 (12)	62·34 (8)	73·13 (5)	76·58 (4)	79·98 (3)	
			5	3	0	23·84 (16)	42·00 (13)	51·04 (11)	66·41 (7)	76·78 (4)	86·72 (1)		
			5	3	1	16·70 (17)	32·31 (15)	42·91 (13)	51·70 (11)	59·53 (9)	73·64 (5)	80·27 (3)	86·77 (1)
			5	3	2	13·26 (18)	23·18 (17)	35·80 (15)	40·80 (14)	53·50 (11)	60·88 (9)	67·76 (7)	71·07 (6)
						74·31 (5)	80·66 (3)	83·79 (2)					
			5	3	3	15·02 (19)	30·21 (17)	48·64 (13)	69·16 (7)	75·28 (5)	87·09 (1)		
			5	4	0	20·44 (18)	43·21 (14)	47·41 (13)	71·80 (6)	81·02 (3)	84·02 (2)		
			5	4	1	12·24 (19)	29·03 (17)	34·62 (16)	39·51 (15)	51·89 (12)	55·54 (11)	65·70 (8)	75·10 (5)
						78·13 (4)	81·12 (3)	87·05 (1)					
			5	4	2	6·38 (20)	26·57 (18)	32·36 (17)	37·34 (16)	60·20 (10)	63·43 (9)	69·65 (7)	78·54 (4)
						84·30 (2)	87·15 (1)	90·00 (0)					
			5	4	3	8·13 (21)	19·47 (20)	26·41 (19)	45·00 (15)	55·55 (12)	64·90 (9)	67·84 (8)	70·73 (7)
						76·37 (5)	81·87 (3)	87·30 (1)	90·00 (0)				
			5	4	4	13·75 (22)	22·00 (21)	51·82 (14)	60·94 (11)	74·64 (6)	77·25 (5)	84·93 (2)	
			5	5	1	11·42 (21)	27·52 (19)	37·49 (17)	52·64 (13)	70·93 (7)	81·95 (3)	87·32 (1)	
			5	5	2	3·68 (22)	30·47 (19)	35·26 (18)	60·07 (11)	65·91 (9)	84·79 (2)	87·40 (1)	
			5	5	3	3·52 (23)	24·31 (21)	42·46 (17)	61·49 (11)	67·01 (9)	82·52 (3)	87·51 (1)	
			5	5	4	10·02 (24)	19·32 (23)	48·97 (16)	57·76 (13)	73·31 (7)	80·55 (4)	82·93 (3)	

h_1	k_1	l_1	h_2	k_2	l_2								
3	1	0	3	1	0	25·84 (9)	36·87 (8)	53·13 (6)	72·54 (3)	84·26 (1)	90·00 (0)		
			3	1	1	17·55 (10)	40·29 (8)	55·10 (6)	67·58 (4)	79·01 (2)	90·00 (0)		
			3	2	0	15·26 (11)	37·87 (9)	52·13 (7)	58·25 (6)	74·74 (3)	79·90 (2)		
			3	2	1	21·62 (11)	32·31 (10)	40·48 (9)	47·46 (8)	53·73 (7)	59·53 (6)	65·00 (5)	75·31 (3)
						85·15 (1)	90·00 (0)						
			3	2	2	32·47 (11)	46·35 (9)	52·15 (8)	57·53 (7)	72·13 (4)	76·70 (3)		
			3	3	1	29·47 (12)	43·49 (10)	54·52 (8)	64·20 (6)	90·00 (0)			

			3	3	2	36·00 (12)	42·13 (11)	52·64 (9)	61·84 (7)	66·14 (6)	78·33 (3)		
			4	1	0	4·40 (13)	23·02 (12)	32·47 (11)	57·53 (7)	72·13 (4)	76·70 (3)	85·60 (1)	
			4	1	1	14·31 (13)	34·93 (11)	58·55 (7)	72·65 (4)	81·43 (2)	85·73 (1)		
			4	2	1	14·96 (14)	26·22 (13)	40·62 (11)	46·36 (10)	61·12 (7)	69·82 (5)	82·07 (2)	86·04 (1)
			4	3	0	18·43 (15)	34·70 (13)	40·63 (12)	55·30 (9)	71·57 (5)	75·35 (4)	79·06 (3)	
			4	3	1	21·52 (15)	36·27 (13)	46·98 (11)	51·67 (10)	56·07 (9)	60·26 (8)	64·27 (7)	68·15 (6)
						71·94 (5)	86·44 (1)	90·00 (0)					
			4	3	2	28·26 (15)	34·70 (14)	40·24 (13)	49·76 (11)	54·04 (10)	58·10 (9)	65·73 (7)	72·93 (5)
						79·85 (3)	83·26 (2)						
			4	3	3	35·56 (15)	45·17 (13)	49·40 (12)	60·78 (9)	71·01 (6)	74·27 (5)		
			4	4	1	28·26 (16)	44·31 (13)	52·73 (11)	63·87 (8)	67·34 (7)	86·84 (1)		
			4	4	3	37·80 (16)	42·20 (15)	50·06 (13)	63·61 (9)	66·73 (8)	75·70 (5)		
			5	1	0	7·13 (16)	21·52 (15)	29·74 (14)	60·26 (8)	71·94 (5)	79·28 (3)	82·87 (2)	86·44 (1)
			5	1	1	13·16 (16)	31·57 (14)	60·87 (8)	75·91 (4)	83·01 (2)			
			5	2	0	3·37 (17)	28·26 (15)	40·24 (13)	49·76 (11)	69·37 (6)	72·93 (5)	83·26 (2)	86·63 (1)
			5	2	1	11·04 (17)	22·52 (16)	36·07 (14)	41·36 (13)	50·57 (11)	62·49 (8)	66·16 (7)	73·22 (5)
						83·37 (2)	86·69 (1)						
			5	2	2	20·64 (17)	44·31 (13)	52·73 (11)	63·87 (8)	77·28 (4)	86·84 (1)		
			5	3	0	12·53 (18)	35·56 (15)	40·60 (14)	49·40 (12)	60·78 (9)	74·27 (5)	77·47 (4)	80·64 (3)
			5	3	1	15·82 (18)	31·21 (16)	41·55 (14)	50·10 (12)	57·69 (10)	64·68 (8)	71·29 (6)	77·65 (4)
						83·86 (2)	90·00 (0)						
			5	3	2	22·57 (18)	29·30 (17)	44·10 (14)	48·17 (13)	52·01 (12)	55·65 (11)	62·50 (9)	68·96 (7)
						78·16 (4)	81·15 (3)	87·06 (1)					
			5	3	3	29·77 (18)	47·53 (14)	54·64 (12)	73·18 (6)	78·88 (4)			
			5	4	0	20·22 (19)	32·91 (17)	42·20 (15)	53·66 (12)	57·09 (11)	69·78 (7)	75·70 (5)	78·61 (4)
			5	4	1	22·01 (19)	33·95 (17)	38·67 (16)	46·91 (14)	50·63 (13)	57·54 (11)	67·02 (8)	70·03 (7)
						84·40 (2)	87·20 (1)						
			5	4	2	26·41 (19)	36·74 (17)	48·70 (14)	52·21 (13)	58·77 (11)	61·87 (10)	70·73 (7)	84·59 (2)
						87·30 (1)							
			5	4	3	31·82 (19)	36·39 (18)	40·51 (17)	47·87 (15)	51·24 (14)	54·45 (13)	57·54 (12)	60·53 (11)
						66·27 (9)	71·76 (7)	77·08 (5)	79·70 (4)				
			5	4	4	37·27 (19)	44·60 (17)	47·92 (16)	62·57 (11)	70·42 (8)	72·95 (7)		
			5	5	1	27·67 (20)	44·89 (16)	51·69 (14)	63·72 (10)	69·25 (8)	84·92 (2)		
			5	5	2	30·61 (20)	42·98 (17)	55·98 (13)	61·75 (11)	64·51 (10)	87·53 (1)		
			5	5	3	34·57 (20)	42·18 (18)	54·80 (14)	60·39 (12)	65·69 (10)	80·52 (4)		
			5	5	4	38·88 (20)	42·30 (19)	48·57 (17)	64·65 (11)	67·09 (10)	74·19 (7)		
h_1	k_1	l_1	h_2	k_2	l_2								
3	1	1	3	1	1	35·10 (9)	50·48 (7)	62·96 (5)	84·78 (1)				
			3	2	0	23·09 (11)	41·18 (9)	54·17 (7)	65·28 (5)	75·47 (3)	85·20 (1)		
			3	2	1	14·76 (12)	36·31 (10)	49·86 (8)	61·09 (6)	71·20 (4)	80·73 (2)		
			3	2	2	18·07 (13)	36·45 (11)	48·84 (9)	59·21 (7)	68·55 (5)	85·81 (1)		
			3	3	1	25·94 (13)	40·46 (11)	51·50 (9)	61·04 (7)	69·77 (5)	78·02 (3)		
			3	3	2	25·85 (14)	39·52 (12)	50·00 (10)	59·05 (8)	67·31 (6)	75·10 (4)	90·00 (0)	
			4	1	0	18·07 (13)	36·45 (11)	59·21 (7)	68·55 (5)	77·33 (3)	85·81 (1)		
			4	1	1	5·77 (14)	31·48 (12)	44·71 (10)	55·35 (8)	64·76 (6)	81·83 (2)	90·00 (0)	
			4	2	1	9·27 (15)	31·20 (13)	43·64 (11)	53·69 (9)	70·79 (5)	78·62 (3)	86·23 (1)	
			4	3	0	25·24 (15)	38·38 (13)	57·13 (9)	65·03 (7)	72·45 (5)	86·54 (1)		
			4	3	1	18·90 (16)	34·12 (14)	44·80 (12)	53·75 (10)	61·77 (8)	69·22 (6)	76·32 (4)	83·21 (2)
			4	3	2	17·86 (17)	32·88 (15)	43·29 (13)	51·98 (11)	66·93 (7)	73·74 (5)	80·33 (3)	86·79 (1)
			4	3	3	21·45 (18)	34·17 (16)	51·65 (12)	58·86 (10)	65·56 (8)	71·93 (6)	84·06 (2)	
			4	4	1	26·84 (17)	38·07 (15)	54·74 (11)	61·81 (9)	68·44 (7)	74·79 (5)	80·94 (3)	
			4	4	3	26·53 (19)	36·82 (17)	52·26 (13)	58·80 (11)	64·93 (9)	76·38 (5)	87·30 (1)	

Table 9 (cont.)

h_1	k_1	l_1	h_2	k_2	l_2								
3 (cont.)	1	1	5	1	0	18·90 (16)	34·12 (14)	61·77 (8)	69·22 (6)	76·32 (4)	83·21 (2)		
			5	1	1	9·45 (17)	29·50 (15)	41·03 (13)	58·52 (9)	66·03 (7)	79·98 (3)	86·67 (1)	
			5	2	0	17·86 (17)	43·29 (13)	51·98 (11)	66·93 (7)	80·33 (3)	86·79 (1)		
			5	2	1	7·75 (18)	28·26 (16)	39·59 (14)	48·66 (12)	56·60 (10)	70·71 (6)	77·28 (4)	83·68 (2)
						90·00 (0)							
			5	2	2	4·26 (19)	38·07 (15)	46·97 (13)	54·74 (11)	61·81 (9)	80·94 (3)	86·99 (1)	
			5	3	0	21·45 (18)	43·62 (14)	51·65 (12)	65·56 (8)	78·06 (4)	84·06 (2)		
			5	3	1	14·46 (19)	29·96 (17)	40·14 (15)	48·51 (13)	55·90 (11)	75·24 (5)	81·21 (3)	87·08 (1)
			5	3	2	11·98 (20)	38·50 (16)	46·78 (14)	54·06 (12)	60·72 (10)	66·97 (8)	72·93 (6)	78·72 (4)
						84·39 (2)							
			5	3	3	15·08 (21)	38·59 (17)	46·39 (15)	59·62 (11)	65·55 (9)	71·22 (7)	87·36 (1)	
			5	4	0	26·53 (19)	36·82 (17)	58·80 (11)	64·93 (9)	70·75 (7)	87·30 (1)		
			5	4	1	21·49 (20)	33·13 (18)	41·89 (16)	56·06 (12)	62·27 (10)	68·15 (8)	73·79 (6)	79·27 (4)
						84·66 (2)							
			5	4	2	19·29 (21)	31·35 (19)	40·17 (17)	47·61 (15)	54·25 (13)	66·14 (9)	71·66 (7)	77·01 (5)
						82·25 (3)							
			5	4	3	20·27 (22)	31·48 (20)	39·87 (18)	46·98 (16)	53·35 (14)	64·76 (10)	70·05 (8)	80·18 (4)
						90·00 (0)							
			5	4	4	23·29 (23)	33·00 (21)	53·20 (15)	58·72 (13)	63·94 (11)	73·77 (7)	83·12 (3)	
			5	5	1	27·55 (21)	36·66 (19)	56·71 (13)	62·33 (11)	67·67 (9)	72·81 (7)	82·72 (3)	
			5	5	2	25·49 (22)	42·39 (18)	48·97 (16)	60·50 (12)	70·84 (8)	75·75 (6)	80·55 (4)	
			5	5	3	25·47 (23)	41·77 (19)	48·14 (17)	59·32 (13)	69·31 (9)	74·05 (7)	87·75 (1)	
			5	5	4	27·04 (24)	35·26 (22)	53·57 (16)	58·70 (14)	63·55 (12)	77·13 (6)	85·74 (2)	

h_1	k_1	l_1	h_2	k_2	l_2								
3	2	0	3	2	0	22·62 (12)	46·19 (9)	62·51 (6)	67·38 (5)	72·08 (4)	90·00 (0)		
			3	2	1	15·50 (13)	27·19 (12)	35·38 (11)	48·15 (9)	53·63 (8)	58·74 (7)	68·25 (5)	72·75 (4)
						77·15 (3)	85·75 (1)	90·00 (0)					
			3	2	2	29·02 (13)	36·18 (12)	47·73 (10)	70·35 (5)	82·27 (2)	90·00 (0)		
			3	3	1	17·36 (15)	45·58 (11)	55·06 (9)	63·55 (7)	79·00 (3)			
			3	3	2	27·51 (15)	39·76 (13)	44·80 (12)	72·80 (5)	79·78 (3)	90·00 (0)		
			4	1	0	19·65 (14)	36·18 (12)	42·27 (11)	47·73 (10)	57·44 (8)	70·35 (5)	78·36 (3)	82·27 (2)
			4	1	1	23·76 (14)	44·02 (11)	49·18 (10)	70·92 (5)	86·25 (1)			
			4	2	1	14·45 (16)	32·08 (14)	48·26 (11)	52·75 (10)	61·04 (8)	64·93 (7)	72·39 (5)	75·99 (4)
						83·05 (2)	86·53 (1)						
			4	3	0	3·18 (18)	19·44 (17)	48·27 (12)	60·05 (9)	63·66 (8)	70·56 (6)	86·82 (1)	
			4	3	1	11·74 (18)	22·38 (17)	40·40 (14)	53·25 (11)	57·05 (10)	60·69 (9)	67·62 (7)	70·95 (6)
						74·22 (5)	80·61 (3)	86·88 (1)					
			4	3	2	22·02 (18)	28·89 (17)	34·51 (16)	43·86 (14)	47·97 (13)	51·83 (12)	65·67 (8)	72·00 (6)
						75·08 (5)	84·09 (2)	87·05 (1)	90·00 (0)				
			4	3	3	31·11 (18)	36·04 (17)	44·48 (15)	73·42 (6)	81·80 (3)	87·27 (1)		
			4	4	1	15·07 (20)	47·47 (14)	57·92 (11)	61·13 (10)	76·03 (5)	78·86 (4)		
			4	4	3	29·97 (20)	38·77 (18)	42·58 (17)	74·94 (6)	80·02 (4)	87·52 (1)		
			5	1	0	22·38 (17)	35·32 (15)	45·00 (13)	57·05 (10)	67·62 (7)	80·61 (3)	83·75 (2)	
			5	1	1	24·85 (17)	46·06 (13)	68·06 (7)	74·52 (5)	86·94 (1)			
			5	2	0	11·89 (19)	34·51 (16)	39·42 (15)	55·49 (11)	59·00 (10)	72·00 (6)	78·11 (4)	
			5	2	1	15·82 (19)	30·59 (17)	35·89 (16)	48·83 (13)	56·15 (11)	66·10 (8)	69·24 (7)	78·31 (4)
						87·10 (1)							
			5	2	2	23·46 (19)	39·42 (16)	57·92 (11)	61·13 (10)	78·86 (4)	84·46 (2)		
			5	3	0	2·73 (21)	25·35 (19)	44·48 (15)	61·60 (10)	64·65 (9)	73·42 (6)	87·27 (1)	
			5	3	1	10·10 (21)	27·03 (19)	37·16 (17)	52·45 (13)	58·96 (11)	65·04 (9)	70·84 (7)	81·91 (3)
						87·31 (1)							

h_1	k_1	l_1	h_2	k_2	l_2								
			5	3	2	19·12 (21)	31·26 (19)	43·96 (16)	54·20 (13)	57·32 (12)	60·34 (11)	66·11 (9)	77·00 (5)
						79·63 (4)	87·42 (1)	90·00 (0)					
			5	3	3	27·35 (21)	36·52 (19)	50·62 (15)	67·63 (9)	82·71 (3)	87·58 (1)		
			5	4	0	4·97 (23)	17·65 (22)	49·48 (15)	58·68 (12)	64·33 (10)	69·73 (8)	72·35 (7)	85·03 (2)
			5	4	1	10·16 (23)	19·69 (22)	43·32 (17)	53·19 (14)	56·20 (13)	61·92 (11)	64·66 (10)	72·57 (7)
						77·64 (5)	85·09 (2)						
			5	4	2	18·02 (23)	24·55 (22)	38·23 (19)	48·58 (16)	54·63 (14)	62·95 (11)	70·69 (8)	73·18 (7)
						80·48 (4)	85·26 (2)						
			5	4	3	25·56 (23)	30·35 (22)	34·54 (21)	41·82 (19)	45·09 (18)	48·18 (17)	69·33 (9)	74·06 (7)
						76·39 (6)	85·50 (2)	87·75 (1)					
			5	4	4	32·34 (23)	36·08 (22)	42·72 (20)	75·10 (7)	81·55 (4)	85·79 (2)		
			5	5	1	13·85 (25)	48·68 (17)	59·68 (13)	74·23 (7)	78·80 (5)			
			5	5	2	19·34 (25)	44·18 (19)	52·85 (16)	65·47 (11)	79·12 (5)	81·32 (4)		
			5	5	3	25·49 (25)	40·69 (21)	46·68 (19)	71·04 (9)	79·60 (5)	87·93 (1)		
			5	5	4	31·41 (25)	38·26 (23)	41·32 (22)	76·17 (7)	80·17 (5)	86·08 (2)		
h_1	k_1	l_1	h_2	k_2	l_2								
3	2	1	3	2	1	21·79 (13)	31·00 (12)	38·21 (11)	44·42 (10)	49·99 (9)	60·00 (7)	64·62 (6)	69·08 (5)
						73·40 (4)	81·79 (2)	85·90 (1)					
			3	2	2	13·52 (15)	24·84 (14)	32·58 (13)	44·52 (11)	49·59 (10)	63·02 (7)	71·09 (5)	78·79 (3)
						82·55 (2)	86·28 (1)						
			3	3	1	11·18 (16)	30·86 (14)	42·63 (12)	52·18 (10)	60·63 (8)	68·41 (6)	75·80 (4)	82·96 (2)
						90·00 (0)							
			3	3	2	14·38 (17)	24·26 (16)	31·27 (15)	42·21 (13)	55·26 (10)	59·15 (9)	62·88 (8)	73·45 (5)
						80·16 (3)	83·46 (2)	86·73 (1)					
			4	1	0	24·84 (14)	32·58 (13)	44·52 (11)	49·59 (10)	54·31 (9)	63·02 (7)	67·11 (6)	71·09 (5)
						82·55 (2)	86·28 (1)						
			4	1	1	19·11 (15)	35·02 (13)	40·89 (12)	46·14 (11)	50·95 (10)	55·46 (9)	67·79 (6)	71·64 (5)
						75·41 (4)	79·11 (3)	86·39 (1)					
			4	2	1	7·49 (17)	21·07 (16)	28·98 (15)	40·70 (13)	45·58 (12)	50·09 (11)	58·34 (9)	62·19 (8)
						65·91 (7)	73·05 (5)	76·51 (4)	79·92 (3)	86·66 (1)	90·00 (0)		
			4	3	0	15·82 (18)	24·68 (17)	36·70 (15)	45·98 (13)	53·99 (11)	57·69 (10)	61·24 (9)	71·29 (6)
						74·50 (5)	83·86 (2)	86·94 (1)					
			4	3	1	5·21 (19)	19·36 (18)	27·00 (17)	33·00 (16)	38·17 (15)	42·79 (14)	47·05 (13)	54·79 (11)
						65·21 (8)	68·48 (7)	74·81 (5)	80·95 (3)	83·98 (2)	87·00 (1)	90·00 (0)	
			4	3	2	6·98 (20)	19·45 (19)	32·47 (17)	37·43 (16)	41·89 (15)	49·82 (13)	56·91 (11)	63·47 (9)
						66·61 (8)	75·63 (5)	78·55 (4)	81·44 (3)	87·16 (1)			
			4	3	3	15·73 (21)	23·55 (20)	29·44 (19)	46·57 (15)	50·08 (14)	59·72 (11)	65·64 (9)	71·29 (7)
						79·44 (4)	82·10 (3)	84·74 (2)	87·37 (1)				
			4	4	1	12·31 (21)	27·88 (19)	33·13 (18)	45·74 (15)	49·36 (14)	62·27 (10)	65·25 (9)	70·99 (7)
						73·79 (6)	76·55 (5)	81·98 (3)	87·33 (1)				
			4	4	3	16·26 (23)	23·33 (22)	28·77 (21)	44·80 (17)	54·24 (14)	57·14 (13)	65·33 (10)	73·01 (7)
						77·95 (5)	82·81 (3)	85·21 (2)	87·61 (1)				
			5	1	0	27·00 (17)	33·00 (16)	42·79 (14)	47·05 (13)	54·79 (11)	61·85 (9)	65·21 (8)	68·48 (7)
						80·95 (3)	83·98 (2)						
			5	1	1	22·21 (18)	34·62 (16)	43·94 (14)	51·89 (12)	59·05 (10)	65·70 (8)	72·02 (6)	78·13 (4)
						90·00 (0)							
			5	2	0	19·45 (19)	32·47 (17)	37·43 (16)	49·82 (13)	53·45 (12)	56·91 (11)	63·47 (9)	66·61 (8)
						78·55 (4)	87·16 (1)						
			5	2	1	12·60 (20)	22·01 (19)	28·56 (18)	33·95 (17)	42·95 (15)	50·63 (13)	54·16 (12)	57·54 (11)
						60·79 (10)	63·95 (9)	72·98 (6)	75·88 (5)	78·74 (4)	81·58 (3)	84·40 (2)	87·20 (1)

Table 9 (cont.)

h_1	k_1	l_1	h_2	k_2	l_2								
3 (cont.)	2	1	5	2	2	12·31 (21)	33·13 (18)	37·73 (17)	45·74 (15)	49·36 (14)	52·78 (13)	65·25 (9)	70·99 (7)
						73·79 (6)	76·55 (5)	81·98 (3)	84·66 (2)				
			5	3	0	15·73 (21)	29·44 (19)	34·41 (18)	50·08 (14)	53·43 (13)	56·63 (12)	59·72 (11)	65·64 (9)
						71·29 (7)	79·44 (4)	87·37 (1)					
			5	3	1	6·35 (22)	25·38 (20)	35·59 (18)	43·71 (16)	50·77 (14)	57·17 (12)	63·14 (10)	68·81 (8)
						74·27 (6)	79·59 (4)	84·82 (2)	90·00 (0)				
			5	3	2	4·31 (23)	17·48 (22)	24·43 (21)	34·54 (19)	38·70 (18)	42·52 (17)	46·08 (16)	52·63 (14)
						55·69 (13)	61·52 (11)	64·31 (10)	69·71 (8)	72·33 (7)	77·48 (5)	82·53 (3)	87·52 (1)
						90·00 (0)							
			5	3	3	11·99 (24)	26·28 (22)	35·40 (20)	42·81 (18)	49·30 (16)	60·72 (12)	65·95 (10)	70·97 (8)
						75·85 (6)	80·62 (4)	85·32 (2)					
			5	4	0	16·26 (23)	23·33 (22)	37·53 (19)	44·80 (17)	54·24 (14)	57·14 (13)	62·67 (11)	73·01 (7)
						75·50 (6)	82·81 (3)	85·21 (2)					
			5	4	1	8·21 (24)	18·47 (23)	24·87 (22)	30·00 (21)	38·41 (19)	45·49 (17)	48·71 (16)	51·79 (15)
						57·58 (13)	63·02 (11)	65·64 (10)	68·21 (9)	70·74 (8)	75·67 (6)	78·10 (5)	82·89 (3)
						87·64 (1)	90·00 (0)						
			5	4	2	5·11 (25)	17·02 (24)	23·60 (23)	33·21 (21)	37·17 (20)	40·80 (19)	53·30 (15)	58·81 (13)
						61·44 (12)	64·01 (11)	68·99 (9)	71·41 (8)	73·81 (7)	78·51 (5)	80·83 (4)	83·14 (3)
						90·00 (0)							
			5	4	3	10·67 (26)	19·11 (25)	29·62 (23)	33·74 (22)	40·89 (20)	44·10 (19)	50·02 (17)	55·46 (15)
						60·57 (13)	63·03 (12)	65·43 (11)	67·79 (10)	76·89 (6)	79·11 (5)	81·30 (4)	83·49 (3)
						87·83 (1)							
			5	4	4	17·10 (27)	23·02 (26)	27·75 (25)	47·73 (19)	50·42 (18)	57·93 (15)	67·08 (11)	71·42 (9)
						77·74 (6)	83·90 (3)	85·94 (2)	87·97 (1)				
			5	5	1	13·34 (26)	26·08 (24)	34·58 (22)	47·65 (18)	63·31 (12)	72·58 (8)	77·02 (6)	81·39 (4)
						85·71 (2)							
			5	5	2	10·89 (27)	29·21 (24)	33·23 (23)	40·20 (21)	54·41 (16)	59·39 (14)	66·42 (11)	70·89 (9)
						75·25 (7)	77·40 (6)	83·74 (3)	87·92 (1)				
			5	5	3	13·03 (28)	25·22 (26)	33·38 (24)	40·05 (22)	56·17 (16)	60·85 (14)	73·84 (8)	77·95 (6)
						82·00 (4)	86·01 (2)						
			5	5	4	17·44 (29)	22·91 (28)	27·35 (27)	46·30 (21)	53·69 (18)	56·00 (17)	66·75 (12)	72·78 (9)
						76·69 (7)	84·34 (3)	86·23 (2)	88·11 (1)				

h_1	k_1	l_1	h_2	k_2	l_2								
3	2	2	3	2	2	19·75 (16)	58·03 (9)	61·93 (8)	76·39 (4)	86·63 (1)			
			3	3	1	18·93 (17)	33·42 (15)	43·67 (13)	59·95 (9)	73·85 (5)	80·39 (3)	86·81 (1)	
			3	3	2	10·75 (19)	21·45 (18)	55·33 (11)	68·78 (7)	71·93 (6)	87·04 (1)		
			4	1	0	34·56 (14)	49·68 (11)	53·97 (10)	69·33 (6)	72·90 (5)			
			4	1	1	23·84 (16)	42·00 (13)	46·69 (12)	59·04 (9)	62·79 (8)	66·41 (7)	80·13 (3)	
			4	2	1	17·70 (18)	32·13 (16)	37·45 (15)	42·19 (14)	50·57 (12)	58·04 (10)	61·55 (9)	68·25 (7)
						71·48 (6)	77·78 (4)	86·97 (1)	90·00 (0)				
			4	3	0	29·18 (18)	34·45 (17)	47·23 (14)	73·08 (6)	84·43 (2)	87·22 (1)		
			4	3	1	17·95 (20)	25·35 (19)	36·04 (17)	40·44 (16)	44·48 (15)	58·45 (11)	67·63 (8)	76·24 (5)
						79·03 (4)	81·80 (3)	87·27 (1)					
			4	3	2	7·77 (22)	18·95 (21)	25·74 (20)	50·91 (14)	54·16 (13)	63·23 (10)	68·88 (8)	76·99 (5)
						79·62 (4)	82·23 (3)	84·83 (2)					
			4	3	3	3·37 (24)	16·93 (23)	60·06 (12)	62·77 (11)	73·07 (7)	78·00 (5)	90·00 (0)	
			4	4	1	21·75 (22)	36·66 (19)	40·54 (18)	56·71 (13)	75·33 (6)	82·72 (3)	85·16 (2)	
			4	4	3	10·00 (26)	18·75 (25)	57·98 (14)	67·74 (10)	70·07 (9)	74·62 (7)	85·66 (2)	
			5	1	0	36·04 (17)	51·80 (13)	55·20 (12)	67·63 (8)	70·55 (7)			
			5	1	1	27·52 (19)	45·56 (15)	59·11 (11)	65·16 (9)	76·50 (5)			
			5	2	0	31·16 (19)	43·90 (16)	50·91 (14)	60·30 (11)	74·32 (6)	79·62 (4)		
			5	2	1	21·58 (21)	37·15 (18)	41·17 (17)	51·69 (14)	54·85 (13)	60·85 (11)	66·51 (9)	74·59 (6)
						82·37 (3)	84·92 (2)						

h_1	k_1	l_1	h_2	k_2	l_2								
			5	2	2	13.82 (23)	32.39 (20)	50.71 (15)	59.56 (12)	70.26 (8)	72.81 (7)	90.00 (0)	
			5	3	0	29.13 (21)	37.79 (19)	48.28 (16)	68.02 (9)	80.42 (4)	87.62 (1)		
			5	3	1	19.45 (23)	30.58 (21)	38.84 (19)	45.82 (17)	57.80 (13)	63.19 (11)	73.32 (7)	82.94 (3)
						87.65 (1)							
			5	3	2	10.39 (25)	25.19 (23)	30.05 (22)	48.02 (17)	53.83 (15)	59.24 (13)	66.83 (10)	78.65 (5)
						83.22 (3)	85.49 (2)						
			5	3	3	3.00 (27)	22.38 (25)	56.30 (15)	61.26 (13)	74.99 (7)	79.34 (5)	83.63 (3)	
			5	4	0	29.40 (23)	33.56 (22)	47.02 (18)	74.62 (7)	85.66 (2)			
			5	4	1	20.67 (25)	26.08 (24)	38.20 (21)	41.54 (20)	55.85 (15)	70.32 (9)	79.22 (5)	81.39 (4)
						87.86 (1)	90.00 (0)						
			5	4	2	12.53 (27)	19.94 (26)	29.81 (24)	46.61 (19)	49.40 (18)	64.29 (12)	66.57 (11)	73.19 (8)
						77.47 (6)	81.68 (4)	83.77 (3)	85.85 (2)				
			5	4	3	5.91 (29)	16.18 (28)	22.17 (27)	54.33 (17)	56.72 (16)	63.52 (13)	67.83 (11)	72.02 (9)
						74.07 (8)	76.11 (7)	82.11 (4)	88.03 (1)				
			5	4	4	5.21 (31)	15.48 (30)	61.19 (15)	63.27 (14)	71.26 (10)	78.89 (6)	88.16 (1)	
			5	5	1	23.51 (27)	38.64 (23)	54.74 (17)	76.25 (7)	84.15 (3)			
			5	5	2	16.84 (29)	30.89 (26)	46.12 (21)	62.48 (14)	72.72 (9)	78.58 (6)	88.11 (1)	
			5	5	3	11.81 (31)	23.70 (29)	53.13 (19)	69.68 (11)	73.49 (9)	88.19 (1)		
			5	5	4	9.87 (33)	17.19 (32)	59.50 (17)	67.16 (13)	69.01 (12)	76.18 (8)	84.86 (3)	
h_1	k_1	l_1	h_2	k_2	l_2								
3	3	1	3	3	1	26.53 (17)	37.86 (15)	61.73 (9)	80.92 (3)	86.98 (1)			
			3	3	2	11.98 (20)	28.31 (18)	38.50 (16)	54.06 (12)	72.93 (6)	84.39 (2)	90.00 (0)	
			4	1	0	33.42 (15)	43.67 (13)	52.26 (11)	59.95 (9)	67.08 (7)	86.81 (1)		
			4	1	1	30.10 (16)	40.80 (14)	57.27 (10)	64.37 (8)	77.51 (4)	83.79 (2)		
			4	2	1	17.98 (19)	31.67 (17)	49.40 (13)	56.59 (11)	69.49 (7)	75.50 (5)	87.13 1)	
			4	3	0	15.52 (21)	46.51 (15)	53.38 (13)	65.61 (9)	76.74 (5)	82.09 (3)		
			4	3	1	8.18 (22)	25.86 (20)	35.92 (18)	43.96 (16)	57.32 (12)	63.26 (10)	68.90 (8)	74.34 (6)
						79.63 (4)	84.84 (2)						
			4	3	2	11.53 (23)	26.54 (21)	35.96 (19)	50.28 (15)	62.06 (11)	67.45 (9)	72.65 (7)	77.70 (5)
						82.66 (3)	87.56 (1)						
			4	3	3	19.22 (24)	30.05 (22)	44.91 (18)	56.58 (14)	76.35 (6)	80.95 (4)	90.00 (0)	
			4	4	1	3.24 (25)	23.29 (23)	40.64 (19)	58.72 (13)	63.94 (11)	78.48 (5)	87.71 (1)	
			4	4	3	14.68 (27)	26.40 (25)	41.20 (21)	52.48 (17)	75.48 (7)	83.83 (3)	87.95 (1)	
			5	1	0	35.92 (18)	43.96 (16)	50.96 (14)	57.32 (12)	68.90 (8)	84.84 (2)		
			5	1	1	32.98 (19)	41.36 (17)	54.97 (13)	60.94 (11)	77.25 (5)	87.47 (1)		
			5	2	0	26.54 (21)	43.60 (17)	56.37 (13)	62.06 (11)	67.45 (9)	87.56 (1)		
			5	2	1	22.86 (22)	33.10 (20)	47.92 (16)	54.10 (14)	65.24 (10)	70.42 (8)	80.36 (4)	85.19 (2)
			5	2	2	23.29 (23)	40.64 (19)	47.24 (17)	63.94 (11)	73.77 (7)	78.48 (5)		
			5	3	0	19.22 (24)	44.91 (18)	56.58 (14)	61.83 (12)	76.35 (6)	80.95 (4)		
			5	3	1	14.20 (25)	26.89 (23)	35.48 (21)	48.76 (17)	54.43 (15)	64.75 (11)	69.57 (9)	74.25 (7)
						78.82 (5)	87.78 (1)						
			5	3	2	14.62 (26)	26.72 (24)	35.04 (22)	41.90 (20)	47.94 (18)	63.47 (12)	68.15 (10)	72.68 (8)
						77.10 (6)	81.44 (4)	85.73 (2)					
			5	3	3	19.16 (27)	36.42 (23)	42.72 (21)	62.95 (13)	71.65 (9)	79.93 (5)	83.98 (3)	
			5	4	0	14.68 (27)	47.10 (19)	52.48 (17)	66.79 (11)	75.48 (7)	83.83 (3)		
			5	4	1	7.61 (28)	23.02 (26)	38.85 (22)	44.93 (20)	55.50 (16)	60.29 (14)	69.27 (10)	73.55 (8)
						81.86 (4)	85.94 (2)						
			5	4	2	7.35 (29)	31.24 (25)	38.13 (23)	54.45 (17)	63.60 (13)	67.90 (11)	80.15 (5)	88.04 (1)
			5	4	3	13.26 (30)	24.71 (28)	32.48 (26)	38.86 (24)	49.54 (20)	58.73 (16)	71.07 (10)	74.96 (8)
						78.77 (6)	86.28 (2)	90.00 (0)					
			5	4	4	19.61 (31)	28.21 (29)	45.66 (23)	54.74 (19)	77.72 (7)	81.26 (5)	88.26 (1)	

Table 9 (cont.)

h_1	k_1	l_1	h_2	k_2	l_2								
3 (cont.)	3	1	5	5	1	5·21 (31)	21·31 (29)	42·37 (23)	56·90 (17)	65·32 (13)	77·00 (7)	88·16 (1)	
			5	5	2	2·53 (32)	29·06 (28)	35·74 (26)	60·03 (16)	64·08 (14)	82·83 (4)	86·42 (2)	
			5	5	3	9·73 (33)	29·99 (29)	36·25 (27)	55·43 (19)	70·82 (11)	84·86 (3)	88·29 (1)	
			5	5	4	16·23 (34)	25·36 (32)	42·76 (26)	51·59 (22)	76·94 (8)	83·51 (4)	86·76 (2)	
h_1	k_1	l_1	h_2	k_2	l_2								
3	3	2	3	3	2	17·34 (21)	50·48 (14)	65·85 (9)	79·52 (4)	82·16 (3)			
			4	1	0	39·14 (15)	43·62 (14)	55·33 (11)	58·86 (10)	62·27 (9)	75·02 (5)		
			4	1	1	31·32 (17)	45·29 (14)	49·21 (13)	56·44 (11)	66·30 (8)	69·40 (7)	84·23 (2)	
			4	2	1	21·49 (20)	27·88 (19)	37·73 (17)	41·89 (16)	52·78 (13)	59·22 (11)	68·15 (8)	76·55 (5)
						79·27 (4)	87·33 (1)						
			4	3	0	26·43 (21)	39·87 (18)	43·54 (17)	75·18 (6)	82·65 (3)	87·56 (1)		
			4	3	1	15·91 (23)	28·59 (21)	33·25 (20)	37·40 (19)	51·16 (15)	54·17 (14)	67·89 (9)	77·93 (5)
						80·37 (4)	82·79 (3)	85·20 (2)	87·60 (1)				
			4	3	2	8·21 (25)	18·16 (24)	24·41 (23)	47·70 (17)	61·64 (12)	64·18 (11)	73·91 (7)	78·58 (5)
						87·73 (1)	90·00 (0)						
			4	3	3	9·17 (27)	18·07 (26)	56·74 (15)	68·55 (10)	70·79 (9)	72·99 (8)	83·70 (3)	
			4	4	1	15·21 (26)	31·39 (23)	35·26 (22)	50·88 (17)	74·94 (7)	85·74 (2)	87·87 (1)	
			4	4	3	2·70 (30)	15·07 (29)	53·18 (18)	64·35 (13)	68·51 (11)	78·48 (6)	80·42 (5)	
			5	1	0	41·18 (18)	44·70 (17)	57·07 (13)	59·88 (12)	72·98 (7)			
			5	1	1	34·85 (20)	48·97 (16)	54·94 (14)	65·78 (10)	80·55 (4)			
			5	2	0	33·76 (21)	41·22 (19)	50·70 (16)	64·18 (11)	69·13 (9)	80·89 (4)		
			5	2	1	26·46 (23)	31·09 (22)	42·30 (19)	51·48 (16)	56·98 (14)	59·60 (13)	64·65 (11)	71·86 (8)
						74·19 (7)	87·77 (1)						
			5	2	2	21·90 (25)	35·26 (22)	50·88 (17)	61·15 (13)	68·21 (10)	79·31 (5)	85·74 (2)	
			5	3	0	28·65 (24)	39·84 (21)	46·00 (19)	70·79 (9)	77·33 (6)	87·90 (1)		
			5	3	1	20·45 (26)	30·13 (24)	37·55 (22)	49·56 (18)	54·79 (16)	64·38 (12)	73·24 (8)	77·51 (6)
						81·71 (4)	85·87 (2)						
			5	3	2	14·44 (28)	20·96 (27)	30·16 (25)	46·23 (20)	58·75 (15)	63·28 (13)	69·77 (10)	75·99 (7)
						80·04 (5)	84·04 (3)	86·03 (2)					
			5	3	3	12·74 (30)	24·45 (28)	54·18 (18)	67·04 (12)	71·03 (10)	74·92 (8)	90·00 (0)	
			5	4	0	25·97 (27)	40·02 (23)	42·90 (22)	76·52 (7)	84·27 (3)	86·18 (2)		
			5	4	1	17·44 (29)	31·20 (26)	34·67 (25)	48·86 (20)	51·31 (19)	70·79 (10)	80·53 (5)	82·44 (4)
						88·11 (1)							
			5	4	2	9·85 (31)	22·83 (29)	27·14 (28)	43·03 (23)	57·30 (17)	59·44 (16)	65·60 (13)	75·27 (8)
						77·15 (7)	82·70 (4)	88·18 (1)					
			5	4	3	5·74 (33)	15·24 (32)	20·82 (31)	50·72 (21)	61·16 (16)	65·03 (14)	66·92 (13)	70·63 (11)
						74·25 (9)	77·82 (7)	84·81 (3)	86·54 (2)				
			5	4	4	8·75 (35)	16·23 (34)	57·55 (19)	66·71 (14)	71·90 (11)	73·60 (10)	81·88 (5)	
			5	5	1	17·19 (32)	33·29 (28)	48·94 (22)	76·18 (8)	86·58 (2)			
			5	5	2	9·45 (34)	25·92 (31)	41·03 (26)	56·55 (19)	71·39 (11)	83·34 (4)	88·34 (1)	
			5	5	3	2·25 (36)	19·31 (34)	48·23 (24)	63·63 (16)	67·13 (14)	80·41 (6)	83·63 (4)	
			5	5	4	4·26 (38)	13·83 (37)	54·74 (22)	63·50 (17)	70·05 (13)	77·88 (8)	79·41 (7)	
h_1	k_1	l_1	h_2	k_2	l_2								
4	1	0	4	1	0	19·75 (16)	28·07 (15)	61·93 (8)	76·39 (4)	86·63 (1)	90·00 (0)		
			4	1	1	13·63 (17)	30·96 (15)	62·79 (8)	73·39 (5)	80·13 (3)	90·00 (0)		
			4	2	1	17·70 (18)	25·88 (17)	37·45 (15)	42·19 (14)	50·57 (12)	61·55 (9)	64·95 (8)	68·25 (7)
						71·48 (6)	77·78 (4)	83·92 (2)	90·00 (0)				
			4	3	0	22·83 (19)	39·09 (16)	50·91 (13)	54·40 (12)	67·17 (8)	78·81 (4)	81·63 (3)	
			4	3	1	25·35 (19)	36·04 (17)	40·44 (16)	44·48 (15)	51·80 (13)	58·45 (11)	67·63 (8)	70·55 (7)
						87·27 (1)	90·00 (0)						

h_1	k_1	l_1	h_2	k_2	l_2								
			4	3	2	31·16 (19)	35·84 (18)	43·90 (16)	50·91 (14)	54·16 (13)	57·29 (12)	60·30 (11)	63·23 (10)
						68·88 (8)	76·99 (5)	79·62 (4)					
			4	3	3	37·79 (19)	48·28 (16)	51·40 (15)	57·27 (13)	68·02 (9)	70·56 (8)		
			4	4	1	32·39 (20)	44·13 (17)	50·71 (15)	59·56 (12)	70·26 (8)	90·00 (0)		
			4	4	3	40·75 (20)	43·97 (19)	52·70 (16)	60·50 (13)	62·97 (12)	72·36 (8)		
			5	1	0	2·73 (21)	17·95 (20)	25·35 (19)	64·65 (9)	76·24 (5)	79·03 (4)	87·27 (1)	
			5	1	1	11·42 (21)	27·52 (19)	65·16 (9)	76·50 (5)	81·95 (3)	87·32 (1)		
			5	2	0	7·77 (22)	25·74 (20)	35·84 (18)	54·16 (13)	68·88 (8)	76·99 (5)	82·23 (3)	84·83 (2)
			5	2	1	13·05 (22)	21·58 (21)	32·72 (19)	37·15 (18)	54·85 (13)	66·51 (9)	71·94 (7)	74·59 (6)
						82·37 (3)	84·92 (2)	87·46 (1)					
			5	2	2	21·75 (22)	40·54 (18)	56·71 (13)	65·03 (10)	75·33 (6)	82·72 (3)		
			5	3	0	16·93 (23)	33·71 (20)	45·00 (17)	60·06 (12)	73·07 (7)	78·00 (5)	82·83 (3)	
			5	3	1	19·45 (23)	30·58 (21)	38·84 (19)	45·82 (17)	57·80 (13)	63·19 (11)	68·35 (9)	73·32 (7)
						87·65 (1)							
			5	3	2	25·19 (23)	30·05 (22)	44·91 (18)	48·02 (17)	56·58 (14)	59·24 (13)	64·36 (11)	66·83 (10)
						74·01 (7)	78·65 (5)	83·22 (3)					
			5	3	3	31·71 (23)	51·04 (17)	56·30 (15)	70·56 (9)	74·99 (7)			
			5	4	0	24·62 (24)	37·30 (21)	40·75 (20)	52·70 (16)	65·38 (11)	79·08 (5)	81·29 (4)	
			5	4	1	26·08 (24)	38·20 (21)	44·68 (19)	50·49 (17)	53·22 (16)	55·85 (15)	65·69 (11)	70·32 (9)
						72·58 (8)	87·86 (1)	90·00 (0)					
			5	4	2	29·81 (24)	37·31 (22)	40·60 (21)	49·40 (18)	54·66 (16)	59·59 (14)	61·96 (13)	64·29 (12)
						66·57 (11)	81·68 (4)	83·77 (3)					
			5	4	3	34·59 (24)	37·92 (23)	43·92 (21)	49·33 (19)	54·33 (17)	56·72 (16)	63·52 (13)	67·83 (11)
						74·07 (8)	76·11 (7)						
			5	4	4	39·56 (24)	47·58 (21)	50·02 (20)	59·07 (16)	67·33 (12)	69·31 (11)		
			5	5	1	31·89 (25)	44·50 (21)	49·81 (19)	59·37 (15)	72·20 (9)	88·05 (1)		
			5	5	2	34·40 (25)	43·44 (22)	53·55 (18)	60·33 (15)	64·59 (13)	84·32 (3)		
			5	5	3	37·87 (25)	43·43 (23)	57·54 (17)	61·73 (15)	77·23 (7)			
			5	5	4	41·72 (25)	44·23 (24)	51·18 (21)	61·47 (16)	63·40 (15)	70·83 (11)		

h_1	k_1	l_1	h_2	k_2	l_2								
4	1	1	4	1	1	27·27 (16)	38·94 (14)	60·00 (9)	67·11 (7)	86·82 (1)			
			4	2	1	12·24 (19)	29·03 (17)	39·51 (15)	48·04 (13)	55·54 (11)	59·05 (10)	72·02 (6)	75·10 (5)
						81·12 (3)	84·10 (2)						
			4	3	0	26·41 (19)	41·04 (16)	52·21 (13)	67·84 (8)	70·73 (7)	87·30 (1)		
			4	3	1	22·41 (20)	33·69 (18)	38·20 (17)	46·10 (15)	49·67 (14)	56·31 (12)	59·44 (11)	65·42 (9)
						71·12 (7)	76·64 (5)	82·03 (3)					
			4	3	2	23·20 (21)	38·02 (18)	41·92 (17)	48·96 (15)	52·21 (14)	61·22 (11)	64·04 (10)	66·80 (9)
						72·16 (7)	74·77 (6)	87·49 (1)					
			4	3	3	27·21 (22)	39·82 (19)	49·70 (16)	58·30 (13)	63·60 (11)	66·16 (10)	78·34 (5)	
			4	4	1	30·50 (21)	38·78 (19)	57·76 (13)	60·50 (12)	63·17 (11)	80·55 (4)		
			4	4	3	32·15 (23)	42·59 (20)	51·26 (17)	56·48 (15)	63·79 (12)	70·65 (9)	81·53 (4)	
			5	1	0	13·90 (21)	28·56 (19)	65·42 (9)	73·90 (6)	79·34 (4)	87·35 (1)		
			5	1	1	3·68 (22)	24·88 (20)	35·26 (18)	63·02 (10)	68·72 (8)	84·79 (2)	90·00 (0)	
			5	2	0	15·65 (22)	38·02 (18)	55·32 (13)	72·16 (7)	82·45 (3)			
			5	2	1	8·21 (23)	25·35 (21)	35·15 (19)	42·98 (17)	52·95 (14)	58·91 (12)	61·75 (11)	72·47 (7)
						80·09 (4)	82·58 (3)	85·06 (2)	87·53 (1)				
			5	2	2	10·02 (24)	34·85 (20)	48·97 (16)	52·01 (15)	63·17 (11)	78·16 (5)	87·65 (1)	
			5	3	0	21·61 (23)	46·59 (17)	71·13 (8)	73·56 (7)	85·36 (2)			
			5	3	1	17·02 (24)	28·78 (22)	44·18 (18)	50·40 (16)	61·44 (12)	71·41 (8)	76·17 (6)	80·83 (4)
						85·43 (2)							
			5	3	2	17·08 (25)	36·59 (21)	43·41 (19)	52·28 (16)	55·00 (15)	67·52 (10)	69·87 (9)	76·74 (6)
						78·98 (5)	90·00 (0)						

Table 9 (cont.)

h_1	k_1	l_1	h_2	k_2	l_2								
4 (cont.)	1	1	5	3	3	20·84 (26)	44·04 (20)	59·79 (14)	68·93 (10)	81·73 (4)			
			5	4	0	27·94 (24)	39·37 (21)	53·92 (16)	66·11 (11)	70·65 (9)	87·89 (1)		
			5	4	1	24·60 (25)	33·23 (23)	36·86 (22)	43·33 (20)	51·81 (17)	56·94 (15)	61·78 (13)	64·12 (12)
						68·67 (10)	79·52 (5)	83·74 (3)					
			5	4	2	24·00 (26)	36·09 (23)	39·38 (22)	48·12 (19)	50·77 (18)	53·32 (17)	60·53 (14)	62·82 (13)
						71·57 (9)	75·76 (7)	87·99 (1)					
			5	4	3	25·84 (27)	36·87 (24)	45·57 (21)	50·70 (19)	53·13 (18)	62·18 (14)	64·32 (13)	68·49 (11
						74·53 (8)	84·26 (3)						
			5	4	4	29·06 (28)	38·69 (25)	51·36 (20)	57·95 (17)	62·08 (15)	68·00 (12)	77·38 (7)	
			5	5	1	30·89 (26)	37·62 (24)	58·12 (16)	62·48 (14)	78·58 (6)	82·41 (4)		
			5	5	2	30·00 (27)	42·46 (23)	54·74 (18)	56·96 (17)	65·36 (13)	75·13 (8)	86·32 (2)	
			5	5	3	30·77 (28)	47·54 (22)	56·47 (18)	68·39 (12)	86·48 (2)			
			5	5	4	32·71 (29)	41·03 (26)	52·46 (21)	56·55 (19)	62·34 (16)	71·39 (11)	79·98 (6)	

h_1	k_1	l_1	h_2	k_2	l_2								
4	2	1	4	2	1	17·75 (20)	25·21 (19)	35·95 (17)	40·37 (16)	44·42 (15)	48·19 (14)	51·75 (13)	55·15 (12)
						58·41 (11)	61·56 (10)	73·40 (6)	79·02 (4)	84·53 (2)	87·27 (1)		
			4	3	0	16·23 (22)	29·21 (20)	33·98 (19)	45·71 (16)	55·43 (13)	61·31 (11)	64·12 (10)	69·56 (8)
						77·40 (5)	79·95 (4)	84·99 (2)					
			4	3	1	10·16 (23)	26·01 (21)	35·60 (19)	39·62 (18)	43·32 (17)	50·06 (15)	53·19 (14)	61·92 (11)
						64·66 (10)	67·35 (9)	72·57 (7)	75·12 (6)	77·64 (5)	82·62 (3)	85·09 (2)	87·55 (1)
			4	3	2	13·46 (24)	21·25 (23)	26·94 (22)	35·86 (20)	39·65 (19)	43·16 (18)	46·46 (17)	52·57 (15)
						58·21 (13)	60·90 (12)	66·10 (10)	68·61 (9)	71·08 (8)	75·93 (6)	80·67 (4)	83·02 (3)
						85·35 (2)							
			4	3	3	20·67 (25)	30·60 (23)	34·58 (22)	44·68 (19)	50·49 (17)	58·40 (14)	60·89 (13)	68·02 (10)
						74·81 (7)	85·71 (2)	87·86 (1)					
			4	4	1	18·26 (25)	29·11 (23)	33·31 (22)	46·86 (18)	52·57 (16)	57·87 (14)	67·67 (10)	70·01 (9)
						72·31 (8)	74·58 (7)	90·00 (0)					
			4	4	3	23·05 (27)	27·62 (26)	35·12 (24)	44·30 (21)	52·16 (18)	56·96 (16)	61·50 (14)	67·98 (11)
						74·18 (8)	78·20 (6)	80·19 (5)	90·00 (0)				
			5	1	0	19·69 (22)	26·01 (21)	35·60 (19)	39·62 (18)	53·19 (14)	61·92 (11)	67·35 (9)	72·57 (7)
						75·12 (6)	82·62 (3)	87·55 (1)					
			5	1	1	15·00 (23)	28·13 (21)	37·07 (19)	44·44 (17)	50·95 (15)	56·91 (13)	62·49 (11)	72·90 (7)
						77·88 (5)	82·76 (3)	87·59 (1)					
			5	2	0	13·46 (24)	26·94 (22)	43·16 (18)	49·58 (16)	58·21 (13)	60·90 (12)	68·61 (9)	71·08 (8)
						83·02 (3)	85·35 (2)	87·68 (1)					
			5	2	1	5·11 (25)	17·02 (24)	23·60 (23)	37·17 (20)	40·80 (19)	47·37 (17)	50·40 (16)	53·30 (15)
						58·81 (13)	61·44 (12)	64·01 (11)	71·41 (8)	78·51 (5)	80·83 (4)	83·14 (3)	87·72 (1)
			5	2	2	9·01 (26)	33·31 (22)	40·56 (20)	46·86 (18)	49·78 (17)	52·57 (16)	57·87 (14)	70·01 (9)
						74·58 (7)	81·26 (4)	87·82 (1)	90·00 (0)				
			5	3	0	13·34 (26)	30·60 (23)	34·58 (22)	50·49 (17)	58·40 (14)	60·89 (13)	65·69 (11)	74·81 (7)
						85·71 (2)	87·86 (1)						
			5	3	1	5·18 (27)	22·76 (25)	31·97 (23)	39·23 (21)	45·51 (19)	51·17 (17)	56·41 (15)	61·35 (13)
						66·06 (11)	70·61 (9)	75·04 (7)	79·37 (5)	83·65 (3)	87·89 (1)		
			5	3	2	7·61 (28)	17·10 (27)	31·83 (24)	41·98 (21)	44·93 (20)	47·73 (19)	55·50 (16)	57·93 (15)
						62·60 (13)	64·86 (12)	67·08 (11)	71·42 (9)	75·65 (7)	79·81 (5)	81·86 (4)	83·90 (3)
						87·97 (1)	90·00 (0)						
			5	3	3	15·19 (29)	33·70 (25)	40·06 (23)	50·78 (19)	55·55 (17)	64·37 (13)	68·53 (11)	80·42 (5)
						88·09 (1)							
			5	4	0	17·40 (28)	27·62 (26)	35·12 (24)	44·30 (21)	56·96 (16)	61·50 (14)	63·70 (13)	65·86 (12)
						67·98 (11)	78·20 (6)	84·13 (3)					
			5	4	1	12·45 (29)	24·61 (27)	28·90 (26)	32·67 (25)	39·24 (23)	42·20 (22)	50·23 (19)	52·69 (18)
						55·08 (17)	61·87 (14)	64·04 (13)	68·26 (11)	70·32 (10)	72·36 (9)	76·37 (7)	80·31 (5)
						86·14 (2)	88·07 (1)						

h_1	k_1	l_1	h_2	k_2	l_2								
			5	4	2	12·60 (30)	24·38 (28)	32·24 (26)	35·59 (25)	38·67 (24)	44·30 (22)	46·91 (21)	49·41 (20)
						56·43 (17)	60·79 (15)	62·91 (14)	67·02 (12)	69·03 (11)	71·02 (10)	74·92 (8)	76·84 (7)
						78·74 (6)	80·64 (5)	82·52 (4)	86·27 (2)				
			5	4	3	16·93 (31)	22·21 (30)	26·50 (29)	33·57 (27)	36·64 (26)	39·51 (25)	44·78 (23)	47·24 (22)
						56·26 (18)	58·36 (17)	62·42 (15)	72·02 (10)	73·87 (9)	79·33 (6)	81·12 (5)	84·69 (3)
						86·46 (2)	88·23 (1)						
			5	4	4	22·34 (32)	29·88 (30)	33·05 (29)	46·08 (24)	50·51 (22)	56·69 (19)	62·45 (16)	67·93 (13)
						73·20 (10)	76·63 (8)	85·03 (3)	86·69 (2)				
			5	5	1	18·69 (31)	27·61 (29)	34·41 (27)	45·35 (23)	54·51 (19)	58·70 (17)	66·59 (13)	70·36 (11)
						74·04 (9)	88·25 (1)						
			5	5	2	18·15 (32)	30·55 (29)	33·75 (28)	46·92 (23)	51·42 (21)	55·65 (19)	67·29 (13)	69·12 (12)
						70·93 (11)	76·26 (8)	78·00 (7)	84·89 (3)				
			5	5	3	20·36 (33)	28·27 (31)	39·91 (27)	53·37 (21)	57·33 (19)	61·12 (17)	68·33 (13)	75·19 (9)
						78·53 (7)	85·11 (3)						
			5	5	4	24·04 (34)	27·58 (33)	33·62 (31)	45·70 (26)	51·84 (23)	55·66 (21)	62·83 (17)	67·91 (14)
						72·81 (11)	79·16 (7)	80·73 (6)	88·46 (1)				
h_1	k_1	l_1	h_2	k_2	l_2								
4	3	0	4	3	0	16·26 (24)	50·21 (16)	61·31 (12)	68·90 (9)	73·74 (7)	90·00 (0)		
			4	3	1	11·31 (25)	19·72 (24)	41·82 (19)	51·13 (16)	53·96 (15)	59·34 (13)	69·33 (9)	71·71 (8)
						74·06 (7)	78·69 (5)	90·00 (0)					
			4	3	2	21·80 (25)	26·96 (24)	35·21 (22)	42·03 (20)	48·05 (18)	50·85 (17)	68·20 (10)	74·93 (7)
						77·12 (6)	81·46 (4)	87·87 (1)	90·00 (0)				
			4	3	3	30·96 (25)	34·59 (24)	43·92 (21)	76·11 (7)	84·09 (3)	90·00 (0)		
			4	4	1	12·88 (28)	48·59 (19)	56·15 (16)	63·09 (13)	73·83 (8)	81·99 (4)		
			4	4	3	29·01 (28)	38·66 (25)	41·44 (24)	77·37 (7)	82·82 (4)	90·00 (0)		
			5	1	0	25·56 (23)	38·33 (20)	41·82 (19)	48·18 (17)	53·96 (15)	64·44 (11)	80·97 (4)	83·24 (3)
			5	1	1	27·72 (23)	43·00 (19)	49·13 (17)	64·95 (11)	74·37 (7)	87·79 (1)		
			5	2	0	15·07 (26)	31·33 (23)	42·03 (20)	56·15 (15)	58·67 (14)	72·72 (8)	74·93 (7)	77·12 (6)
			5	2	1	18·31 (26)	32·88 (23)	46·07 (19)	51·63 (17)	59·26 (14)	66·32 (11)	68·58 (10)	75·19 (7)
						79·48 (5)	85·81 (2)						
			5	2	2	25·15 (26)	36·80 (23)	60·83 (14)	75·89 (7)	86·01 (2)			
			5	3	0	5·91 (29)	22·17 (27)	46·69 (20)	59·04 (15)	65·69 (12)	67·83 (11)	72·02 (9)	84·09 (3)
			5	3	1	11·37 (29)	24·11 (27)	38·96 (23)	50·04 (19)	54·92 (17)	59·53 (15)	63·93 (13)	68·17 (11)
						72·29 (9)	80·27 (5)	84·18 (3)					
			5	3	2	19·80 (29)	28·84 (27)	32·48 (26)	41·74 (23)	54·27 (18)	56·53 (17)	62·99 (14)	69·09 (11)
						76·87 (7)	78·77 (6)	84·41 (3)	88·14 (1)				
			5	3	3	27·81 (29)	34·56 (27)	50·17 (21)	70·40 (11)	84·75 (3)			
			5	4	0	1·79 (32)	14·47 (31)	51·34 (20)	60·02 (16)	62·06 (15)	67·99 (12)	75·53 (8)	88·21 (1)
			5	4	1	9·05 (32)	16·93 (31)	44·78 (23)	54·10 (19)	58·36 (17)	60·41 (16)	66·35 (13)	70·16 (11)
						75·71 (8)	88·23 (1)						
			5	4	2	17·44 (32)	22·45 (31)	39·18 (26)	46·71 (23)	49·01 (22)	53·40 (20)	65·33 (14)	72·65 (10)
						76·20 (8)	77·95 (7)	83·15 (4)	88·29 (1)				
			5	4	3	25·16 (32)	28·74 (31)	34·89 (29)	40·21 (27)	45·00 (25)	47·25 (24)	71·87 (11)	76·92 (8)
						78·58 (7)	85·13 (3)	88·38 (1)	90·00 (0)				
			5	4	4	32·04 (32)	34·79 (31)	42·12 (28)	77·76 (8)	83·92 (4)	88·48 (1)		
			5	5	1	11·42 (35)	49·90 (23)	57·85 (19)	61·57 (17)	72·06 (11)	81·95 (5)		
			5	5	2	17·72 (35)	44·96 (26)	51·25 (23)	67·60 (14)	79·02 (7)	82·18 (5)		
			5	5	3	24·31 (35)	40·97 (29)	45·33 (27)	73·36 (11)	82·52 (5)	85·52 (3)		
			5	5	4	30·50 (35)	38·02 (32)	40·26 (31)	78·64 (8)	82·93 (5)	88·59 (1)		

Table 9 (cont.)

h_1	k_1	l_1	h_2	k_2	l_2								
4	3	1	4	3	1	15·94 (25)	22·62 (24)	27·80 (23)	32·20 (22)	43·05 (19)	49·17 (17)	52·02 (16)	60·00 (13)
						64·97 (11)	67·38 (10)	69·75 (9)	72·08 (8)	76·66 (6)	78·91 (5)	87·80 (1)	
			4	3	2	10·49 (27)	18·76 (26)	24·43 (25)	33·11 (23)	36·76 (22)	40·11 (21)	46·22 (19)	51·75 (17)
						59·35 (14)	61·74 (13)	68·64 (10)	70·87 (9)	75·23 (7)	79·51 (5)	83·73 (3)	85·82 (2)
						87·91 (1)							
			4	3	3	19·65 (28)	24·75 (27)	32·77 (25)	42·27 (22)	45·06 (21)	55·13 (17)	70·35 (10)	76·38 (7)
						82·27 (4)	84·21 (3)	88·07 (1)					
			4	4	1	8·09 (29)	22·81 (27)	38·26 (23)	46·94 (20)	54·52 (17)	59·20 (15)	65·82 (12)	72·11 (9)
						80·17 (5)	82·15 (4)	84·12 (3)					
			4	4	3	18·29 (31)	27·35 (29)	30·95 (28)	40·03 (25)	49·97 (21)	52·22 (20)	70·31 (11)	77·62 (7)
						82·96 (4)	84·73 (3)	88·24 (1)					
			5	1	0	27·80 (23)	36·13 (21)	43·05 (19)	49·17 (17)	52·02 (16)	57·42 (14)	64·97 (11)	69·75 (9)
						72·08 (8)	85·59 (2)	87·80 (1)					
			5	1	1	25·07 (24)	33·87 (22)	40·99 (20)	47·21 (18)	52·85 (16)	63·07 (12)	67·83 (10)	76·91 (6)
						81·32 (4)	85·67 (2)						
			5	2	0	18·76 (26)	33·11 (23)	36·76 (22)	49·04 (18)	51·75 (17)	59·35 (14)	61·74 (13)	66·38 (11)
						75·23 (7)	83·73 (3)	87·91 (1)					
			5	2	1	14·82 (27)	26·47 (25)	30·76 (24)	38·03 (22)	41·24 (21)	47·13 (19)	52·50 (17)	55·05 (16)
						57·51 (15)	62·26 (13)	69·02 (10)	71·20 (9)	73·35 (8)	75·48 (7)	77·59 (6)	79·69 (5)
						83·83 (3)	90·00 (0)						
			5	2	2	17·08 (28)	31·41 (25)	34·98 (24)	44·20 (21)	49·56 (19)	56·89 (16)	65·82 (12)	72·11 (9)
						76·17 (7)	80·17 (5)	84·12 (3)					
			5	3	0	12·74 (29)	24·75 (27)	39·32 (23)	52·74 (18)	55·13 (17)	61·91 (14)	66·20 (12)	68·29 (11)
						76·38 (7)	82·27 (4)	84·21 (3)					
			5	3	1	6·02 (30)	21·85 (28)	30·47 (26)	43·17 (22)	48·47 (20)	53·37 (18)	57·97 (16)	62·35 (14)
						66·56 (12)	70·64 (10)	74·62 (8)	82·38 (4)	86·20 (2)	90·00 (0)		
			5	3	2	9·52 (31)	22·69 (29)	30·80 (27)	34·19 (26)	37·31 (25)	42·97 (23)	45·58 (22)	50·48 (20)
						57·26 (17)	65·57 (13)	67·56 (12)	69·52 (11)	71·45 (10)	73·36 (9)	79·00 (6)	80·85 (5)
						82·69 (4)	88·18 (1)						
			5	3	3	16·86 (32)	26·20 (30)	38·96 (26)	44·13 (24)	61·41 (16)	65·25 (14)	76·16 (8)	79·66 (6)
						86·57 (2)	90·00 (0)						
			5	4	0	11·45 (32)	18·29 (31)	42·69 (24)	49·97 (21)	54·41 (19)	58·62 (17)	60·66 (16)	70·31 (11)
						75·82 (8)	77·62 (7)	88·24 (1)					
			5	4	1	3·00 (33)	14·45 (32)	20·26 (31)	24·79 (30)	35·21 (27)	43·43 (24)	45·89 (23)	50·54 (21)
						54·90 (19)	57·00 (18)	63·00 (15)	64·93 (14)	66·83 (13)	70·56 (11)	74·20 (9)	75·99 (8)
						77·77 (7)	84·79 (3)	86·53 (2)	90·00 (0)				
			5	4	2	6·28 (34)	15·26 (33)	28·71 (30)	32·02 (29)	37·87 (27)	43·04 (25)	49·97 (22)	56·26 (19)
						58·25 (18)	60·20 (17)	63·99 (15)	71·24 (11)	73·00 (10)	74·74 (9)	79·90 (6)	81·59 (5)
						84·97 (3)	88·32 (1)						
			5	4	3	13·90 (35)	19·44 (34)	23·76 (33)	30·71 (31)	33·69 (30)	36·46 (29)	39·05 (28)	46·10 (25)
						50·36 (23)	56·31 (20)	58·20 (19)	65·42 (15)	70·56 (12)	72·24 (11)	78·81 (7)	82·03 (5)
						83·63 (4)	86·82 (2)	88·41 (1)					
			5	4	4	20·75 (36)	24·61 (35)	30·99 (33)	43·34 (28)	45·46 (27)	53·31 (23)	71·84 (12)	76·48 (9)
						82·54 (5)	84·04 (4)	85·53 (3)	88·51 (1)				
			5	5	1	8·65 (36)	20·98 (34)	39·74 (28)	48·77 (24)	52·83 (22)	60·38 (18)	63·94 (16)	67·39 (14)
						70·76 (12)	80·52 (6)	83·69 (4)					
			5	5	2	9·08 (37)	28·27 (33)	34·17 (31)	41·65 (28)	55·91 (21)	59·53 (19)	61·29 (18)	71·32 (12)
						76·10 (9)	79·23 (7)	85·41 (3)	86·94 (2)				
			5	5	3	14·02 (38)	29·76 (34)	35·21 (32)	52·21 (24)	55·83 (22)	65·89 (16)	78·21 (8)	81·19 (6)
						87·07 (2)							
			5	5	4	19·70 (39)	26·72 (37)	29·65 (36)	41·55 (31)	49·32 (27)	51·12 (26)	71·71 (13)	77·45 (9)
						81·67 (6)	84·46 (4)	85·85 (3)	88·62 (1)				

h_1	k_1	l_1	h_2	k_2	l_2								
4	3	2	4	3	2	15·09 (28)	26·29 (26)	30·45 (25)	43·60 (21)	46·40 (20)	56·51 (16)	61·13 (14)	67·71 (11)
						69·83 (10)	71·92 (9)	76·03 (7)	82·07 (4)	84·06 (3)	86·05 (2)		
			4	3	3	9·16 (31)	17·18 (30)	22·55 (29)	52·77 (19)	55·02 (18)	65·54 (13)	69·49 (11)	78·98 (6)
						80·84 (5)	88·18 (1)						
			4	4	1	14·13 (30)	29·22 (27)	32·81 (26)	39·12 (24)	47·25 (21)	58·85 (16)	69·17 (11)	75·01 (8)
						78·82 (6)	80·70 (5)	86·29 (2)	90·00 (0)				
			4	4	3	9·59 (34)	16·86 (33)	21·87 (32)	50·36 (22)	60·46 (17)	62·35 (16)	64·21 (15)	73·14 (10)
						76·58 (8)	86·67 (2)	88·34 (1)					
			5	1	0	33·11 (23)	36·76 (22)	46·22 (19)	49·04 (18)	51·75 (17)	59·35 (14)	61·74 (13)	66·38 (11)
						75·23 (7)	77·38 (6)						
			5	1	1	26·69 (25)	41·37 (21)	47·23 (19)	52·59 (17)	57·58 (15)	62·32 (13)	66·85 (11)	71·24 (9)
						83·85 (3)							
			5	2	0	26·29 (26)	34·15 (24)	37·52 (23)	49·07 (19)	51·63 (18)	56·51 (16)	61·13 (14)	67·71 (11)
						76·03 (7)	82·07 (4)	86·05 (2)					
			5	2	1	18·32 (28)	23·74 (27)	32·05 (25)	35·54 (24)	38·76 (23)	44·60 (21)	47·31 (20)	49·90 (19)
						57·15 (16)	59·43 (15)	63·85 (13)	65·99 (12)	72·23 (9)	74·26 (8)	76·27 (7)	80·24 (5)
						88·06 (1)	90·00 (0)						
			5	2	2	14·13 (30)	29·22 (27)	39·12 (24)	44·67 (22)	52·11 (19)	54·42 (18)	67·18 (12)	69·17 (11)
						71·14 (10)	75·01 (8)	82·57 (4)	84·43 (3)				
			5	3	0	22·55 (29)	30·70 (27)	34·11 (26)	45·52 (22)	48·03 (21)	52·77 (19)	63·52 (14)	69·49 (11)
						73·34 (9)	84·52 (3)	86·35 (2)	88·18 (1)				
			5	3	1	13·34 (31)	24·46 (29)	32·06 (27)	38·31 (25)	43·79 (23)	53·39 (19)	57·75 (17)	61·91 (15)
						65·92 (13)	69·80 (11)	73·59 (9)	80·97 (5)	84·60 (3)	88·20 (1)		
			5	3	2	6·24 (33)	15·43 (32)	20·96 (31)	29·12 (29)	32·49 (28)	35·58 (27)	41·14 (25)	46·14 (23)
						52·95 (20)	59·20 (17)	61·19 (16)	63·14 (15)	66·95 (13)	70·65 (11)	74·27 (9)	76·06 (8)
						77·83 (7)	83·08 (4)	88·27 (1)					
			5	3	3	7·63 (35)	20·85 (33)	28·61 (31)	49·36 (23)	53·51 (21)	61·22 (17)	68·40 (13)	71·85 (11)
						75·23 (9)	78·57 (7)	81·86 (5)	85·13 (3)				
			5	4	0	21·87 (32)	25·97 (31)	35·71 (28)	41·06 (26)	48·16 (23)	50·36 (22)	69·63 (12)	76·58 (8)
						78·29 (7)	79·98 (6)	86·67 (2)	88·34 (1)				
			5	4	1	13·04 (34)	18·99 (33)	27·35 (31)	30·73 (30)	33·80 (29)	39·32 (27)	41·84 (26)	44·25 (25)
						48·77 (23)	57·02 (19)	58·95 (18)	64·55 (15)	71·63 (11)	73·35 (10)	75·06 (9)	80·10 (6)
						85·07 (3)	86·71 (2)	88·36 (1)					
			5	4	2	4·76 (36)	14·34 (35)	19·75 (34)	27·65 (32)	30·89 (31)	33·85 (30)	39·19 (28)	41·63 (27)
						52·48 (22)	56·38 (20)	60·11 (18)	65·47 (15)	67·20 (14)	70·60 (12)	73·93 (10)	80·44 (6)
						82·04 (5)	83·64 (4)	85·24 (3)	88·41 (1)	90·00 (0)			
			5	4	3	3·69 (38)	13·67 (37)	23·20 (35)	26·76 (34)	46·94 (26)	48·96 (25)	56·53 (21)	60·07 (19)
						63·48 (17)	66·80 (15)	68·43 (14)	73·21 (11)	74·77 (10)	79·41 (7)	82·45 (5)	85·48 (3)
						86·99 (2)							
			5	4	4	10·32 (40)	16·41 (39)	20·83 (38)	53·82 (24)	55·55 (23)	63·72 (18)	66·83 (16)	69·86 (14)
						77·21 (9)	80·09 (7)	81·51 (6)	90·00 (0)				
			5	5	1	15·83 (37)	30·90 (33)	41·06 (29)	45·41 (27)	56·90 (21)	70·24 (13)	76·47 (9)	79·51 (7)
						85·53 (3)	88·51 (1)						
			5	5	2	9·76 (39)	24·53 (36)	33·50 (33)	38·43 (31)	52·66 (24)	64·56 (17)	66·15 (16)	70·82 (13)
						76·85 (9)	84·20 (4)	85·65 (3)	88·55 (1)				
			5	5	3	7·61 (41)	19·47 (39)	26·56 (37)	45·49 (29)	59·49 (21)	62·66 (19)	65·73 (17)	71·68 (13)
						74·58 (11)	80·26 (7)	88·61 (1)					
			5	5	4	10·62 (43)	16·26 (42)	20·42 (41)	51·89 (27)	59·81 (22)	61·31 (21)	65·70 (18)	72·71 (13)
						75·44 (11)	78·13 (9)	86·07 (3)	87·38 (2)				

J

Table 9 (cont.)

h_1	k_1	l_1	h_2	k_2	l_2								
4	3	3	4	3	3	13·93 (33)	61·93 (16)	63·82 (15)	74·65 (9)	86·63 (2)			
			4	4	1	22·26 (31)	33·29 (28)	41·72 (25)	53·34 (20)	77·94 (7)	83·14 (4)	88·29 (1)	
			4	4	3	7·70 (37)	15·38 (36)	59·41 (19)	69·62 (13)	71·25 (12)	82·30 (5)		
			5	1	0	39·32 (23)	50·28 (19)	52·74 (18)	55·13 (17)	66·20 (12)	68·29 (11)		
			5	1	1	30·89 (26)	43·44 (22)	48·69 (20)	58·12 (16)	62·48 (14)	74·69 (8)		
			5	2	0	34·11 (26)	42·91 (23)	48·03 (21)	63·52 (14)	73·34 (9)	77·12 (7)		
			5	2	1	24·77 (29)	35·50 (26)	38·48 (25)	43·93 (23)	51·23 (20)	57·84 (17)	65·98 (13)	69·85 (11)
						71·75 (10)	80·99 (5)	82·81 (4)					
			5	2	2	17·19 (32)	30·03 (29)	53·34 (20)	59·50 (17)	67·16 (13)	76·18 (8)	88·29 (1)	
			5	3	0	31·47 (29)	37·43 (27)	45·10 (24)	71·12 (11)	79·84 (6)	84·94 (3)		
			5	3	1	21·93 (32)	29·58 (30)	35·74 (28)	41·09 (26)	45·91 (24)	54·57 (20)	66·06 (14)	73·15 (10)
						76·59 (8)	79·98 (6)	86·68 (2)	90·00 (0)				
			5	3	2	13·16 (35)	23·35 (33)	27·09 (32)	50·22 (23)	54·25 (21)	61·77 (17)	63·57 (16)	67·08 (14)
						75·50 (9)	82·00 (5)	85·21 (3)	86·81 (2)				
			5	3	3	6·37 (38)	19·69 (36)	58·46 (20)	61·92 (18)	71·71 (12)	80·97 (6)	87·00 (2)	
			5	4	0	31·01 (32)	33·87 (31)	43·68 (27)	77·63 (8)	85·39 (3)	88·47 (1)		
			5	4	1	22·15 (35)	25·88 (34)	34·88 (31)	39·88 (29)	42·19 (28)	52·51 (23)	73·08 (11)	79·32 (7)
						82·40 (5)	83·92 (4)	86·97 (2)	88·48 (1)				
			5	4	2	13·71 (38)	18·93 (37)	26·52 (35)	48·34 (26)	50·27 (25)	60·94 (19)	69·03 (14)	73·67 (11)
						79·69 (7)	82·66 (5)	87·07 (2)					
			5	4	3	6·07 (41)	14·04 (40)	18·93 (39)	56·09 (23)	57·75 (22)	65·65 (17)	68·67 (15)	75·96 (10)
						77·39 (9)	78·81 (8)	88·61 (1)					
			5	4	4	1·84 (44)	12·37 (43)	62·98 (20)	64·43 (19)	72·82 (13)	75·53 (11)	84·79 (4)	
			5	5	1	24·14 (38)	35·26 (34)	39·78 (32)	51·36 (26)	78·92 (8)	84·49 (4)	87·25 (2)	
			5	5	2	16·89 (41)	27·52 (38)	47·41 (29)	59·11 (22)	75·12 (11)	79·24 (8)	88·66 (1)	
			5	5	3	10·77 (44)	20·32 (42)	54·51 (26)	66·30 (18)	71·79 (14)	74·46 (12)	84·88 (4)	
			5	5	4	7·17 (47)	13·82 (46)	60·95 (23)	68·97 (17)	70·26 (16)	72·81 (14)	81·50 (7)	

h_1	k_1	l_1	h_2	k_2	l_2								
4	4	1	4	4	1	20·05 (31)	43·34 (24)	61·00 (16)	75·97 (8)	88·26 (1)			
			4	4	3	17·91 (35)	29·55 (32)	37·96 (29)	49·27 (24)	77·44 (8)	85·32 (3)	90·00 (0)	
			5	1	0	34·98 (24)	44·20 (21)	49·56 (19)	56·89 (16)	72·11 (9)	88·04 (1)		
			5	1	1	33·12 (25)	39·60 (23)	55·28 (17)	59·83 (15)	64·18 (13)	80·36 (5)	84·23 (3)	
			5	2	0	25·16 (28)	44·67 (22)	54·42 (18)	65·15 (13)	67·18 (12)	84·43 (3)		
			5	2	1	22·83 (29)	30·89 (27)	34·28 (26)	45·64 (22)	55·10 (18)	57·30 (17)	63·58 (14)	65·60 (13)
						69·54 (11)	73·38 (9)	77·15 (7)	88·18 (1)				
			5	2	2	24·62 (30)	38·01 (26)	50·48 (21)	64·90 (14)	70·53 (11)	72·36 (10)	81·29 (5)	
			5	3	0	17·19 (32)	46·63 (23)	59·50 (17)	76·18 (8)	77·94 (7)			
			5	3	1	13·83 (33)	24·19 (31)	37·40 (27)	51·84 (21)	56·01 (19)	67·51 (13)	71·12 (11)	74·64 (9)
						78·11 (7)	84·94 (3)						
			5	3	2	16·23 (34)	28·91 (31)	32·09 (30)	45·09 (25)	64·94 (15)	73·60 (10)	75·28 (9)	80·25 (6)
						88·38 (1)							
			5	3	3	21·70 (35)	39·66 (29)	59·71 (19)	73·02 (11)	82·37 (5)			
			5	4	0	11·84 (36)	49·27 (24)	55·19 (21)	64·22 (16)	72·60 (11)	83·76 (4)		
			5	4	1	6·36 (37)	19·93 (35)	41·23 (28)	47·82 (25)	57·51 (20)	62·83 (17)	66·24 (15)	71·20 (12)
						79·16 (7)	82·28 (5)	85·38 (3)					
			5	4	2	9·56 (38)	28·08 (34)	33·86 (32)	41·19 (29)	51·48 (24)	60·46 (19)	65·47 (16)	70·28 (13)
						78·02 (8)	81·04 (6)	85·54 (3)	87·03 (2)				
			5	4	3	16·24 (39)	27·59 (36)	35·67 (33)	46·42 (28)	55·51 (23)	72·82 (12)	77·20 (9)	80·08 (7)
						84·35 (4)	88·59 (1)						
			5	4	4	22·74 (40)	31·45 (37)	42·45 (32)	51·50 (27)	79·37 (8)	83·38 (5)	90·00 (0)	

h_1	k_1	l_1	h_2	k_2	l_2								
			5	5	1	1·98 (41)	18·07 (39)	45·02 (29)	59·21 (21)	62·41 (19)	74·45 (11)	88·60 (1)	
			5	5	2	5·77 (42)	25·82 (38)	38·58 (33)	56·99 (23)	66·25 (17)	80·45 (7)	87·28 (2)	
			5	5	3	12·96 (43)	33·01 (37)	52·27 (27)	72·87 (13)	86·10 (3)			
			5	5	4	19·47 (44)	28·53 (41)	39·52 (36)	48·38 (31)	78·88 (9)	85·08 (4)	88·77 (1)	
h_1	k_1	l_1	h_2	k_2	l_2								
4	4	3	4	4	3	12·68 (40)	55·88 (23)	67·03 (16)	77·32 (9)	78·75 (8)			
			5	1	0	42·69 (24)	45·21 (23)	54·41 (19)	58·62 (17)	60·66 (16)	70·31 (11)		
			5	1	1	35·76 (27)	46·27 (23)	50·86 (21)	55·18 (19)	63·20 (15)	67·00 (13)	77·85 (7)	
			5	2	0	35·71 (28)	41·06 (26)	48·16 (23)	66·05 (14)	69·63 (12)	78·29 (7)		
			5	2	1	27·88 (31)	31·20 (30)	39·66 (27)	44·53 (25)	51·15 (22)	57·20 (19)	59·12 (18)	64·68 (15)
						71·72 (11)	73·43 (10)	75·13 (9)	85·09 (3)				
			5	2	2	22·43 (34)	32·57 (31)	53·27 (22)	60·70 (18)	65·93 (15)	80·61 (6)	88·44 (1)	
			5	3	0	31·01 (32)	39·04 (29)	43·68 (27)	72·87 (11)	77·63 (8)	85·39 (3)		
			5	3	1	22·49 (35)	29·41 (33)	35·08 (31)	40·04 (29)	48·70 (25)	52·62 (23)	66·67 (15)	73·12 (11)
						79·35 (7)	82·42 (5)	88·49 (1)					
			5	3	2	15·69 (38)	20·38 (37)	27·54 (35)	48·80 (26)	57·86 (21)	61·23 (19)	69·23 (14)	73·82 (11)
						82·72 (5)	85·64 (3)	87·10 (2)					
			5	3	3	12·45 (41)	21·75 (39)	56·79 (23)	66·12 (17)	69·07 (15)	77·62 (9)	88·64 (1)	
			5	4	0	28·59 (36)	38·69 (32)	40·88 (31)	78·75 (8)	84·40 (4)	88·60 (1)		
			5	4	1	19·98 (39)	29·83 (36)	32·50 (35)	37·32 (33)	47·57 (28)	49·41 (27)	73·19 (12)	80·29 (7)
						83·08 (5)	84·47 (4)	85·85 (3)	88·62 (1)				
			5	4	2	12·09 (42)	21·37 (40)	24·78 (39)	45·70 (30)	56·03 (24)	57·62 (23)	68·13 (16)	76·54 (10)
						77·91 (9)	80·62 (7)	87·33 (2)	90·00 (0)				
			5	4	3	6·34 (45)	13·64 (44)	18·25 (43)	53·39 (27)	63·79 (20)	65·19 (19)	73·31 (13)	75·94 (11)
						83·66 (5)	84·93 (4)						
			5	4	4	6·82 (48)	13·53 (47)	60·23 (24)	69·41 (17)	70·67 (16)	71·92 (15)	80·47 (8)	
			5	5	1	19·89 (43)	31·47 (39)	35·99 (37)	47·32 (31)	78·65 (9)	86·24 (3)	88·75 (1)	
			5	5	2	12·15 (46)	23·96 (43)	43·73 (34)	54·98 (27)	73·96 (13)	82·67 (6)	86·34 (3)	
			5	5	3	4·95 (49)	17·14 (47)	50·93 (31)	62·12 (23)	69·78 (17)	79·46 (9)	81·82 (7)	
			5	5	4	1·59 (52)	11·36 (51)	57·43 (28)	66·19 (21)	68·58 (19)	76·66 (12)	77·79 (11)	
h_1	k_1	l_1	h_2	k_2	l_2								
5	1	0	5	1	0	15·94 (25)	22·62 (24)	67·38 (10)	78·91 (5)	87·80 (1)	90·00 (0)		
			5	1	1	11·10 (26)	25·07 (24)	67·83 (10)	76·91 (6)	81·32 (4)	90·00 (0)		
			5	2	0	10·49 (27)	24·43 (25)	33·11 (23)	56·89 (15)	68·64 (10)	79·51 (5)	85·82 (2)	
			5	2	1	14·82 (27)	21·42 (26)	30·76 (24)	34·56 (23)	57·51 (15)	66·80 (11)	69·02 (10)	71·20 (9)
						75·48 (7)	79·69 (5)	83·83 (3)	90·00 (0)				
			5	2	2	22·81 (27)	38·26 (23)	59·20 (15)	65·82 (12)	74·15 (8)	80·17 (5)		
			5	3	0	19·65 (28)	32·77 (25)	42·27 (22)	47·73 (20)	59·70 (15)	70·35 (10)	80·32 (5)	84·21 (3)
			5	3	1	21·85 (28)	30·47 (26)	37·29 (24)	43·17 (22)	48·47 (20)	57·97 (16)	62·35 (14)	70·64 (10)
						74·62 (8)	86·20 (2)	90·00 (0)					
			5	3	2	27·03 (28)	30·80 (27)	42·97 (23)	45·58 (22)	50·48 (20)	57·26 (17)	61·50 (15)	65·57 (13)
						71·45 (10)	77·13 (7)	80·85 (5)					
			5	3	3	33·13 (28)	48·86 (22)	53·26 (20)	57·43 (18)	68·97 (12)	72·60 (10)		
			5	4	0	27·35 (29)	40·03 (25)	49·97 (21)	52·22 (20)	62·25 (15)	81·19 (5)	82·96 (4)	
			5	4	1	28·65 (29)	38·11 (26)	40·84 (25)	43·43 (24)	50·54 (21)	54·90 (19)	63·00 (15)	72·39 (10)
						74·20 (9)	88·27 (1)	90·00 (0)					
			5	4	2	32·02 (29)	37·87 (27)	43·04 (25)	47·75 (23)	49·97 (22)	52·13 (21)	58·25 (18)	63·99 (15)
						65·84 (14)	79·90 (6)	81·59 (5)					

Table 9 (cont.)

h_1	k_1	l_1	h_2	k_2	l_2								
5 (cont.)	1	0	5	4	3	36·46 (29)	39·05 (28)	46·10 (25)	50·36 (23)	52·40 (22)	54·38 (21)	56·31 (20)	58·20 (19)
						61·87 (17)	65·42 (15)	72·24 (11)	73·90 (10)				
			5	4	4	41·12 (29)	49·50 (25)	51·43 (24)	56·94 (21)	65·44 (16)	67·07 (15)		
			5	5	1	34·53 (30)	44·44 (26)	48·77 (24)	56·69 (20)	74·06 (10)	90·00 (0)		
			5	5	2	36·81 (30)	43·90 (27)	52·13 (23)	57·74 (20)	66·40 (15)	82·33 (5)		
			5	5	3	40·01 (30)	44·36 (28)	55·83 (22)	59·29 (20)	75·21 (10)			
			5	5	4	43·60 (30)	45·57 (29)	52·88 (25)	59·54 (21)	61·13 (20)	68·77 (15)		

h_1	k_1	l_1	h_2	k_2	l_2								
5	1	1	5	1	1	22·19 (25)	31·59 (23)	65·96 (11)	70·53 (9)	87·88 (1)			
			5	2	0	15·23 (27)	34·72 (23)	57·58 (15)	75·51 (7)	79·71 (5)	83·85 (3)		
			5	2	1	10·32 (28)	24·00 (26)	32·51 (24)	39·38 (22)	55·79 (16)	60·53 (14)	65·06 (12)	73·67 (8)
						77·83 (6)	81·92 (4)	85·97 (2)					
			5	2	2	13·70 (29)	33·12 (25)	45·29 (21)	55·28 (17)	64·18 (13)	76·44 (7)	84·23 (3)	
			5	3	0	22·46 (28)	43·44 (22)	48·69 (20)	70·73 (10)	74·69 (8)	86·22 (2)		
			5	3	1	19·37 (29)	28·56 (27)	41·57 (23)	46·91 (21)	51·82 (19)	64·98 (13)	69·03 (11)	72·98 (9)
						76·84 (7)	84·40 (3)						
			5	3	2	20·51 (30)	35·74 (26)	41·47 (24)	46·62 (22)	51·36 (20)	55·81 (18)	68·00 (12)	75·54 (8)
						86·42 (2)							
			5	3	3	24·52 (31)	42·80 (25)	47·54 (23)	56·11 (19)	60·07 (17)	67·57 (13)	78·14 (7)	
			5	4	0	29·35 (29)	41·29 (25)	50·86 (21)	63·20 (15)	74·31 (9)	88·28 (1)		
			5	4	1	27·02 (30)	33·75 (28)	39·46 (26)	44·55 (24)	49·21 (22)	53·56 (20)	61·63 (16)	65·43 (14)
						79·74 (6)	83·18 (4)						
			5	4	2	27·21 (31)	39·23 (27)	48·71 (23)	56·97 (19)	60·81 (17)	68·10 (13)	71·60 (11)	75·04 (9)
						88·36 (1)							
			5	4	3	29·43 (32)	40·35 (28)	44·96 (26)	49·22 (24)	53·22 (22)	60·67 (18)	64·19 (16)	67·60 (14)
						70·94 (12)	80·60 (6)						
			5	4	4	32·73 (33)	42·33 (29)	50·41 (25)	57·64 (21)	61·03 (19)	64·32 (17)	73·72 (11)	
			5	5	1	33·34 (31)	38·60 (29)	55·53 (21)	59·20 (19)	66·16 (15)	82·26 (5)		
			5	5	2	33·06 (32)	42·84 (28)	54·82 (22)	58·41 (20)	61·87 (18)	74·82 (10)	90·00 (0)	
			5	5	3	34·23 (33)	47·43 (27)	51·22 (25)	54·81 (23)	64·79 (17)	67·92 (15)	82·80 (5)	
			5	5	4	36·35 (34)	44·71 (30)	51·98 (26)	55·35 (24)	61·72 (20)	67·73 (16)	76·30 (10)	

h_1	k_1	l_1	h_2	k_2	l_2								
5	2	0	5	2	0	30·45 (25)	43·60 (21)	46·40 (20)	69·83 (10)	82·07 (4)	90·00 (0)		
			5	2	1	10·52 (29)	23·74 (27)	38·76 (23)	44·60 (21)	47·31 (20)	59·43 (15)	65·99 (12)	72·23 (9)
						74·26 (8)	80·24 (5)	88·06 (1)	90·00 (0)				
			5	2	2	20·37 (29)	47·25 (21)	49·72 (20)	63·09 (14)	78·82 (6)	90·00 (0)		
			5	3	0	9·16 (31)	37·23 (25)	52·77 (19)	61·46 (15)	71·43 (10)	78·98 (6)	80·84 (5)	
			5	3	1	13·34 (31)	32·06 (27)	38·31 (25)	43·79 (23)	53·39 (19)	57·75 (17)	61·91 (15)	65·92 (13)
						69·80 (11)	80·97 (5)	88·20 (1)					
			5	3	2	20·96 (31)	29·12 (29)	41·14 (25)	50·76 (21)	52·95 (20)	55·09 (19)	61·19 (16)	70·65 (11)
						81·34 (5)	83·08 (4)	90·00 (0)					
			5	3	3	28·61 (31)	44·93 (25)	53·51 (21)	57·45 (19)	75·23 (9)	81·86 (5)		
			5	4	0	16·86 (33)	29·54 (30)	43·53 (25)	54·55 (20)	60·46 (17)	73·14 (10)	76·58 (8)	
			5	4	1	18·99 (33)	30·73 (30)	39·32 (27)	48·77 (23)	50·92 (22)	58·95 (18)	60·85 (17)	64·55 (15)
						68·13 (13)	73·35 (10)	81·76 (5)	85·07 (3)				
			5	4	2	24·91 (33)	33·85 (30)	36·60 (29)	48·37 (24)	54·46 (21)	56·38 (20)	60·11 (18)	61·93 (17)
						63·71 (16)	73·93 (10)	86·83 (2)	90·00 (0)				
			5	4	3	29·93 (33)	35·50 (31)	38·02 (30)	46·94 (26)	48·96 (25)	52·84 (23)	60·07 (19)	63·48 (17)
						68·43 (14)	74·77 (10)	79·41 (7)	82·45 (5)				
			5	4	4	35·74 (33)	42·45 (30)	46·47 (28)	65·28 (17)	72·83 (12)	75·76 (10)		

			h_2	k_2	l_2								
			5	5	1	24·48 (35)	45·41 (27)	53·27 (23)	67·04 (15)	82·53 (5)			
			5	5	2	27·82 (35)	42·88 (29)	57·95 (21)	59·64 (20)	67·73 (15)	90·00 (0)		
			5	5	3	32·21 (35)	41·46 (31)	52·82 (25)	62·66 (19)	68·74 (15)	83·06 (5)		
			5	5	4	36·87 (35)	41·04 (33)	46·71 (30)	67·13 (17)	69·95 (15)	76·79 (10)		

h_1	k_1	l_1	h_2	k_2	l_2								
5	2	1	5	2	1	14·84 (29)	21·04 (28)	33·56 (25)	42·83 (22)	45·57 (21)	48·19 (20)	50·70 (19)	55·48 (17)
						62·18 (14)	64·32 (13)	76·51 (7)	78·46 (6)	82·34 (4)	84·26 (3)	88·09 (1)	
			5	2	2	9·85 (31)	30·89 (27)	43·03 (23)	45·64 (22)	52·85 (19)	55·10 (18)	69·54 (11)	73·38 (9)
						86·36 (2)	88·18 (1)						
			5	3	0	13·92 (31)	28·75 (28)	38·48 (25)	46·46 (22)	51·23 (20)	53·49 (19)	65·98 (13)	69·85 (11)
						71·75 (10)	77·34 (7)	80·99 (5)	88·21 (1)				
			5	3	1	9·05 (32)	22·21 (30)	36·64 (26)	42·21 (24)	47·24 (22)	51·89 (20)	56·26 (18)	60·41 (16)
						68·26 (12)	75·71 (8)	79·33 (6)	82·91 (4)	86·46 (2)			
			5	3	2	12·21 (33)	18·60 (32)	30·81 (29)	36·90 (27)	39·64 (26)	44·70 (24)	47·06 (23)	51·54 (21)
						57·78 (18)	59·77 (17)	61·71 (16)	65·50 (14)	70·99 (11)	74·54 (9)	78·03 (7)	79·76 (6)
						84·90 (3)	88·30 (1)						
			5	3	3	18·80 (34)	38·78 (28)	43·62 (26)	52·23 (22)	63·55 (16)	67·06 (14)	77·13 (8)	83·61 (4)
						86·81 (2)							
			5	4	0	19·79 (33)	31·20 (30)	34·22 (29)	44·53 (25)	53·22 (21)	61·01 (17)	64·68 (15)	66·47 (14)
						68·24 (13)	73·43 (10)	80·15 (6)	85·09 (3)				
			5	4	1	16·70 (34)	25·64 (32)	29·15 (31)	35·22 (29)	40·48 (27)	49·61 (23)	57·64 (19)	59·53 (18)
						61·39 (17)	63·21 (16)	68·52 (13)	71·95 (11)	75·31 (9)	76·98 (8)	86·77 (2)	88·39 (1)
			5	4	2	17·72 (35)	26·08 (33)	29·43 (32)	32·47 (31)	37·88 (29)	40·35 (28)	47·12 (25)	49·22 (24)
						51·25 (23)	55·14 (21)	58·86 (19)	62·44 (17)	64·19 (16)	65·91 (15)	69·28 (13)	70·94 (12)
						72·58 (11)	77·42 (8)	79·02 (7)	83·75 (4)	85·32 (3)			
			5	4	3	21·64 (36)	25·35 (35)	31·56 (33)	36·83 (31)	39·23 (30)	41·52 (29)	43·70 (28)	45·80 (27)
						53·57 (23)	57·17 (21)	58·91 (20)	60·62 (19)	62·31 (18)	67·21 (15)	68·81 (14)	70·39 (13)
						71·95 (12)	76·56 (9)	79·59 (7)	87·04 (2)	88·52 (1)			
			5	4	4	26·52 (37)	34·69 (34)	37·06 (33)	45·47 (29)	51·04 (26)	56·21 (23)	59·48 (21)	65·73 (17)
						70·21 (14)	71·68 (13)	80·25 (7)	81·66 (6)				
			5	5	1	23·02 (36)	29·63 (34)	35·11 (32)	44·29 (28)	55·78 (22)	59·25 (20)	62·60 (18)	65·85 (16)
						69·03 (14)	75·19 (10)	90·00 (0)					
			5	5	2	23·18 (37)	32·36 (34)	34·93 (33)	49·76 (26)	51·60 (25)	53·40 (24)	65·02 (17)	66·58 (16)
						68·12 (15)	71·16 (13)	82·86 (5)					
			5	5	3	25·41 (38)	31·16 (36)	40·48 (32)	44·51 (30)	51·83 (26)	55·22 (24)	61·62 (20)	64·67 (18)
						70·56 (14)	73·43 (12)	76·25 (10)	90·00 (0)				
			5	5	4	28·78 (39)	31·35 (38)	38·13 (35)	45·84 (31)	51·00 (28)	55·82 (25)	60·37 (22)	64·72 (19)
						70·30 (15)	74·35 (12)	75·69 (11)	83·55 (5)				

h_1	k_1	l_1	h_2	k_2	l_2								
5	2	2	5	2	2	40·75 (25)	43·34 (24)	58·99 (17)	61·00 (16)	83·04 (4)			
			5	3	0	22·26 (31)	41·72 (25)	55·44 (19)	61·47 (16)	81·42 (5)	83·14 (4)		
			5	3	1	13·83 (33)	31·43 (29)	37·40 (27)	47·41 (23)	51·84 (21)	59·99 (17)	71·12 (11)	74·64 (9)
						78·11 (7)	84·94 (3)	88·31 (1)					
			5	3	2	8·75 (35)	35·02 (29)	40·32 (27)	42·76 (26)	49·50 (23)	53·63 (21)	64·94 (15)	66·71 (14)
						75·28 (9)	80·25 (6)	88·38 (1)					
			5	3	3	10·82 (37)	34·62 (31)	48·42 (25)	59·71 (19)	69·81 (13)	73·02 (11)	88·48 (1)	
			5	4	0	26·21 (33)	35·35 (30)	60·70 (18)	62·47 (17)	74·22 (10)	86·88 (2)		
			5	4	1	19·93 (35)	30·73 (32)	33·62 (31)	41·23 (28)	51·84 (23)	59·31 (19)	66·24 (15)	69·56 (13)
						71·20 (12)	77·59 (8)	79·16 (7)	85·38 (3)				

Table 9 (cont.)

h_1	k_1	l_1	h_2	k_2	l_2								
5 (cont.)	2	2	5	4	2	16·23 (37)	28·08 (34)	41·19 (29)	43·40 (28)	47·57 (26)	56·98 (21)	68·70 (14)	70·28 (13)
						71·86 (12)	78·02 (8)	81·04 (6)					
			5	4	3	16·24 (39)	27·59 (36)	35·67 (33)	48·34 (27)	53·78 (24)	55·51 (23)	65·26 (17)	66·80 (16)
						71·33 (13)	74·29 (11)	84·35 (4)	85·76 (3)				
			5	4	4	19·03 (41)	28·82 (38)	54·80 (25)	59·52 (22)	65·48 (18)	78·02 (9)	87·36 (2)	
			5	5	1	25·59 (37)	36·45 (33)	52·45 (25)	65·52 (17)	68·55 (15)	71·53 (13)	83·00 (5)	
			5	5	2	22·50 (39)	42·75 (31)	44·71 (30)	63·25 (19)	74·90 (11)	76·30 (10)		
			5	5	3	21·69 (41)	37·51 (35)	48·91 (29)	61·58 (21)	70·13 (15)	78·23 (9)	83·49 (5)	
			5	5	4	22·87 (43)	31·01 (40)	54·65 (27)	60·47 (23)	64·62 (20)	81·37 (7)	90·00 (0)	

h_1	k_1	l_1	h_2	k_2	l_2								
5	3	0	5	3	0	28·07 (30)	42·67 (25)	61·93 (16)	63·82 (15)	74·65 (9)	90·00 (0)		
			5	3	1	9·73 (34)	29·58 (30)	35·74 (28)	50·38 (22)	54·57 (20)	58·55 (18)	62·37 (16)	66·06 (14)
						69·64 (12)	73·15 (10)	83·34 (4)	90·00 (0)				
			5	3	2	18·93 (34)	30·41 (31)	33·42 (30)	45·93 (25)	54·25 (21)	58·09 (19)	63·57 (16)	75·50 (9)
						82·00 (5)	88·41 (1)	90·00 (0)					
			5	3	3	27·23 (34)	38·32 (30)	51·12 (24)	65·26 (16)	80·97 (6)	90·00 (0)		
			5	4	0	7·70 (37)	20·38 (35)	47·96 (25)	57·61 (20)	66·31 (15)	69·62 (13)	71·25 (12)	82·30 (5)
			5	4	1	11·73 (37)	22·15 (35)	42·19 (28)	52·51 (23)	54·40 (22)	58·04 (20)	63·26 (17)	69·88 (13)
						74·66 (10)	79·32 (7)	82·40 (5)					
			5	4	2	18·93 (37)	26·52 (35)	37·58 (31)	48·34 (26)	50·27 (25)	55·78 (22)	60·94 (19)	69·03 (14)
						70·59 (13)	82·66 (5)	87·07 (2)					
			5	4	3	26·18 (37)	31·91 (35)	34·45 (34)	43·31 (30)	45·30 (29)	49·09 (27)	67·17 (16)	71·62 (13)
						74·53 (11)	83·03 (5)	85·83 (3)	90·00 (0)				
			5	4	4	32·81 (37)	37·34 (35)	43·37 (32)	72·82 (13)	79·53 (8)	83·48 (5)		
			5	5	1	16·14 (40)	47·75 (28)	58·11 (22)	61·30 (20)	76·10 (10)			
			5	5	2	21·01 (40)	43·66 (31)	54·31 (25)	63·68 (19)	76·50 (10)	83·30 (5)		
			5	5	3	26·74 (40)	40·61 (34)	47·95 (30)	69·07 (16)	77·10 (10)	90·00 (0)		
			5	5	4	32·39 (40)	38·64 (37)	42·37 (35)	74·07 (13)	77·81 (10)	83·94 (5)		

h_1	k_1	l_1	h_2	k_2	l_2								
5	3	1	5	3	1	19·46 (33)	27·66 (31)	34·05 (29)	44·42 (25)	48·92 (23)	57·12 (19)	60·94 (17)	64·62 (15)
						68·20 (13)	75·10 (9)	78·46 (7)	88·36 (1)				
			5	3	2	9·20 (36)	21·20 (34)	28·66 (32)	39·85 (28)	44·53 (26)	48·85 (24)	52·90 (22)	60·42 (18)
						63·98 (16)	67·43 (14)	77·33 (8)	80·53 (6)	83·70 (4)	86·86 (2)		
			5	3	3	17·49 (37)	31·72 (33)	36·96 (31)	41·62 (29)	45·89 (27)	60·68 (19)	70·42 (13)	73·53 (11)
						85·56 (3)	88·52 (1)						
			5	4	0	12·38 (37)	22·49 (35)	40·04 (29)	48·70 (25)	56·33 (21)	59·90 (19)	63·34 (17)	66·67 (15)
						69·93 (13)	73·12 (11)	79·35 (7)	82·42 (5)				
			5	4	1	7·64 (38)	20·12 (36)	27·53 (34)	33·42 (32)	43·09 (28)	47·30 (26)	51·25 (24)	54·98 (22)
						62·00 (18)	65·33 (16)	68·58 (14)	71·76 (12)	81·00 (6)	84·01 (4)	87·01 (2)	
			5	4	2	10·67 (39)	21·20 (37)	28·13 (35)	33·74 (33)	38·64 (31)	43·05 (29)	47·13 (27)	54·58 (23)
						58·05 (21)	61·40 (19)	64·64 (17)	67·79 (15)	73·91 (11)	76·89 (9)	79·84 (7)	85·66 (3)
						88·56 (1)							
			5	4	3	17·02 (40)	24·72 (38)	35·63 (34)	40·10 (32)	44·18 (30)	51·57 (26)	54·99 (24)	58·27 (22)
						61·44 (20)	67·51 (16)	73·33 (12)	76·17 (10)	78·98 (8)	81·75 (6)	84·51 (4)	87·26 (2)
			5	4	4	23·37 (41)	29·17 (39)	34·07 (37)	42·37 (33)	46·05 (31)	52·81 (27)	67·63 (17)	73·08 (13)
						78·38 (9)	86·15 (3)	88·72 (1)					
			5	5	1	13·97 (41)	22·62 (39)	38·64 (33)	50·28 (27)	53·72 (25)	57·02 (23)	66·27 (17)	69·20 (15)
						74·91 (11)	77·70 (9)	83·20 (5)					
			5	5	2	14·96 (42)	29·06 (38)	34·10 (36)	46·36 (30)	53·27 (26)	56·49 (24)	62·61 (20)	71·21 (14)
						73·98 (12)	76·70 (10)	79·40 (8)	90·00 (0)				

h_1	k_1	l_1	h_2	k_2	l_2								
			5	5	3	18.87 (43)	30.88 (39)	35.49 (37)	39.63 (35)	50.34 (29)	56.62 (25)	62.48 (21)	73.38 (13)
						75.99 (11)	81.14 (7)	83.68 (5)					
			5	5	4	23.73 (44)	29.09 (42)	33.67 (40)	41.49 (36)	48.26 (32)	51.38 (30)	68.01 (18)	73.06 (14)
						77.99 (10)	80.42 (8)	82.83 (6)	90.00 (0)				

h_1	k_1	l_1	h_2	k_2	l_2								
5	3	2	5	3	2	13.17 (37)	26.53 (34)	35.33 (31)	37.86 (30)	40.26 (29)	46.83 (26)	48.86 (25)	58.24 (20)
						60.00 (19)	69.99 (13)	71.59 (12)	73.17 (11)	76.30 (9)	83.96 (4)	88.49 (1)	
			5	3	3	8.29 (40)	27.05 (36)	32.74 (34)	46.16 (28)	53.58 (24)	57.03 (22)	66.68 (16)	69.74 (14)
						75.68 (10)	81.46 (6)	84.32 (4)					
			5	4	0	20.38 (37)	27.54 (35)	33.28 (33)	40.53 (30)	54.36 (23)	56.13 (22)	64.49 (17)	70.77 (13)
						75.32 (10)	79.78 (7)	82.72 (5)	87.10 (2)				
			5	4	1	12.52 (39)	22.16 (37)	25.69 (36)	28.83 (35)	34.31 (33)	41.33 (30)	45.50 (28)	47.48 (27)
						59.96 (20)	63.22 (18)	64.82 (17)	67.95 (15)	69.49 (14)	71.01 (13)	72.52 (12)	74.02 (11)
						79.91 (7)	85.69 (3)	87.13 (2)					
			5	4	2	7.48 (41)	19.42 (39)	29.47 (36)	37.06 (33)	39.30 (32)	41.44 (31)	49.24 (27)	54.52 (24)
						56.21 (23)	65.73 (17)	67.24 (16)	71.68 (13)	73.13 (12)	74.57 (11)	77.43 (9)	78.85 (8)
						84.45 (4)	85.84 (3)	88.61 (1)					
			5	4	3	9.43 (43)	15.52 (42)	19.85 (41)	26.53 (39)	29.33 (38)	31.91 (37)	44.67 (31)	48.29 (29)
						53.38 (26)	56.59 (24)	59.69 (22)	61.20 (21)	64.16 (19)	67.05 (17)	72.65 (13)	75.38 (11)
						79.42 (8)	80.76 (7)	88.69 (1)					
			5	4	4	14.78 (45)	22.49 (43)	25.52 (42)	51.46 (29)	54.54 (27)	61.79 (22)	63.18 (21)	67.25 (18)
						73.78 (13)	83.83 (5)	86.30 (3)	87.54 (2)				
			5	5	1	17.44 (42)	30.32 (38)	43.37 (32)	47.04 (30)	62.98 (20)	65.87 (18)	74.18 (12)	76.87 (10)
						79.53 (8)	90.00 (0)						
			5	5	2	13.75 (44)	25.16 (41)	37.37 (36)	39.41 (35)	50.19 (29)	62.38 (21)	70.66 (15)	72.00 (14)
						78.54 (9)	82.39 (6)	83.66 (5)					
			5	5	3	13.71 (46)	21.68 (44)	32.35 (40)	44.11 (34)	56.69 (26)	59.54 (24)	65.01 (20)	70.25 (16)
						77.81 (10)	82.72 (6)	85.15 (4)					
			5	5	4	16.57 (48)	20.20 (47)	26.03 (45)	50.28 (32)	57.38 (27)	60.05 (25)	62.66 (23)	68.94 (18)
						72.57 (15)	84.27 (5)	86.57 (3)	87.71 (2)				

h_1	k_1	l_1	h_2	k_2	l_2								
5	3	3	5	3	3	24.91 (39)	54.45 (25)	60.77 (21)	77.92 (9)	80.63 (7)			
			5	4	0	28.21 (37)	33.53 (35)	49.98 (27)	71.96 (13)	83.16 (5)	85.90 (3)		
			5	4	1	19.74 (40)	26.60 (38)	36.86 (34)	41.15 (32)	58.82 (22)	67.88 (16)	76.39 (10)	79.15 (8)
						87.30 (2)							
			5	4	2	12.17 (43)	21.24 (41)	32.74 (37)	45.19 (31)	48.76 (29)	64.41 (19)	67.27 (17)	72.81 (13)
						75.52 (11)	80.84 (7)	88.70 (1)					
			5	4	3	7.22 (46)	18.39 (44)	25.07 (42)	52.85 (28)	55.89 (26)	61.68 (22)	67.16 (18)	72.43 (14)
						75.00 (12)	85.05 (4)						
			5	4	4	8.21 (49)	18.31 (47)	59.67 (25)	62.32 (23)	69.92 (17)	81.87 (7)	88.84 (1)	
			5	5	1	23.33 (43)	37.81 (37)	41.63 (35)	57.73 (25)	73.88 (13)	81.40 (7)	83.87 (5)	
			5	5	2	17.33 (46)	33.89 (40)	45.12 (34)	65.48 (20)	70.61 (16)	78.02 (10)	85.24 (4)	
			5	5	3	13.39 (49)	26.69 (45)	52.01 (31)	67.84 (19)	72.67 (15)	88.86 (1)		
			5	5	4	12.55 (52)	20.19 (50)	58.29 (28)	65.61 (22)	67.95 (20)	79.18 (10)	87.85 (2)	

Table 9 (cont.)

h_1	k_1	l_1	h_2	k_2	l_2								
5	4	0	5	4	0	12·68 (40)	52·43 (25)	60·80 (20)	67·03 (16)	77·32 (9)	90·00 (0)		
			5	4	1	8·88 (41)	15·44 (40)	45·67 (29)	52·95 (25)	54·67 (24)	59·60 (21)	67·32 (16)	68·81 (15)
						74·63 (11)	77·47 (9)	90·00 (0)					
			5	4	2	17·35 (41)	21·37 (40)	39·80 (33)	45·70 (30)	49·32 (28)	52·75 (26)	66·69 (17)	73·78 (12)
						76·54 (10)	77·91 (9)	81·97 (6)	90·00 (0)				
			5	4	3	25·10 (41)	27·94 (40)	35·20 (37)	39·37 (35)	45·03 (32)	46·79 (31)	73·31 (13)	78·53 (9)
						79·82 (8)	83·66 (5)	88·73 (1)	90·00 (0)				
			5	4	4	31·99 (41)	34·16 (40)	41·87 (36)	79·27 (9)	85·25 (4)	90·00 (0)		
			5	5	1	10·23 (45)	50·64 (29)	56·86 (25)	62·66 (21)	70·85 (15)	83·72 (5)		
			5	5	2	16·99 (45)	45·47 (33)	50·39 (30)	68·82 (17)	77·73 (10)	83·90 (5)		
			5	5	3	23·80 (45)	41·21 (37)	44·63 (35)	74·67 (13)	84·17 (5)			
			5	5	4	30·11 (45)	37·99 (41)	39·74 (40)	80·04 (9)	84·48 (5)	90·00 (0)		

h_1	k_1	l_1	h_2	k_2	l_2								
5	4	1	5	4	1	12·53 (41)	17·75 (40)	21·79 (39)	38·21 (33)	46·33 (29)	51·75 (26)	53·47 (25)	60·00 (21)
						63·10 (19)	66·12 (17)	67·61 (16)	74·82 (11)	76·23 (10)	79·02 (8)	81·79 (6)	88·64 (1)
			5	4	2	8·47 (43)	14·96 (42)	26·22 (39)	29·06 (38)	31·67 (37)	38·55 (34)	40·62 (33)	44·51 (31)
						48·16 (29)	53·27 (26)	58·06 (23)	61·12 (21)	66·98 (17)	71·21 (14)	72·60 (13)	75·34 (11)
						80·73 (7)	82·07 (6)	87·36 (2)	88·68 (1)				
			5	4	3	16·23 (44)	20·23 (43)	26·53 (41)	31·67 (39)	33·98 (38)	36·16 (37)	38·23 (36)	43·94 (33)
						47·43 (31)	53·90 (27)	55·43 (26)	68·22 (17)	73·52 (13)	74·82 (12)	78·67 (9)	82·48 (6)
						84·99 (4)	86·25 (3)	88·75 (1)					
			5	4	4	23·12 (45)	25·94 (44)	33·07 (41)	40·87 (37)	42·63 (36)	50·69 (31)	74·59 (13)	79·40 (9)
						84·13 (5)	85·31 (4)	88·83 (1)					
			5	5	1	6·33 (46)	18·07 (44)	42·72 (34)	49·59 (30)	55·82 (26)	58·76 (24)	64·40 (20)	69·77 (16)
						77·52 (10)	82·55 (6)	85·04 (4)					
			5	5	2	9·28 (47)	25·46 (43)	37·07 (38)	42·70 (35)	53·99 (28)	58·34 (25)	62·49 (22)	71·64 (15)
						75·41 (12)	81·55 (7)	83·97 (5)	86·39 (3)				
			5	5	3	15·37 (48)	32·46 (42)	36·53 (40)	50·00 (32)	52·94 (30)	68·80 (18)	78·41 (10)	80·75 (8)
						87·70 (2)	90·00 (0)						
			5	5	4	21·46 (49)	29·11 (46)	31·27 (45)	38·86 (41)	46·86 (36)	48·34 (35)	74·58 (14)	80·16 (9)
						84·55 (5)	85·64 (4)	88·91 (1)					

h_1	k_1	l_1	h_2	k_2	l_2								
5	4	2	5	4	2	12·10 (44)	24·34 (41)	32·39 (38)	34·69 (37)	36·87 (36)	56·25 (25)	60·73 (22)	66·42 (18)
						69·17 (16)	73·21 (13)	78·46 (9)	83·62 (5)	84·90 (4)	87·45 (2)		
			5	4	3	7·76 (47)	14·13 (46)	18·43 (45)	24·97 (43)	27·69 (42)	30·19 (41)	42·45 (35)	44·21 (34)
						52·31 (29)	55·30 (27)	62·37 (22)	63·72 (21)	67·70 (18)	71·57 (15)	74·09 (13)	76·59 (11)
						79·06 (9)	82·73 (6)	83·95 (5)	86·37 (3)	87·58 (2)			
			5	4	4	14·65 (49)	18·60 (48)	24·73 (46)	49·34 (33)	50·81 (32)	59·11 (26)	70·39 (17)	73·95 (14)
						80·91 (8)	83·20 (6)	88·87 (1)					
			5	5	1	11·16 (47)	26·16 (43)	35·50 (39)	43·06 (35)	49·68 (31)	58·54 (25)	66·63 (19)	71·75 (15)
						76·73 (11)	81·60 (7)	84·01 (5)	86·41 (3)				
			5	5	2	6·27 (49)	29·27 (43)	33·72 (41)	35·76 (40)	56·79 (27)	62·19 (23)	66·06 (20)	79·48 (9)
						81·84 (7)	88·84 (1)	90·00 (0)					
			5	5	3	8·20 (51)	24·20 (47)	29·15 (45)	40·81 (39)	58·40 (27)	60·98 (25)	63·49 (23)	73·08 (15)
						77·67 (11)	84·43 (5)	86·66 (3)	88·89 (1)				
			5	5	4	13·46 (53)	20·64 (51)	23·44 (50)	47·24 (37)	55·33 (31)	56·60 (30)	69·60 (19)	76·20 (13)
						79·43 (10)	86·84 (3)	88·95 (1)					

h_1	k_1	l_1	h_2	k_2	l_2								
5	4	3	5	4	3	11·48 (49)	19·95 (47)	23·07 (46)	50·21 (32)	51·68 (31)	60·00 (25)	62·61 (23)	68·90 (18)
						70·12 (17)	71·34 (16)	73·74 (14)	79·63 (9)	81·95 (7)	88·85 (1)	90·00 (0)	
			5	4	4	6·89 (53)	13·08 (52)	17·19 (51)	57·10 (29)	58·37 (28)	66·84 (21)	69·15 (19)	77·01 (12)
						78·11 (11)	86·78 (3)						
			5	5	1	18·10 (48)	29·39 (44)	33·72 (42)	37·62 (40)	44·53 (36)	53·55 (30)	73·90 (14)	78·58 (10)
						80·88 (8)	83·18 (6)	87·73 (2)	90·00 (0)				
			5	5	2	11·04 (51)	22·52 (48)	30·00 (45)	41·36 (39)	51·99 (32)	61·24 (25)	69·73 (18)	73·22 (15)
						77·78 (11)	84·48 (5)	87·79 (2)	88·90 (1)				
			5	5	3	6·16 (54)	16·78 (52)	22·99 (50)	48·49 (36)	58·97 (28)	66·11 (22)	68·39 (20)	75·06 (14)
						79·39 (10)	85·78 (4)	87·89 (2)					
			5	5	4	7·14 (57)	12·88 (56)	16·78 (55)	54·94 (33)	63·09 (26)	64·20 (25)	65·31 (24)	72·79 (17)
						74·86 (15)	83·00 (7)	84·00 (6)					

h_1	k_1	l_1	h_2	k_2	l_2							
5	4	4	5	4	4	10·75 (56)	63·99 (25)	65·10 (24)	73·70 (16)	82·95 (7)		
			5	5	1	24·66 (49)	33·42 (45)	40·50 (41)	49·52 (35)	80·39 (9)	84·68 (5)	88·94 (1)
			5	5	2	17·20 (53)	25·68 (50)	48·17 (37)	57·27 (30)	76·45 (13)	79·62 (10)	86·90 (3)
			5	5	3	10·61 (57)	18·48 (55)	55·32 (33)	64·46 (25)	72·95 (17)	75·01 (15)	83·07 (7)
			5	5	4	5·99 (61)	11·98 (60)	61·78 (29)	69·98 (21)	70·97 (20)	79·67 (11)	

h_1	k_1	l_1	h_2	k_2	l_2							
5	5	1	5	5	1	16·10 (49)	46·66 (35)	60·65 (25)	72·90 (15)	88·88 (1)		
			5	5	2	7·74 (52)	23·84 (48)	40·34 (40)	55·13 (30)	67·60 (20)	79·01 (10)	87·82 (2)
			5	5	3	14·94 (53)	31·04 (47)	34·88 (45)	50·35 (35)	74·13 (15)	84·77 (5)	86·86 (3)
			5	5	4	21·45 (54)	30·48 (50)	37·55 (46)	46·41 (40)	80·07 (10)	86·05 (4)	90·00 (0)

h_1	k_1	l_1	h_2	k_2	l_2							
5	5	2	5	5	2	31·59 (46)	33·56 (45)	62·42 (25)	84·69 (5)	85·75 (4)		
			5	5	3	7·20 (56)	27·65 (50)	38·78 (44)	57·89 (30)	69·25 (20)	83·90 (6)	90·00 (0)
			5	5	4	13·70 (58)	22·89 (55)	45·29 (42)	54·11 (35)	75·45 (15)	82·30 (8)	85·20 (5)

h_1	k_1	l_1	h_2	k_2	l_2							
5	5	3	5	5	3	21·22 (55)	45·98 (41)	64·93 (25)	81·23 (9)	85·14 (5)		
			5	5	4	6·51 (62)	15·95 (60)	52·49 (38)	61·27 (30)	71·31 (20)	78·91 (12)	80·78 (10)

h_1	k_1	l_1	h_2	k_2	l_2					
5	5	4	5	5	4	9·99 (65)	58·99 (34)	67·74 (25)	75·97 (16)	76·86 (15)

11 Standard diffraction patterns

Figs. 32 and 33 show standard diffraction diagrams for cubic and hexagonal crystals in certain orientations. The symmetry of these diagrams should enable crystals belonging to one of the above two systems and possibly to the tetragonal system to be recognized. Rotation of the crystal about an axis so that two or more of the standard patterns can be obtained, may enable a satisfactory identification of the diffracting medium to be made without recording the patterns. However, it is advisable not to rely completely on such judgement. Some notes on these diagrams are given:

Cubic crystals (Fig. 32)

The patterns in each column have the same fundamental network but different reflections occur according to the extinction rules.

A body-centred cubic structure gives a face-centred reciprocal lattice and vice versa.

The diamond cube is essentially the same as a face-centred cube except that reflections with Σh^2 = odd multiples of four are weak or absent. However, in this case these reflections could occur by *double reflection.*

Though double reflection does not affect the other cases, it does tend to even out the intensities of spots.

For the [111] projection, hexagonal and rectangular symmetries can be discerned.

[123] represents the more general case for [*hkl*], with $h \neq k \neq l$. The diagram for such a case requires the smallest whole numbers which satisfy $hu+kv+lw = 0$, and careful choice of the correct angle for the particular set of indices with the correct signs.

Hexagonal crystals (Fig. 33)

The diagrams are to scale for the axial ratio $c/a = 1{\cdot}633$ (spherical close packing).

The fundamental basal plane pattern is treated separately and is shown in relation to the crystal lattice in the top panel. The stereographic projection has already been given (Fig. 25) and so has one possible projection with [00.1] in the plane of projection (see Fig. 26.)

The choice of different indices for the same spacings is indicated and follows from the treatment in §9.2.

There are two types of axes about which the crystal has *twofold* symmetry. These pass through the edges of the hexagon or are perpendicular to its faces. Rotation about either axis gives the series of patterns shown. These are chosen to include points of short ($1/d$) and long d, i.e. near the centre of the pattern and having low indices.

Two out of the three possible sets of *planar* index choices are shown and can easily be recognized by referring to the basal diagram. In every case (*hk.l*) represents ($hk\,\overline{k+h}\,l$) in the four-index system and the alternative merely involves another choice of two out of the three.

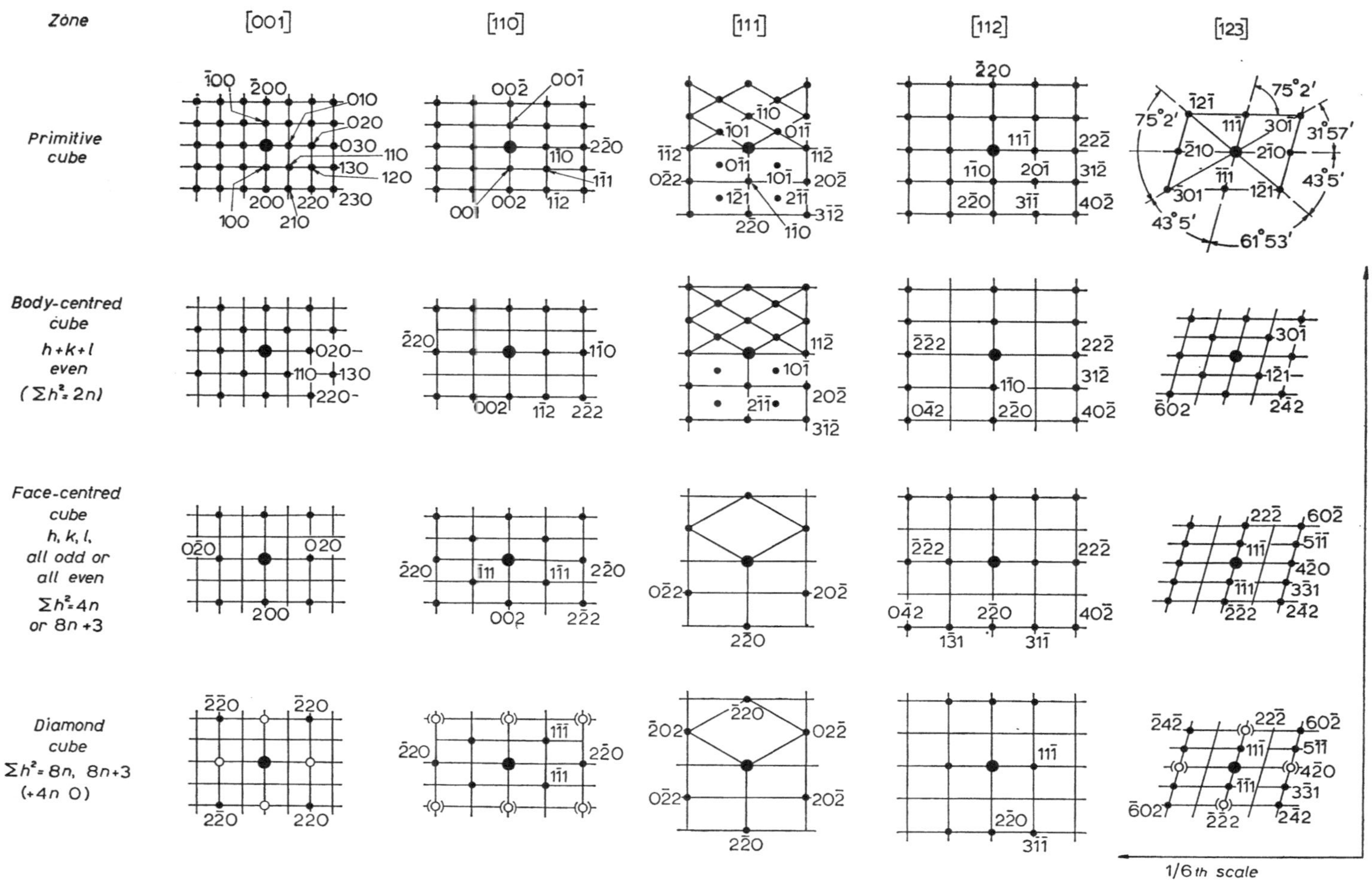

FIG. 32. Standard electron diffraction patterns for cubic crystals.
○ = additional weaker spot. ( ) = spot that could also occur by double reflection.

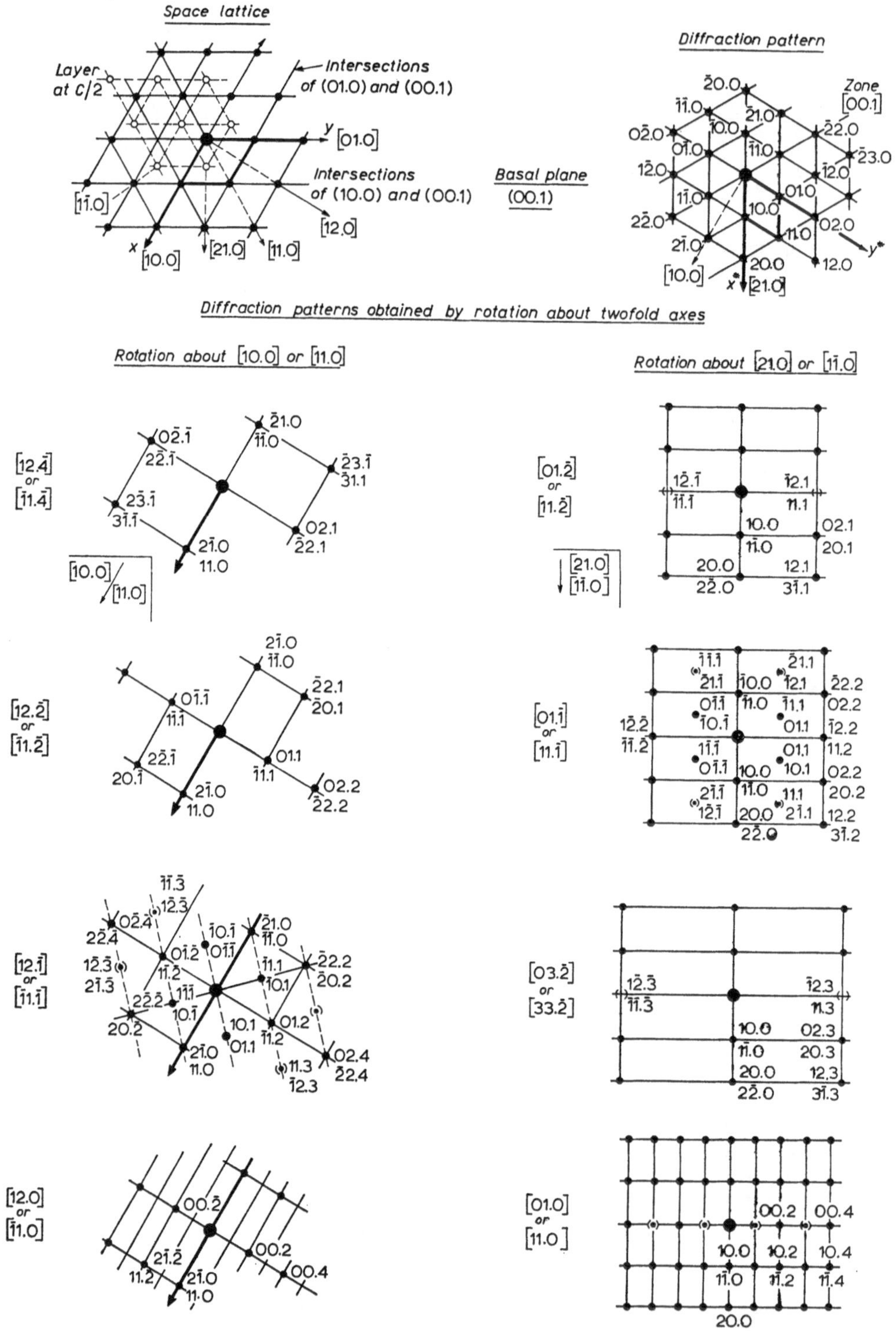

FIG. 33. Standard electron diffraction patterns for close-packed hexagonal crystals ($c/a = 1{\cdot}633$) indexed using the three-index system. ( ) = could occur by double reflection.

The zone axis symbols used here are *strictly confined to three* owing to the ambiguity that may arise, so the four-index addition law should not be used. The zone axes are found by cross multiplication using only three indices. These axes can be recognized as passing through lattice points and are at right angles to the rotation axes.

Extension to other cases

The cubic projection [001] applies equally well to [100] or [010] for a *tetragonal* lattice.

With a change in scale of one edge relative to the other, the pattern can apply to one of the same projections of an *orthorhombic* lattice or to [001] according to choice of axes.

In *tetragonal* and *orthorhombic* crystals the first and second projections will always be rectangular, as in Fig. 32.

In the other three projections, departure from cubic symmetry to *tetragonal* or *orthorhombic* alters all the angles and the relative scale of the distances of spots from the centre (except in the tetragonal case for $d_{hkl} = d_{khl}$).

Distortion of a cubic lattice to rhombohedral will alter the relative magnitudes and angles in every diagram. Only in the [111] case, keeping this as the threefold axis, will the symmetry of the pattern be unaltered.

Rhombohedral cases otherwise behave like hexagonal cases and should be treated as such.

Orthorhombic crystals have three axes about which rectangular symmetry is retained on rotation.

A *hexagonal* lattice behaves like a centred orthorhombic lattice and the only real distinction is the threefold symmetry corresponding to a ratio of two of the orthorhombic axes of exactly $\sqrt{3}$. This double symmetry is also shown for the cubic case in Fig. 32.

12 Twins in cubic crystals

Kelly[28] has pointed out that whilst face-centred cubic metals twin on {111} planes and body-centred cubic metals on {112}, they produce equivalent results. The reason is indicated in Fig. 34. The body-centred twin plane, e.g. $(11\bar{2})$, always has a face-centred twin plane (111) perpendicular to it. The line of intersection is a ⟨110⟩ direction, in

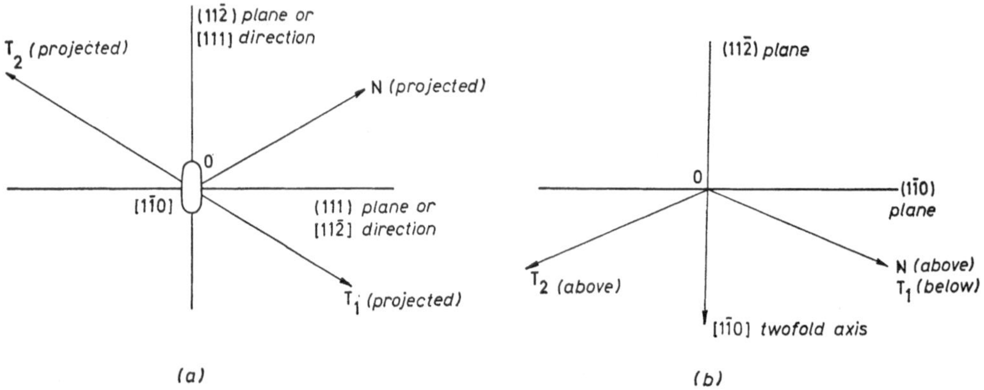

FIG. 34. Equivalence of two types of cubic twins. (*a*) Projection on $(1\bar{1}0)$. (*b*) (111) and $[1\bar{1}0]$ in plane of diagram.

this case $[1\bar{1}0]$ which is a twofold axis. The diagram shows that the normal ON to a plane (direction [*uvw*]) is translated into OT_1 by twinning on (111). Twinning on $(11\bar{2})$ would translate ON into OT_2. The planes normal to OT_1, OT_2, or these directions themselves, will be of the same type because they are rotated into each other by the twofold axis $[1\bar{1}0]$. They will therefore only differ in the sign or permutation of their indices. The stereographic projection (Fig. 35), kindly provided by Dr P. M. Kelly, is thus sufficient to interpret twins in either structure. A simple analytical proof of this relationship is as follows:

Let the face-centred cubic twins be on (111) and the body-centred cubic twins on $(11\bar{2})$. These planes intersect along $[1\bar{1}0]$.
Therefore for all $(h_1k_1l_1)$

$$h_2 = -k_3, \quad k_2 = -h_3, \quad l_2 = -l_3$$

which proves the point that $(h_2k_2l_2)$ and $(h_3k_3l_3)$ belong to the same set $\{hkl\}$. Secondly, the indices add vectorially to $[1\bar{1}0]$ apart from a common factor, which proves that they make equal angles with this direction.

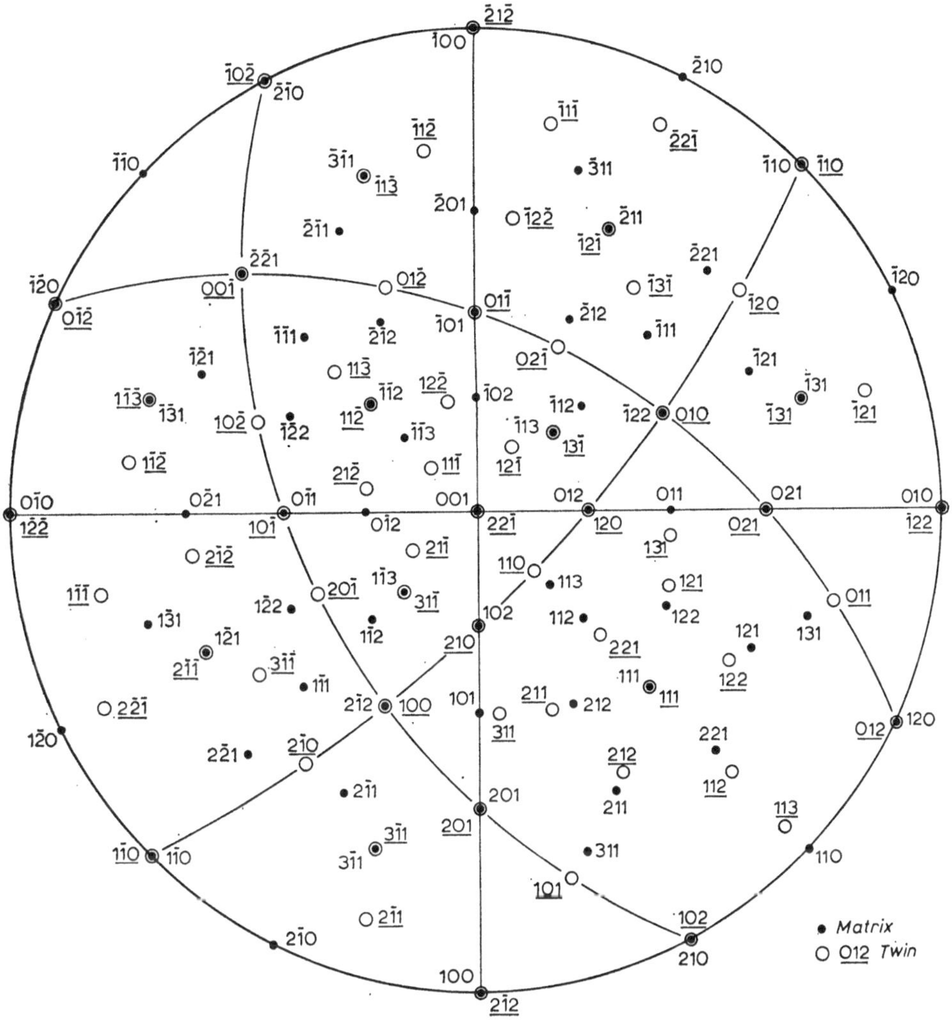

FIG. 35. Stereographic projection representing twin relationships in body- and face-centred cubic crystals. (After Kelly.[28])

Table 10 shows the location of cubic twin spots in reciprocal space due to these relationships, and is a modification of a table originally compiled by Meieran and Richman.[29] A further guide to the position of twin points in reciprocal space is given

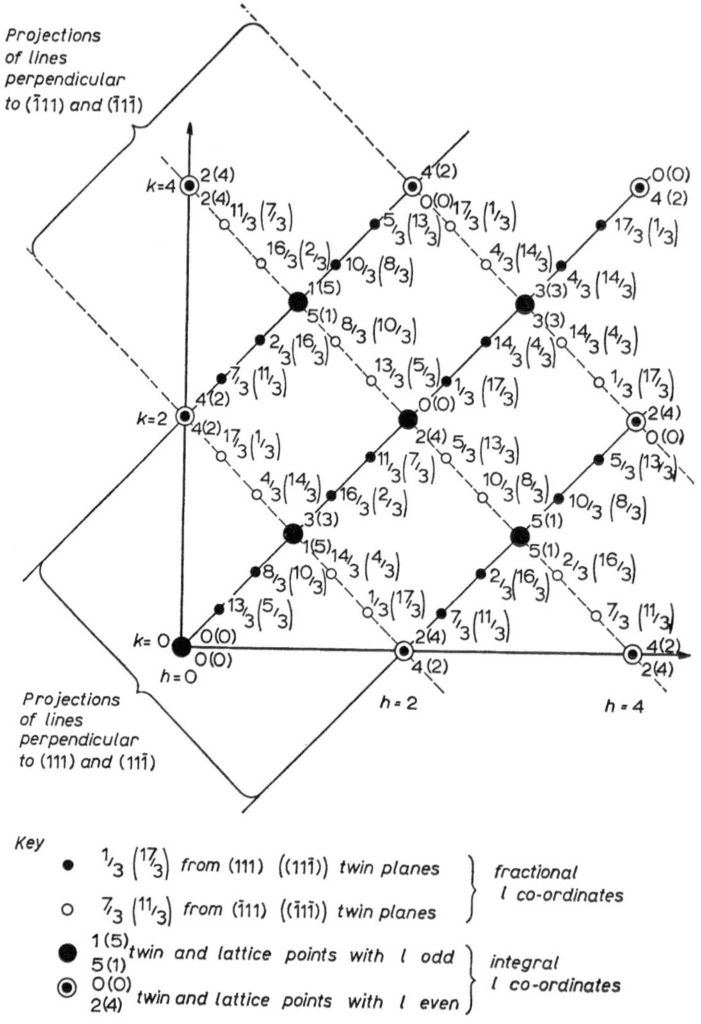

FIG. 36. Projection on (001) of the reciprocal lattice showing twin reflections from a face-centred cubic unit cell. Twinning on {111} planes. (After Menzer.[43])

Indices before twinning	Indices after twinning on (111)		Indices after twinning on (112̄)		Sum (of h_2+h_3, etc.)
h_1	h_2	$h_1-\frac{2}{3}(h_1+k_1+l_1)$ $=\frac{1}{3}(h_1-2k_1-2l_1)$	h_3	$h_1-\frac{1}{3}(h_1+k_1-2l_1)$ $=\frac{1}{3}(2h_1-k_1+2l_1)$	h_1-k_1
k_1	k_2	$\frac{1}{3}(-2h_1+k_1-2l_1)$	k_3	$\frac{1}{3}(-h_1+2k_1+2l_1)$	$-(h_1-k_1)$
l_1	l_2	$\frac{1}{3}(-2h_1-2k_1+l_1)$	l_3	$\frac{1}{3}(2h_1+2k_1-l_1)$	0

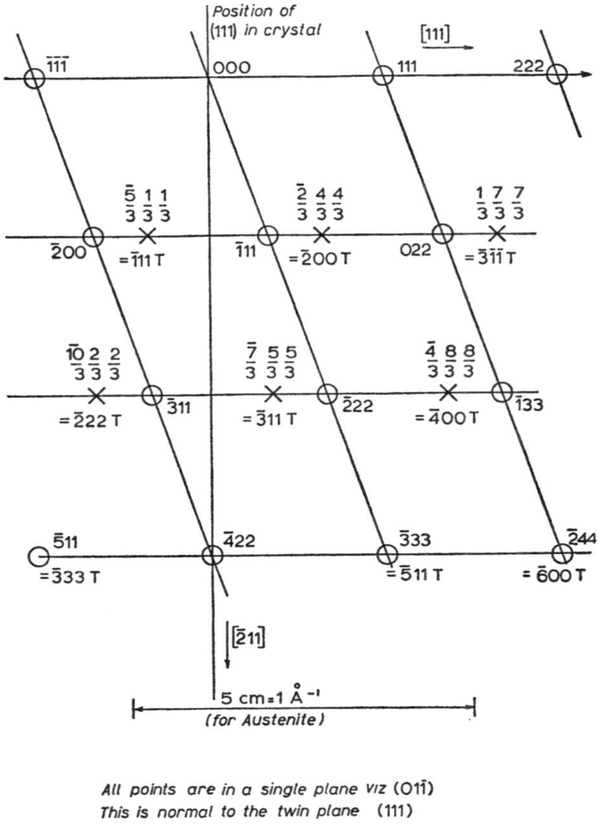

Fig. 37. Effect of cubic twinning on (111) ((011̄) in plane of diagram).

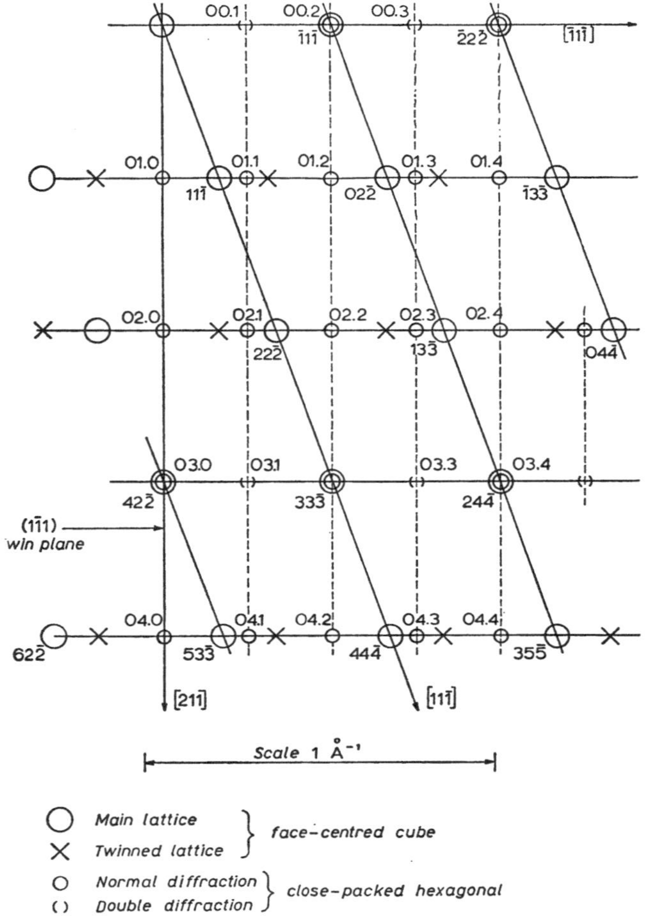

Fig. 38. Combined effects of cubic twinning and hexagonal phase.

in Fig. 36 for face-centred twins on {111} planes. It is a simple matter to determine which spots will lie in any chosen zone. This diagram is due to Menzer.[43] Fig. 37 shows a {011} layer and was originally used to study patterns from austenites containing twins. A more specific pattern (Fig. 38) shows where additional spots form if a hexagonal phase also occurs, as it does with some stainless steels and cobalt.

TABLE 10. Location of twin spots in reciprocal space

× denotes the $h_1k_1l_1$ and $h_2k_2l_2$ reflections which do not occur in b.c.c. structures.	Twinning across (21$\bar{1}$) (i.e. as in b.c.c. structures) →						← Twinning across (1$\bar{1}$1) (i.e. as in f.c.c. structures)			× denotes the $h_1k_1l_1$ and $h_2k_2l_2$ reflections which do not occur in f.c.c. structures.
	h_1	k_1	l_1	h_2	k_2	l_2	h_1	k_1	l_1	
	0	1	1	0	1	1	0	1	1	×
	0	-1	1	4/3	-1/3	1/3	0	1	-1	×
	0	1	-1	-4/3	1/3	-1/3	0	-1	1	×
	1	0	1	1/3	-1/3	4/3	-1	1	0	×
	-1	0	1	1	1	0	1	1	0	×
	1	0	-1	-1	-1	0	-1	-1	0	×
	1	1	0	-1	0	1	-1	0	1	×
	-1	1	0	-1/3	4/3	-1/3	1	0	1	×
	1	-1	0	1/3	-4/3	1/3	-1	0	-1	×
×	1	1	1	-1/3	1/3	5/3	-1	1	1	
×	-1	1	1	1/3	5/3	1/3	1	1	1	
×	1	-1	1	1	-1	1	-1	1	-1	
×	1	1	-1	-5/3	-1/3	1/3	-1	-1	1	
	0	0	2	4/3	2/3	4/3	0	2	0	
	0	2	0	-4/3	4/3	2/3	0	0	2	
	2	0	0	-2/3	-4/3	4/3	-2	0	0	
	1	1	2	1/3	2/3	7/3	-1	2	1	×
	-1	1	2	1	2	1	1	2	1	×
	1	-1	2	5/3	-2/3	5/3	-1	2	-1	×
	1	1	-2	-7/3	-2/3	-1/3	-1	-2	1	×
	1	2	1	-1	1	2	-1	1	2	×
	-1	2	1	-1/3	7/3	2/3	1	1	2	×
	1	-2	1	5/3	-5/3	2/3	-1	1	-2	×
	1	2	-1	-7/3	1/3	2/3	-1	-1	2	×
	2	1	1	-2/3	-1/3	7/3	2	-1	-1	×
	2	-1	1	2/3	-5/3	5/3	-2	1	-1	×
	-2	1	1	2/3	7/3	-1/3	2	1	1	×
	2	1	-1	-2	-1	1	-2	-1	1	×
	0	2	2	0	2	2	0	2	2	
	0	2	-2	-8/3	2/3	-2/3	0	-2	2	
	2	0	2	2/3	-2/3	8/3	-2	2	0	
	2	0	-2	-2	-2	0	-2	-2	0	
	2	2	0	-2	0	2	-2	0	2	

× denotes the $h_1k_1l_1$ and $h_2k_2l_2$ reflections which do not occur in b.c.c. structures.	Twinning across (21$\bar{1}$) (i.e. as in b.c.c. structures) →						← Twinning across (1$\bar{1}$1) (i.e. as in f.c.c. structures)			× denotes the $h_1k_1l_1$ and $h_2k_2l_2$ reflections which do not occur in f.c.c. structures.
	h_1	k_1	l_1	h_2	k_2	l_2	h_1	k_1	l_1	
	2	-2	0	2/3	-8/3	2/3	-2	0	-2	
	0	1	3	4/3	5/3	7/3	0	3	1	×
	0	-1	3	8/3	1/3	5/3	0	3	-1	×
	0	3	1	-4/3	7/3	5/3	0	1	3	×
	0	-3	1	8/3	-5/3	-1/3	0	1	-3	×
	3	0	1	-1/3	-5/3	8/3	-3	1	0	×
	-3	0	1	5/3	7/3	-4/3	3	1	0	×
	1	0	3	5/3	1/3	8/3	-1	3	0	×
	-1	0	3	7/3	5/3	4/3	1	3	0	×
	1	3	0	-7/3	4/3	5/3	-1	0	3	×
	-1	3	0	-5/3	8/3	1/3	1	0	3	×
	3	1	0	-5/3	-4/3	7/3	-3	0	1	×
	-3	1	0	1/3	8/3	-5/3	3	0	1	×
×	1	1	3	1	1	3	-1	3	1	
×	-1	1	3	5/3	7/3	5/3	1	3	1	
×	1	-1	3	7/3	-1/3	7/3	-1	3	-1	
×	1	1	-3	-3	-1	-1	-1	-3	1	
×	1	3	1	-5/3	7/3	5/3	-1	1	3	
×	-1	3	1	-1	3	1	1	1	3	
×	1	-3	1	7/3	-7/3	1/3	-1	1	-3	
×	1	3	-1	-3	1	1	-1	-1	3	
×	3	1	1	-1	-1	3	-3	1	1	
×	-3	1	1	1	3	-1	3	1	1	
×	3	-1	1	1/3	-7/3	7/3	-3	1	-1	
×	3	1	-1	-7/3	-5/3	5/3	-3	-1	1	
	2	2	2	-2/3	2/3	10/3	-2	2	2	
	-2	2	2	2/3	10/3	2/3	2	2	2	
	2	-2	2	2	-2	2	-2	2	-2	
	2	2	-2	-10/3	-2/3	2/3	-2	-2	2	
	1	2	3	1/3	5/3	10/3	-1	3	2	×
	-1	2	3	1	3	2	1	3	2	×
	1	-2	3	3	-1	2	-1	3	-2	×
	1	2	-3	-11/3	-1/3	-2/3	-1	-3	2	×

TABLE 10 (cont.)

× denotes the $h_1k_1l_1$ and $h_2k_2l_2$ reflections which do not occur in b.c.c. structures.	Twinning across (21$\bar{1}$) (i.e. as in b.c.c. structures) →						← Twinning across (1$\bar{1}$1) (i.e. as in f.c.c. structures)			× denotes the $h_1k_1l_1$ and $h_2k_2l_2$ reflections which do not occur in f.c.c. structures.
	h_1	k_1	l_1	h_2	k_2	l_2	h_1	k_1	l_1	
	2	1	3	2/3	1/3	11/3	−2	3	1	×
	−2	1	3	2	3	1	2	3	1	×
	2	−1	3	2	−1	3	−2	3	−1	×
	2	1	−3	−10/3	−5/3	−1/3	−2	−3	1	×
	3	2	1	−5/3	−1/3	10/3	−3	1	2	×
	−3	2	1	1/3	11/3	−2/3	3	1	2	×
	3	−2	1	−1	3	−2	3	−1	2	×
	3	2	−1	−3	−1	2	−3	−1	2	×
	2	3	1	−2	−1	3	−2	1	3	×
	−2	3	1	−2/3	11/3	1/3	2	1	3	×
	2	−3	1	2	−3	1	−2	1	−3	×
	2	3	−1	−10/3	1/3	5/3	−2	−1	3	×
	3	1	2	−1/3	−2/3	11/3	−3	2	1	×
	−3	1	2	5/3	10/3	−1/3	3	2	1	×
	3	−1	2	1	−2	3	−3	2	−1	×
	3	1	−2	−3	−2	1	−3	−2	1	×
	1	3	2	−1	2	3	−1	2	3	×
	−1	3	2	−1/3	10/3	5/3	1	2	3	×
	1	−3	2	3	−2	1	−1	2	−3	×
	1	3	−2	−11/3	2/3	1/3	−1	−2	3	×
	0	0	4	8/3	4/3	8/3	0	4	0	
	0	4	0	−8/3	8/3	4/3	0	0	4	
	4	0	0	−4/3	−8/3	8/3	−4	0	0	
	1	1	4	5/3	4/3	11/3	−1	4	1	×
	−1	1	4	7/3	8/3	7/3	1	4	1	×
	1	−1	4	3	0	3	−1	4	−1	×
	1	1	−4	−11/3	−4/3	−5/3	−1	−4	1	×
	1	4	1	−7/3	7/3	8/3	−1	1	4	×
	−1	4	1	−5/3	11/3	4/3	1	1	4	×
	1	−4	1	3	−3	0	−1	1	−4	×
	1	4	−1	−11/3	5/3	−4/3	−1	−1	4	×
	4	1	1	−4/3	−5/3	11/3	−4	1	1	×
	−4	1	1	4/3	11/3	−5/3	4	1	1	×

× denotes the $h_1k_1l_1$ and $h_2k_2l_2$ reflections which do not occur in b.c.c. structures.	Twinning across (21$\bar{1}$) (i.e. as in b.c.c. structures) →						← Twinning across (1$\bar{1}$1) (i.e. as in f.c.c. structures)			× denotes the $h_1k_1l_1$ and $h_2k_2l_2$ reflections which do not occur in f.c.c. structures.
	h_1	k_1	l_1	h_2	k_2	l_2	h_1	k_1	l_1	
	4	−1	1	0	−3	3	−4	1	−1	×
	4	1	−1	−8/3	−7/3	7/3	−4	−1	1	×
	0	3	3	0	3	3	0	3	3	×
	0	3	−3	−4	1	−1	0	−3	3	×
	3	0	3	1	−1	4	−3	3	0	×
	3	0	−3	−3	−3	0	−3	−3	0	×
	3	3	0	−3	0	3	−3	0	3	×
	3	−3	0	1	−4	1	−3	0	−3	×
×	3	3	1	−7/3	1/3	11/3	−3	1	3	
×	−3	3	1	−1/3	13/3	−1/3	3	1	3	
×	3	−3	1	5/3	−11/3	5/3	−3	1	−3	
×	3	3	−1	−11/3	−1/3	7/3	−3	−1	3	
	3	1	3	1/3	−1/3	13/3	−3	3	1	
×	−3	1	3	7/3	11/3	1/3	3	3	1	
×	3	−1	3	5/3	−5/3	11/3	−3	3	−1	
×	3	1	−3	−11/3	−7/3	1/3	−3	−3	1	
×	1	3	3	−1/3	7/3	11/3	−1	3	3	
×	−1	3	3	1/3	11/3	7/3	1	3	3	
×	1	−3	3	11/3	−5/3	5/3	−1	3	−3	
×	1	3	−3	−13/3	1/3	−1/3	−1	−3	3	
	0	2	4	4/3	8/3	10/3	0	4	2	
	0	−2	4	4	0	2	0	4	−2	
	0	4	2	−4/3	10/3	8/3	0	2	4	
	0	−4	2	4	−2	0	0	2	−4	
	4	2	0	−8/3	−4/3	10/3	−4	0	2	
	4	−2	0	0	−4	2	−4	0	−2	
	2	4	0	−10/3	4/3	8/3	−2	0	4	
	2	−4	0	2	−4	0	−2	0	−4	
	4	0	2	0	−2	4	−4	2	0	
	−4	0	2	8/3	10/3	−4/3	4	2	0	
	2	0	4	2	0	4	−2	4	0	
	2	0	−4	−10/3	−8/3	−4/3	−2	−4	0	

13 Superimposed patterns and projections (for martensite transformations etc.)

In addition to the appearance of extra spots due to twinning etc., patterns are also obtained from specimens in which two phases with some particular mutual orientation relationship are present. One phase may have formed directly from the other by a martensitic type of transformation or the pattern may be from a precipitate and matrix. The case where a close-packed hexagonal phase can form from a face-centred cubic phase is closely connected with twinning (and stacking fault formation) in the latter. The corresponding positions of diffraction spots for face-centred cubic twin planes {111} and hexagonal (00.1) normal to the diffraction pattern have already been shown in Figs. 37 and 38. It is useful to consider what kinds of diffraction pattern can be formed when the two cubic phases, i.e. face-centred or body-centred, are present together or, alternatively, when either or both occur with the hexagonal phase.

The co-existence of body-centred and face-centred cubic structure is of interest in connection with the martensite transformation in steels and other alloys. The formation of the hexagonal ε-martensite, by heat treatment or deformation of some austenitic steels, and the cobalt transformation involve the two close-packed phases. Other systems involve body-centred cubic phases and close-packed hexagonal phases. The three phases are sometimes found together.

Fig. 39 shows the individual patterns to be expected initially for those cases where the planes of closest packing are parallel to the plane of the diffraction pattern. In order to bring out some essential details, the upper row of patterns provides a representation of the atomic arrangements on the planes of the crystal and the lower row the corresponding diffraction patterns or reciprocal lattice nets with absences. The patterns are in pairs, (*a*) with (*b*) etc. The first two structures give reciprocal plane patterns with sixfold symmetry, but the third has only rectangular (twofold) symmetry. The last two patterns (*g*) and (*h*) bring out the point that the body-centred cube can also show the sixfold reciprocal symmetry in another orientation. In this case, however, the atoms shown in (*h*) are not all co-planar. This is a cube viewed along a diagonal. These diffraction patterns have already appeared in Figs. 32 and 33.

The simplest way of accounting for the patterns which may arise when two or three of the phases occur together is to consider that the close-packed planes (*a*) and (*c*) coincide completely. In the body-centred cube (*e*), however, the atoms in (011) are the closest packed, but spheres touch along only two rows out of three, i.e. [$1\bar{1}1$] and [$11\bar{1}$], but not [100]. A choice follows:

Either a close-packed row of the body-centred cube coincides with a close-packed row of the face-centred cube or a close-packed hexagonal phase. The second close-packed row will then be $70°32' - 60° = 10°32'$ away from the corresponding row of the other planes. This is the condition actually shown in Fig. 39 and it results in the body-centred cubic pattern being tilted with respect to the other two. This relationship corresponds to the mutual orientation in the Kurdjumov and Sachs, K–S, transformation relationship for martensite.[44]

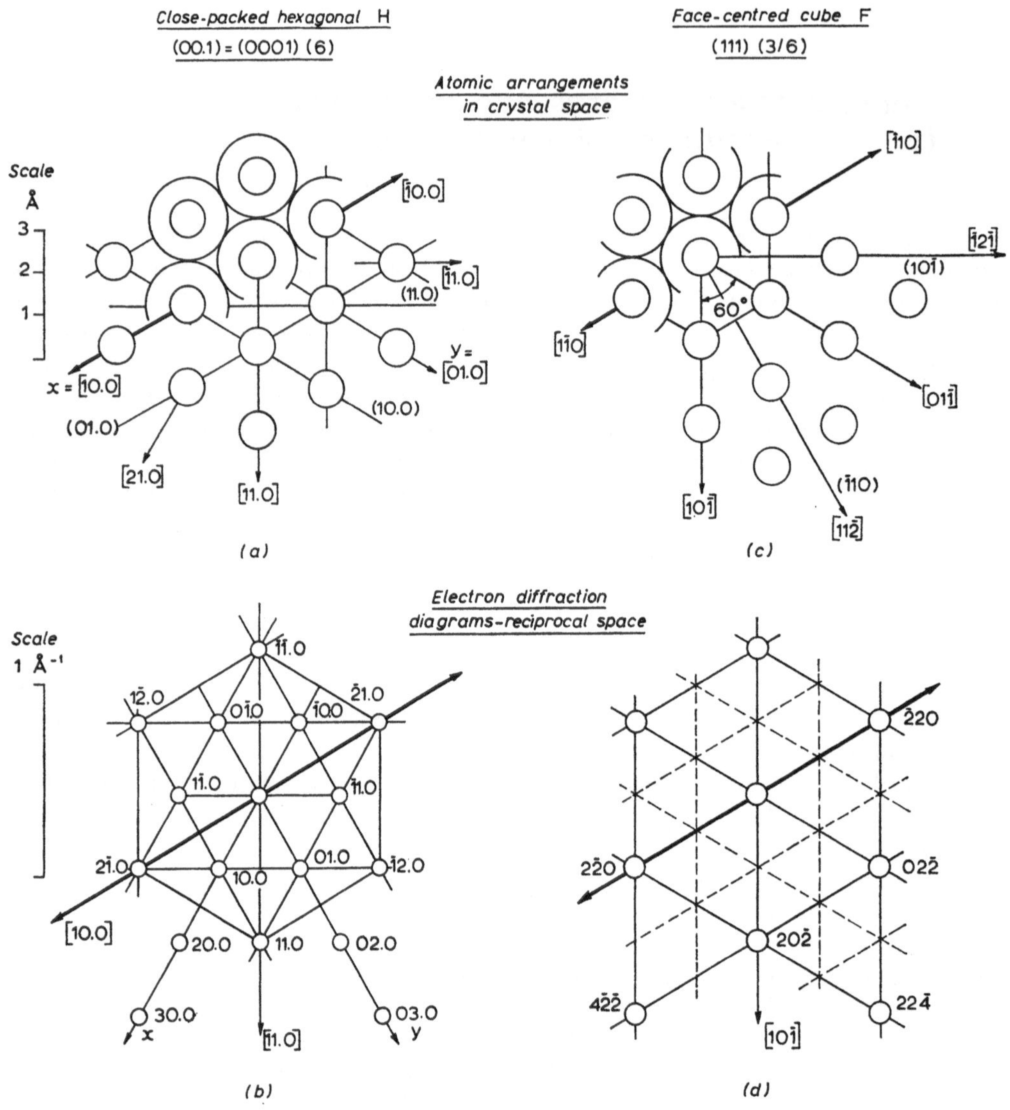

FIG. 39. Relationships between the three structures in the two spaces, i.e. crystal

Or the angle of 10°32′ is equally divided so that the two close-packed body-centred cubic rows make 5°16′ with two of the corresponding rows in the other phases. This is the mutual orientation found by Nishiyama[45] and Wassermann,[46] N–W. All the patterns are now symmetrical with respect to each other.

The superimposed diffraction patterns are shown in Fig. 40. These diagrams refer to the quite general case of the three phases I (body-centred cubic), F (face-centred cubic) and H (close-packed hexagonal) occuring together. The relative distances indicated have, however, been chosen for the phases occurring in steels.

It is not expected that actual patterns will always happen to fall in this orientation. In face-centred metal strip, for example, {110} planes are frequently found in the

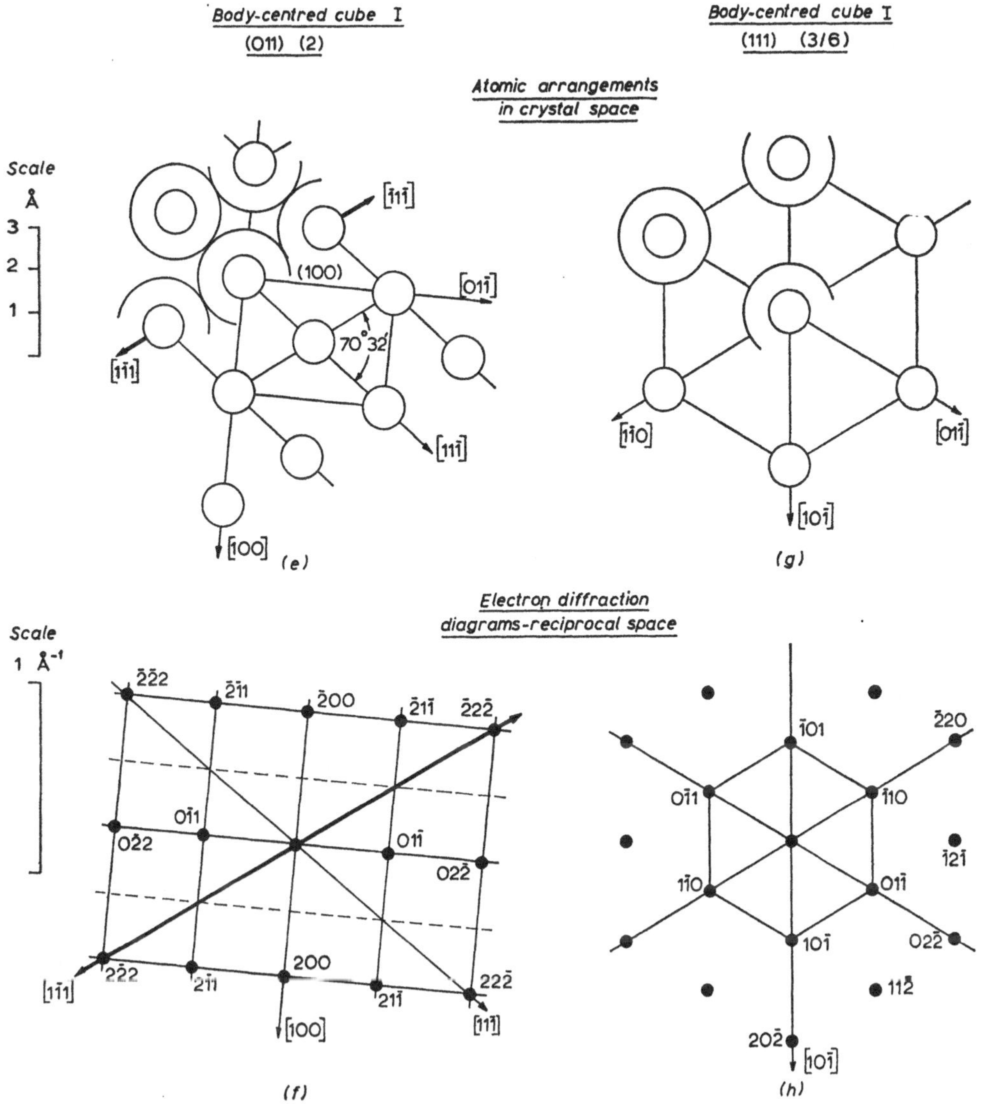

and diffraction patterns. All planes marked are perpendicular to the diagram.

plane of the strip. It is therefore useful to consider the case of a plane of this kind in the plane of the pattern. This involves rotating Figs. 39 and 40 through 90°.† Fig. 41(a) shows a face-centred cubic pattern not shown in Fig. 39, and corresponds to the body-centred cubic orientation of Fig. 39(*f*) but has different systematic absences. This pattern then forms the basis of Fig. 41(*b*) to which the body-centred cubic and close-packed hexagonal reflections have been added. The grouping and spacings of reflections along lines parallel to $[00.1]_H$ should provide a clear indication of the phases present.

† In addition there has also been a change in indices based on one of the alternative choices of axes referred to on p. 141 and conforming with the orientations in Fig. 42. This does not affect the conclusions in any way.

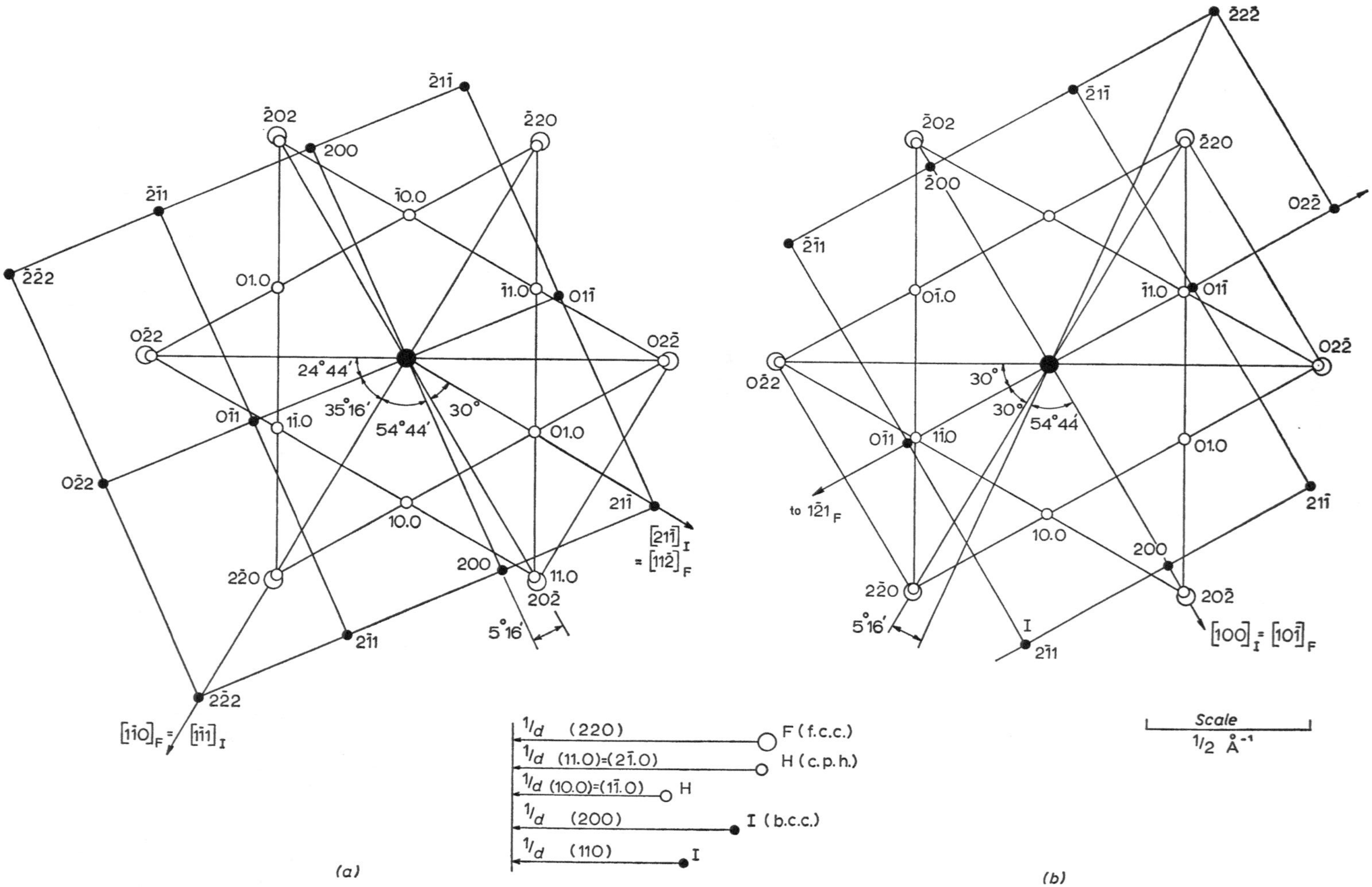

FIG. 40. Superimposed patterns in which (111)F//(011)I//(00.1)H. (*a*) K–S orientation. (*b*) N–W orientation. Note that the only difference between the two orientations is a rotation of 5°16′. (Dimensions approximately the same as those for α-, γ-, and ϵ-iron alloys.)

With the interplanar spacings so close, reflections may merge into each other but will separate at greater distances.

If now the face-centred cubic and close-packed hexagonal patterns are held in the plane of the diagram and the body-centred cubic pattern rotated out of this plane by 5°16′, as in Fig. 41(*c*), the pattern of Fig. 39(*d*) remains. Alternatively, if the tilt is made the other way a pattern with sixfold symmetry [Fig. 39(*h*)] appears. In favourable circumstances it should therefore be possible to distinguish between different orientation relationships of this kind by using a tilting stage.

It is appreciated that the mutual relationships can occur in a number of different ways and interpretation would become very difficult if more than one combination appeared on one pattern. Fig. 42 shows the stereographic projection for the orientation of Fig. 41 with that of Fig. 40 represented by the diameter normal to $(00.1)_{H_1}$; a projection on this plane is given for a specific body-centred cubic/close-packed hexagonal case in Fig. 52. The other possible hexagonal face-centred cubic relationships are represented by H_2 (broken line), also normal to the plane, and by H_3 and H_4 for which the (00.1) poles are shown. It is thus possible for the hexagonal and body-centred phases to be related to the face-centred phase through different {111} planes of the latter. The alternative choices of face-centred cubic and body-centred cubic axes are easily envisaged, since there is a total of six ways of achieving the relationship of Figs. 39(*c*) and (*e*), i.e. one close-packed row parallel for one particular {111} face-centred cubic plane, making twenty-four possibilities in all. In the symmetrical N–W case these numbers are halved.

The interpretation in the above diagrams has been confined to the simplest case in which a single {111} face-centred cubic plane is involved and the other two phases are then related through it. The stereographic projection in Fig. 42 represents the following mutual orientation relationships referred to the axes usually accepted. The indices are given without brackets as they are true for planes ( ) or zones [ ].

K–S (F) f.c.c	K–S (I) b.c.c	From K–S to N–W rotation 5°16′ ↔	N–W (F) f.c.c	N–W (I) b.c.c	N–W to Bain rotation ~9° ↔	Bain (F) f.c.c	Bain (I) b.c.c
011	111		$\bar{1}12$	011		$\bar{1}10$	010
$21\bar{1}$	$2\bar{1}\bar{1}$		110	100		110	100
$1\bar{1}1$	$0\bar{1}1$		$1\bar{1}1$	$0\bar{1}1$		001	001

The third system of axes has been added because it is of interest in connection with Bain's original hypothesis[47] for the martensite transformation, and because patterns based on {001} or {011} in either phase may be encountered. It is useful to recognize, from Fig. 42 for example, that, whereas the $(011)_F$ and $(\bar{1}10)_F$ planes do have the same face-centred cubic pattern, the pattern for the body-centred cubic phase is different. This type of pattern and orientation could again be recognized by tilting the specimen.

To assist in the interpretation of superimposed patterns from both body-centred and face-centred cubic lattices the three double projections in Figs. 43, 44 and 45 are provided. These are based on the three relationships given in the above table (but the specific choice of axes is different).

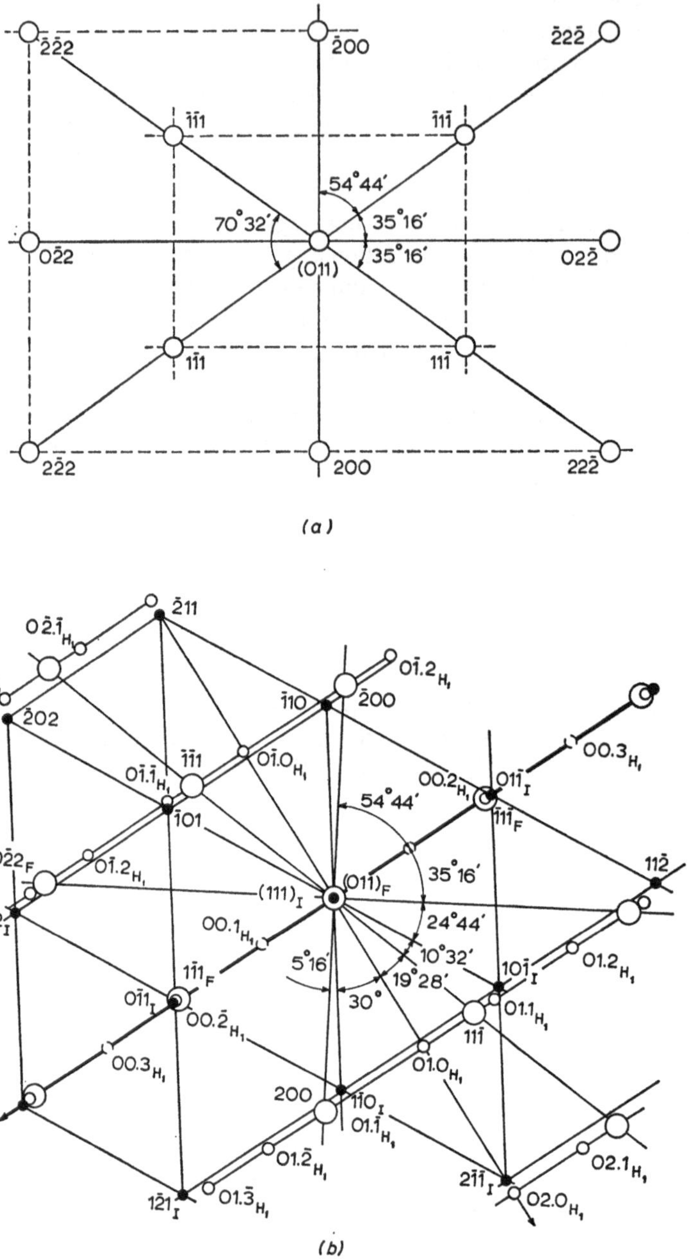

FIG. 41. Superimposed patterns in which $(011)_F//(111)_I//(\bar{2}1.0)_H$. (*a*) Face-centred cubic pattern alone. (*b*) Superimposed patterns for K–S orientation.

Fig. 45 is thus also useful with orientations based on the K–S and N–W relationships when a cube face of the body-centred cubic phase, or one of the alternative face-centred cubic {110} planes of Fig. 41(*b*), is selected for the plane of the diffraction pattern. One would then require a small relative rotation of the superimposed projections

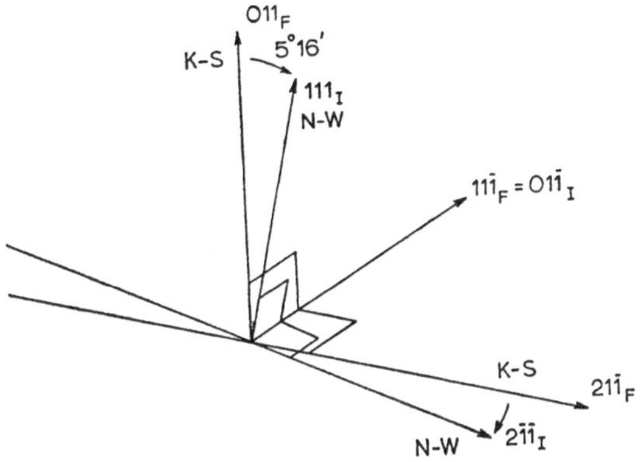

(c)

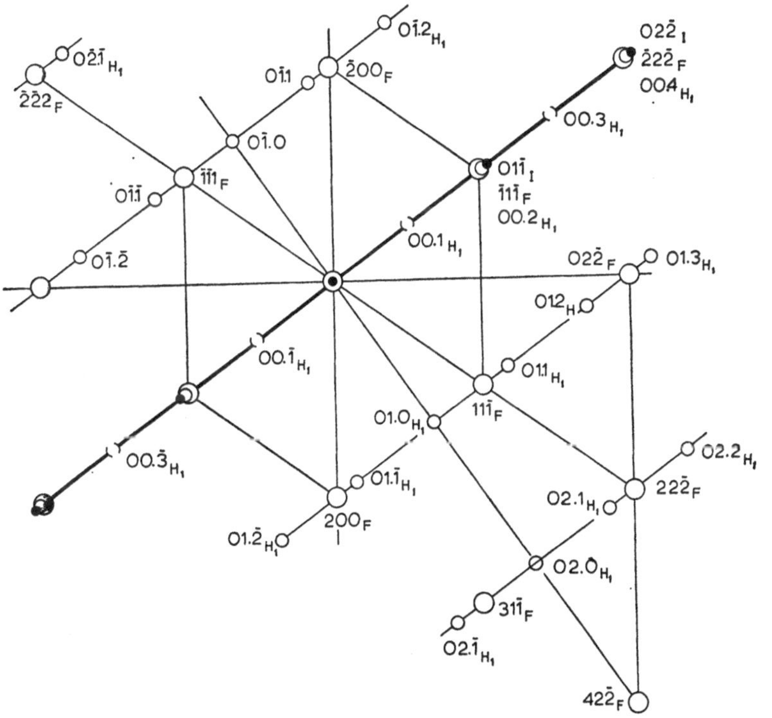

(d)

FIG. 41 (cont). (c) Perspective drawing of axes showing rotation of 5°16′ from K–S to N–W orientation. (d) N–W orientation with (001)F in plane.

in Fig. 45 in order to obtain a K–S or N–W orientation. On the other hand, the Bain relationship may occur for some transformations and for epitaxial oxide layers[48] or for coherent precipitations.[49]

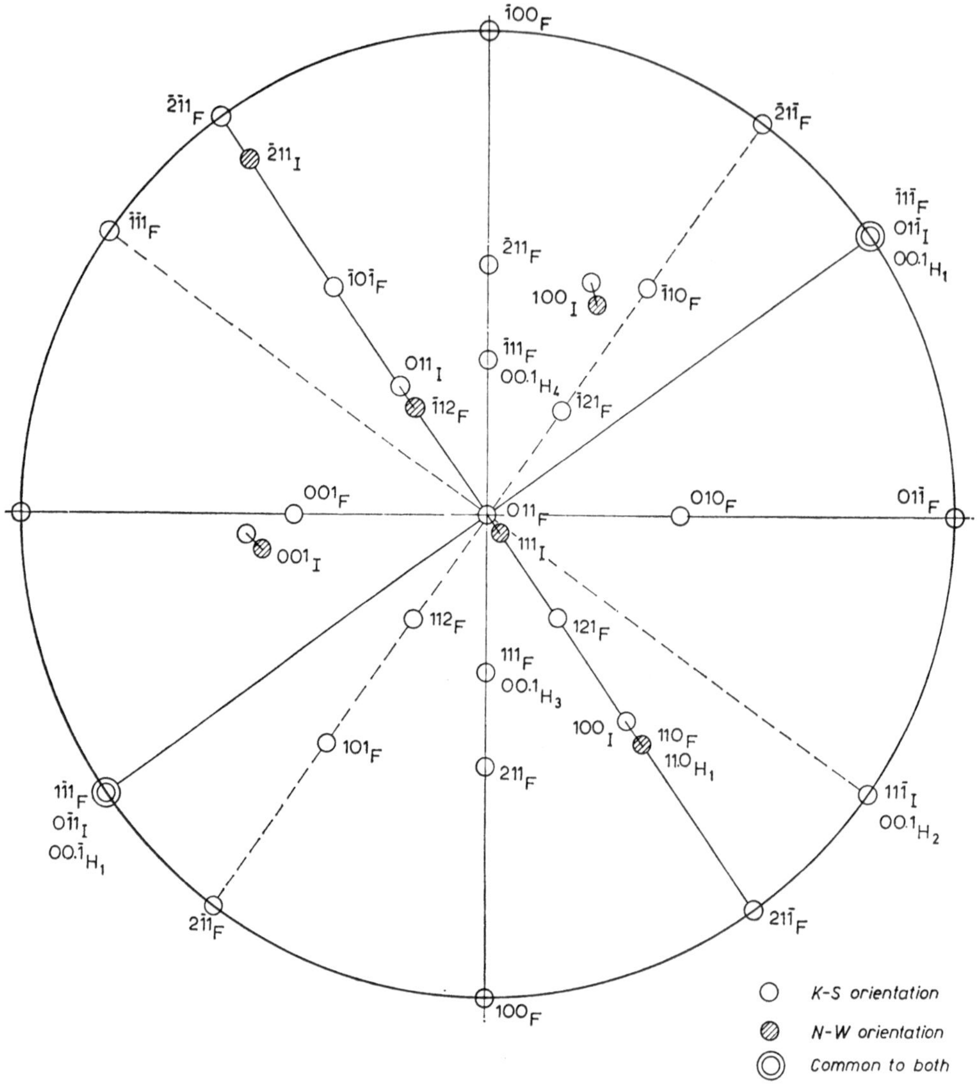

Fig. 42. Stereographic projection for Fig. 41.

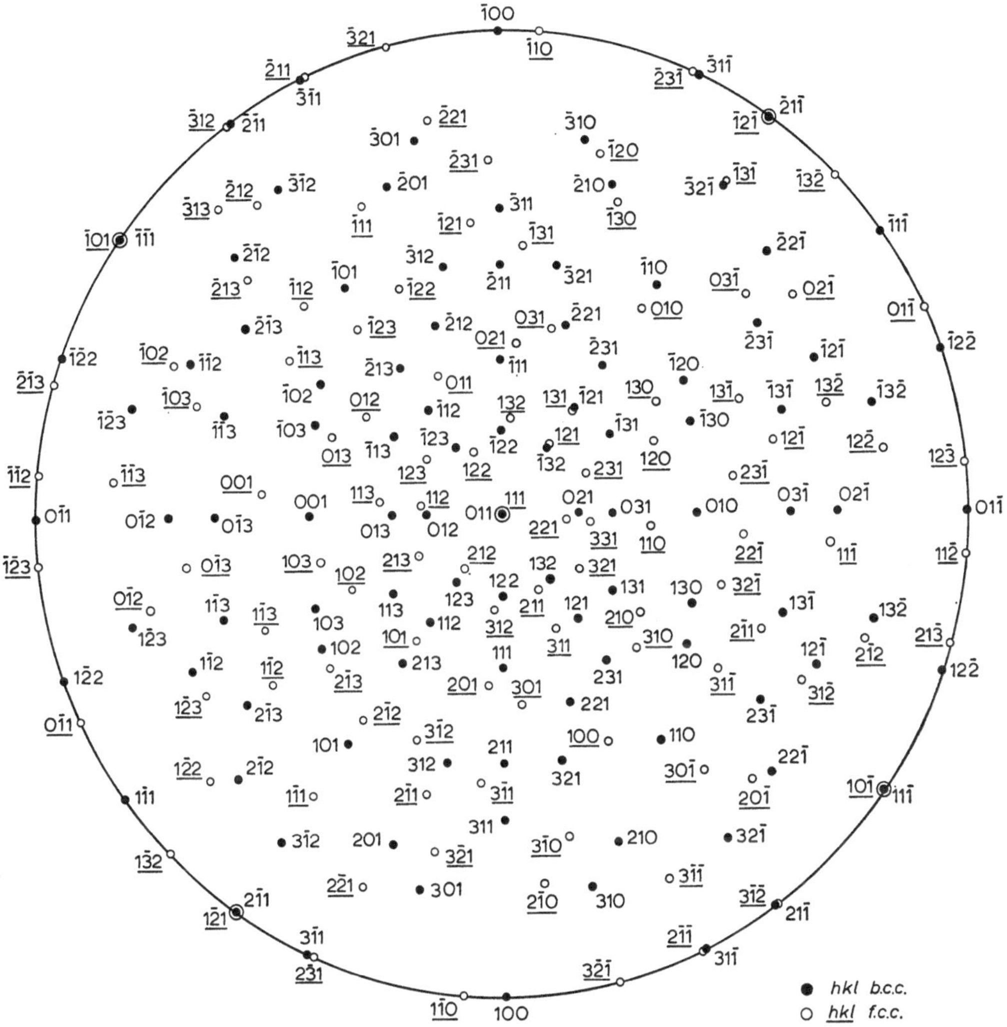

FIG. 43. Stereographic projection representing the K–S orientation relationship between body- and face-centred cubic materials.
(011) b.c.c. //(111) f.c.c.
$(11\bar{1})$ b.c.c. //$(10\bar{1})$ f.c.c.
$(\bar{2}1\bar{1})$ b.c.c. //$(\bar{1}2\bar{1})$ f.c.c.

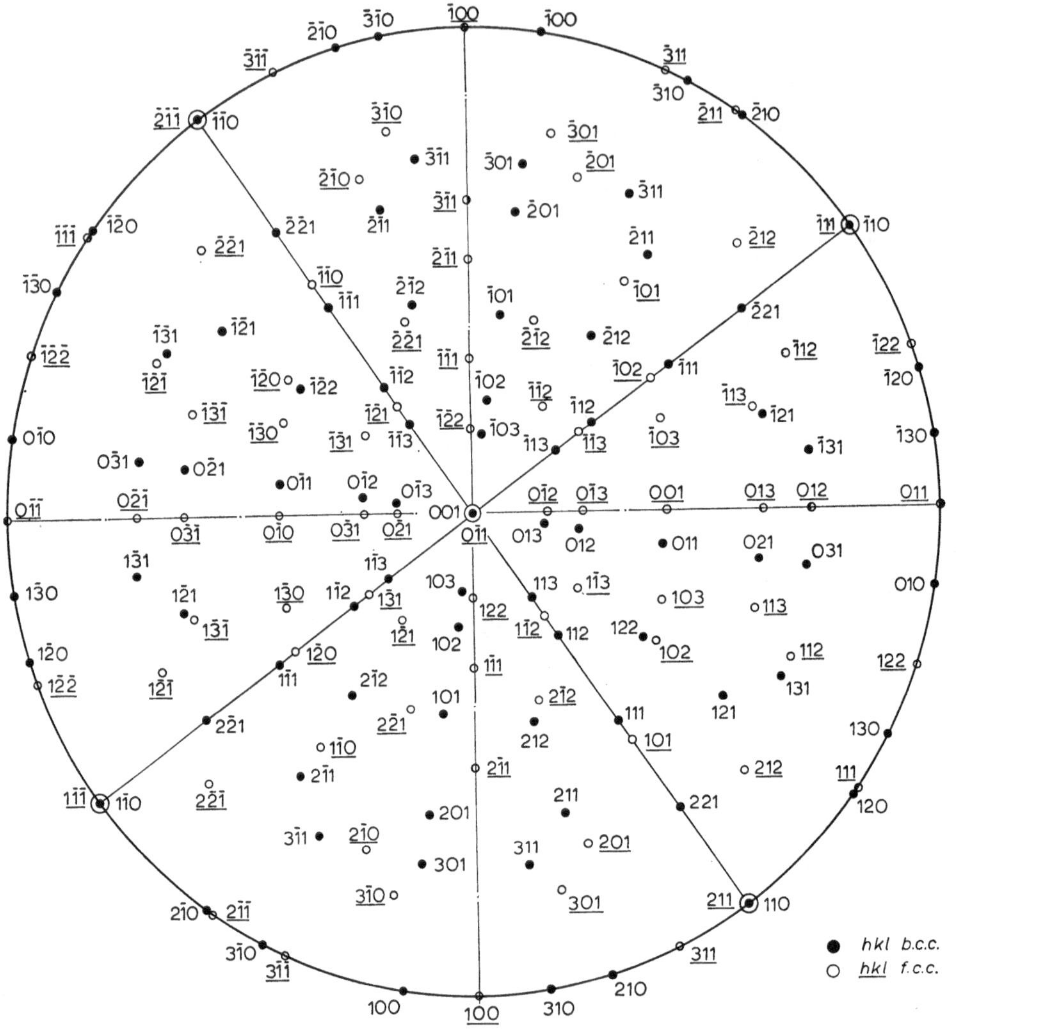

Fig. 44. Stereographic projection representing the N–W orientation relationship between body- and face-centred cubic materials.
(001) b.c.c. //$(0\bar{1}\bar{1})$ f.c.c.
$(\bar{1}10)$ b.c.c. //$(\bar{1}11)$ f.c.c.
(110) b.c.c. //(211) f.c.c.

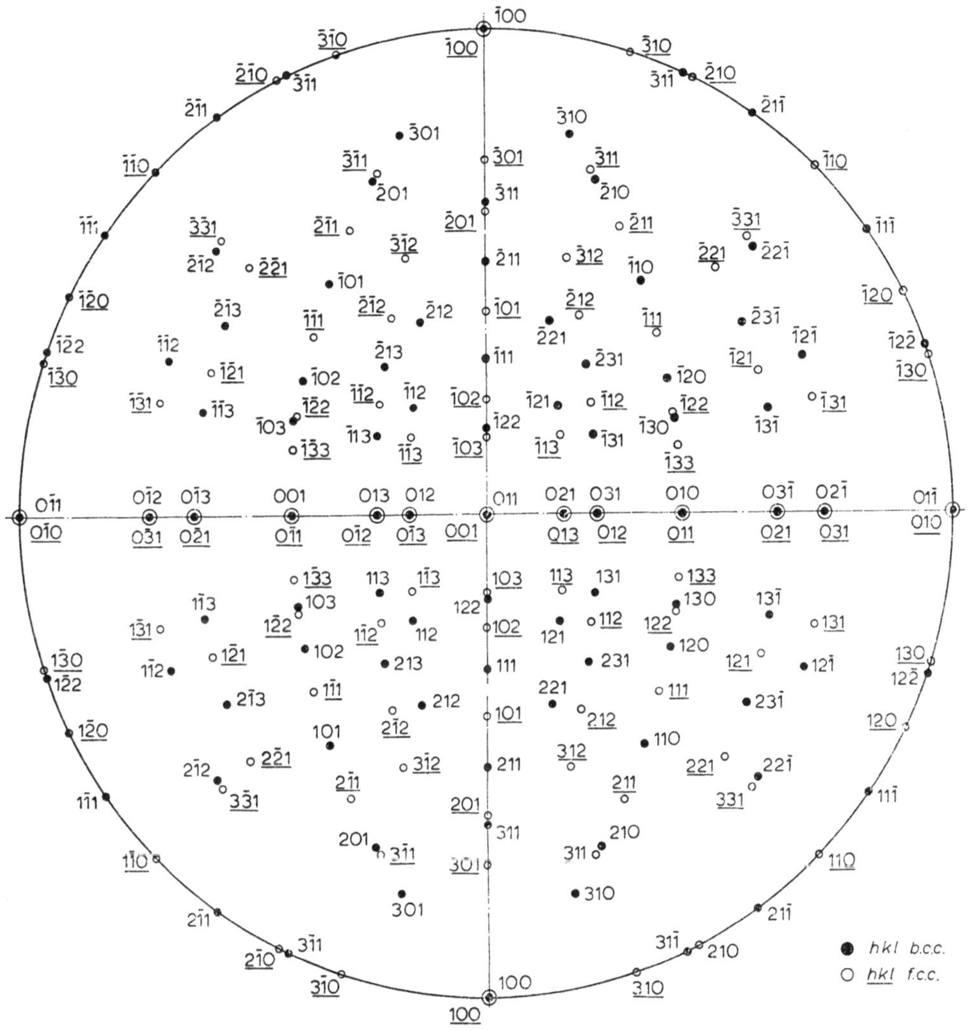

FIG. 45. Stereographic projection representing the Bain relationship between body- and face-centred cubic materials.
(100) b.c.c. //(100) f.c.c.
$(01\bar{1})$ b.c.c. //(010) f.c.c.
(011) b.c.c. //(001) f.c.c.

Part 3

DATA FOR SPECIFIC MATERIALS

14 General

The data given in this part show a bias towards ferrous metallurgy. Apart from their specific practical application they also demonstrate:

(*a*) Crystal systems embraced by commonly occurring elements and compounds.

(*b*) The large number of *d*-spacings possible from a given unit cell, particularly illustrating the need for diffraction patterns with small spot separation (equivalent to large *d*-spacing) for unambiguous interpretation.

It is expected that many phases occur in cubic, hexagonal or tetragonal systems which are already conveniently covered by data and diagrams in Part 2.

TABLE 11. Interplanar spacings of cementite, Fe_3C (see §15.1.1)

Crystal structure: Orthorhombic

Unit cell: $a_0 = 4{\cdot}5241_1$ Å, $b_0 = 5{\cdot}0883_3$ Å, $c_0 = 6{\cdot}7416_6$ Å

hkl	*d*-spacing (Å)	*hkl*	*d*-spacing (Å)	*hkl*	*d*-spacing (Å)	*hkl*	*d*-spacing (Å)
001	6·742	023	1·684	320 }	1·297	240	1·109
010	5·088	031	1·645	133 }		043	1·107
100	4·524	221	1·640	105	1·292	410	1·104
011	4·061	014	1·600	321	1·274	314 }	1·097
101	3·757	203	1·594	040	1·272	016 }	
110	3·381	130	1·588	232	1·259	241	1·094
002	3·371	104	1·579	115 }	1·252	106	1·091
111	3·022	123	1·578	303 }		411	1·090
012	2·810	131	1·546	041	1·250	143	1·075
102	2·703	213	1·521	140	1·225	403	1·072
020	2·544	032	1·515	313	1·216	332	1·069
112	2·387	222	1·511	322	1·211	116	1·066
021	2·380	114 }	1·508	141	1·205	234	1·057
200	2·262	300 }		034	1·196	035	1·056
003	2·247	301	1·472	224	1·194	225	1·054
120	2·218	310	1·446	025	1·191	242	1·053
201	2·145	132	1·437	043	1·190	412	1·049
121	2·107	311	1·414	233	1·162	420	1·034
210	2·067	024	1·405	205	1·158	324 }	
013	2·056	302	1·377	134	1·156	135 }	1·028
022	2·031	230	1·357	125	1·152	026 }	
103	2·013	033	1·354	142	1·151	421	1·022
211	1·976	204	1·352	400	1·131	050	1·018
202	1·878	223	1·351	215	1·129	044	1·015
113	1·872	005	1·348	330	1·127	403	1·010
122	1·853	124	1·342	304 }		333	1·007
212	1·762	231	1·330	006 }	1·124	051 }	1·006
030	1·696	312	1·329	323 }		206 }	
220	1·691	214	1·306	401	1·115	305	1·005
004	1·685	015	1·303	331	1·112	126	1·002

15 Carbides

The carbides presented here are commonly found in low-alloy steels. Where the symbol M occurs in the chemical formula this represents a fixed total of metal atoms of different kinds. In this respect therefore it is possible to find carbides with the same metal–carbon ratio but with different proportions of metal atoms. Differences in the lattice parameters are therefore found for such carbides. Thus, although lattice parameters are given and M is not specified, these are included to demonstrate the range of *d*-spacings and are only very slightly different from the *d*-spacings of any other carbide which fits the particular chemical formula.

15.1 CEMENTITE, Fe_3C

15.1.1 *Interplanar spacings* (Table 11)

Interplanar spacings have been calculated at the authors' laboratory using lattice parameters intermediate between those of Lipson and Petch[50] and those of Hume-Rothery, Raynor and Little[51]; they represent the possible reflections from *d*-spacings greater than 1·00 Å.

15.1.2 *Interplanar angles* (Table 12)

Interplanar angles have been calculated by Drs Swann, Warlimont and Whitmore of The United States Steel Corporation. They have used the lattice parameters of Lipson and Petch, viz. $a_0 = 4{\cdot}523_5$ Å, $b_0 = 5{\cdot}089_0$ Å, and $c_0 = 6{\cdot}743_3$ Å.

15.1.3 *Interzonal angles* (Table 13)

Interzonal angles have been supplied by Professor J. Nutting of Leeds University. He used the parameters $a_0 = 4{\cdot}524$ Å, $b_0 = 5{\cdot}088$ Å, and $c_0 = 6{\cdot}742$ Å.

The variation in the three parameters has negligible effect on the combined results from the three sources. Figs. 46 and 47 are stereographic projections of this phase on (100) and (001).

As a result of the study of the orientation of cementite with ferrite in both pearlite and bainite structures, two different relationships have been strongly supported.[52] The double projection shown in Fig. 48 corresponds to the orientation originally found by Bagaryatskii[53] and Fig. 49 corresponds to that of Pitsch's.[54] These double projections were kindly provided by Drs Kelly and Shackleton of Leeds University.

Table 12

Angles between crystallographic planes in cementite (in degrees)

In this table only, the indices of the planes are given by HKL and hkl instead of by $h_1k_1l_1$ and $h_2k_2l_2$. This convention was used in the original paper and is retained here for convenience as is θ for the angler instead of ϕ.

HKL	hkl	θ_{hkl}	$\theta_{hk\bar{l}}$	$\theta_{\bar{h}\bar{k}l}$	$\theta_{\bar{h}\bar{k}\bar{l}}$
001	001	00·00	180·00	00·00	180·00
	010	90·00	90·00	90·00	90·00
	011	52·96	127·04	52·96	127·04
	012	33·53	146·47	33·53	146·47
	013	23·83	156·17	23·83	156·17
	021	69·33	110·67	69·33	110·67
	023	41·46	138·54	41·46	138·54
	031	75·88	104·12	75·88	104·12
	032	63·29	116·71	63·29	116·71
	100	90·00	90·00	90·00	90·00
	101	56·15	123·85	56·15	123·85
	102	36·70	143·30	36·70	143·30
	103	26·42	153·58	26·42	153·58
	110	90·00	90·00	90·00	90·00
	111	63·37	116·63	63·37	116·63
	112	44·92	135·08	44·92	135·08
	113	33·62	146·38	33·62	146·38
010	010	00·00	00·00	180·00	180·00
	011	37·04	37·04	142·96	142·96
	012	56·47	56·47	123·53	123·53
	013	66·17	66·17	113·83	113·83
	021	20·67	20·67	159·33	159·33
	023	48·54	48·54	131·46	131·46
	031	14·12	14·12	165·88	165·88
	032	26·71	26·71	153·29	153·29
	100	90·00	90·00	90·00	90·00
	101	90·00	90·00	90·00	90·00
	102	90·00	90·00	90·00	90·00
	103	90·00	90·00	90·00	90·00
	110	48·37	48·37	131·63	131·63
	111	53·57	53·57	126·43	126·43
	112	62·02	62·02	117·98	117·98
	113	68·42	68·42	111·58	111·58

HKL	hkl	θ_{hkl}	$\theta_{hk\bar{l}}$	$\theta_{\bar{h}\bar{k}l}$	$\theta_{\bar{h}\bar{k}\bar{l}}$
001	120	90·00	90·00	90·00	90·00
	121	71·80	108·20	71·80	108·20
	122	56·67	123·33	56·67	123·33
	123	45·39	134·61	45·39	134·61
	130	90·00	90·00	90·00	90·00
	131	76·75	103·25	76·75	103·25
	132	64·78	115·22	64·78	115·22
	133	54·75	125·25	54·75	125·25
	201	71·46	108·54	71·46	108·54
	203	44·82	135·18	44·82	135·18
	210	90·00	90·00	90·00	90·00
	211	72·96	107·04	72·96	107·04
	212	58·49	121·51	58·49	121·51
	213	47·40	132·60	47·40	132·60
	221	75·93	104·07	75·93	104·07
	223	53·05	126·95	53·05	126·95
	230	90·00	90·00	90·00	90·00
010	120	29·36	29·36	150·64	150·64
	121	34·11	34·11	145·89	145·89
	122	43·27	43·27	136·73	136·73
	123	51·65	51·65	128·35	128·35
	130	20·56	20·56	159·44	159·44
	131	24·30	24·30	155·70	155·70
	132	32·11	32·11	147·89	147·89
	133	40·12	40·12	139·88	139·88
	201	90·00	90·00	90·00	90·00
	203	90·00	90·00	90·00	90·00
	210	66·04	66·04	113·96	113·96
	211	67·15	67·15	112·85	112·85
	212	69·74	69·74	110·26	110·26
	213	72·60	72·60	107·40	107·40
	221	49·88	49·88	130·12	130·12
	223	57·93	57·93	122·07	122·07

HKL	hkl	θ_{hkl}	$\theta_{hk\bar{l}}$	$\theta_{\bar{h}\bar{k}l}$	$\theta_{\bar{h}\bar{k}\bar{l}}$
001	231	78·62	101·38	78·62	101·38
	232	68·08	111·92	68·08	111·92
	233	58·88	121·12	58·88	121·12
	301	77·40	102·60	77·40	102·60
	302	65·91	114·09	65·91	114·09
	310	90·00	90·00	90·00	90·00
	311	77·90	102·10	77·90	102·10
	312	66·79	113·21	66·79	113·21
	313	57·25	122·75	57·25	122·75
	320	90·00	90·00	90·00	90·00
	321	79·11	100·89	79·11	100·89
	322	68·96	111·04	68·96	111·04
	323	60·01	119·99	60·01	119·99
	331	80·51	99·49	80·51	99·49
	332	71·52	108·48	71·52	108·48
010	230	36·87	36·87	143·13	143·13
	231	38·35	38·35	141·65	141·65
	232	42·09	42·09	137·91	137·91
	233	46·78	46·78	133·22	133·22
	301	90·00	90·00	90·00	90·00
	302	90·00	90·00	90·00	90·00
	310	73·50	73·50	106·50	106·50
	311	73·87	73·87	106·13	106·13
	312	74·86	74·86	105·14	105·14
	313	76·18	76·18	103·82	103·82
	320	59·35	59·35	120·65	120·65
	321	59·96	59·96	120·04	120·04
	322	61·59	61·59	118·41	118·41
	323	63·80	63·80	116·20	116·20
	331	49·06	49·06	130·94	130·94
	332	50·94	50·94	129·06	129·06

Table 12 (cont.)

HKL	hkl	θ_{hkl}	$\theta_{hk\bar{l}}$	$\theta_{h\bar{k}l}$	$\theta_{h\bar{k}\bar{l}}$
011	011	00·00	74·08	105·92	180·00
	012	19·43	93·51	86·49	160·57
	013	29·13	103·21	76·79	150·87
	021	16·37	57·71	122·29	163·63
	023	11·50	85·58	94·42	168·50
	031	22·92	51·16	128·84	157·08
	032	10·33	63·75	116·25	169·67
	100	90·00	90·00	90·00	90·00
	101	70·39	109·61	70·39	109·61
	102	61·12	118·88	61·12	118·88
	103	57·35	122·65	57·35	122·65
	110	57·97	57·97	122·03	122·03
	111	41·92	78·23	101·77	138·08
	112	36·77	92·98	87·02	143·23
	113	37·32	102·01	77·99	142·68
	120	45·92	45·92	134·08	134·08
012	012	00·00	112·95	67·05	180·00
	013	9·70	122·64	57·36	170·30
	021	35·80	77·15	102·85	144·20
	023	7·93	105·02	74·98	172·07
	031	42·35	70·59	109·41	137·65
	032	29·77	83·18	96·82	150·23
	100	90·00	90·00	90·00	90·00
	101	62·33	117·67	62·33	117·67
	102	48·06	131·94	48·06	131·94
	103	41·71	138·29	41·71	138·29
	110	68·47	68·47	111·53	111·53
	111	45·44	92·61	87·39	134·56
	112	31·85	109·34	70·66	148·15
	113	26·19	119·41	60·59	153·81
	120	61·22	61·22	118·78	118·78
	121	44·13	78·65	101·35	135·87
013	013	00·00	132·34	47·66	180·00
	021	45·50	86·84	93·16	134·50
	023	17·63	114·71	65·29	162·37
	031	52·05	80·29	99·71	127·95
	032	39·46	92·88	87·12	140·54
	100	90·00	90·00	90·00	90·00
	101	59·36	120·64	59·36	120·64
	102	42·83	137·17	42·83	137·17
	103	35·00	145·00	35·00	145·00
	110	74·43	74·43	105·57	105·57
	111	49·46	99·79	80·21	130·54
	112	33·15	117·27	62·73	146·85
	113	24·44	127·82	52·18	155·56
	120	69·38	69·38	110·62	110·62
	121	51·66	87·21	92·79	128·34

HKL	hkl	θ_{hkl}	$\theta_{hk\bar{l}}$	$\theta_{h\bar{k}l}$	$\theta_{h\bar{k}\bar{l}}$
011	121	31·89	61·79	118·21	148·11
	122	24·18	75·51	104·49	155·82
	123	23·32	85·86	94·14	156·68
	130	41·64	41·64	138·36	138·36
	131	30·05	53·89	126·11	149·95
	132	21·12	65·20	114·80	158·88
	133	16·66	74·77	105·23	163·34
	201	78·96	101·04	78·96	101·04
	203	64·71	115·29	64·71	115·29
	210	71·08	71·08	108·92	108·92
	211	60·89	82·33	97·67	119·11
	212	53·76	92·20	87·80	126·24
	213	49·73	99·73	80·27	130·27
	221	48·63	68·41	111·59	131·37
	223	38·20	86·46	93·54	141·80
	230	50·31	50·31	129·69	129·69
012	122	30·65	93·21	86·79	149·35
	123	21·85	104·05	75·95	158·15
	130	58·86	58·86	121·14	121·14
	131	46·01	71·80	108·20	133·99
	132	34·60	83·54	96·46	145·40
	133	25·39	93·37	86·63	154·61
	201	74·63	105·37	74·63	105·37
	203	53·75	126·25	53·75	126·25
	210	77·04	77·04	102·96	102·96
	211	62·69	91·71	88·29	117·31
	212	51·18	104·15	75·85	128·82
	213	43·17	113·52	66·48	136·83
	221	56·04	81·19	98·81	123·96
	223	37·41	101·99	78·01	142·59
	230	63·78	63·78	116·22	116·22
013	122	37·17	102·03	77·97	142·83
	123	26·73	113·06	66·94	153·27
	130	67·77	67·77	112·23	112·23
	131	54·69	80·88	99·12	125·31
	132	42·94	92·73	87·27	137·06
	133	33·19	102·65	77·35	146·81
	201	73·09	106·91	73·09	106·91
	203	49·55	130·45	49·55	130·45
	210	80·56	80·56	99·44	99·44
	211	64·85	96·38	83·62	115·15
	212	51·83	109·77	70·23	128·17
	213	42·27	119·89	60·11	137·73
	221	61·13	87·83	92·17	118·87
	223	40·15	109·59	70·41	139·85
	230	71·14	71·14	108·86	108·86

HKL	hkl	θ_{hkl}	$\theta_{hk\bar{l}}$	$\theta_{h\bar{k}l}$	$\theta_{h\bar{k}\bar{l}}$
011	231	41·85	59·52	120·48	138·15
	232	35·18	68·44	111·56	144·82
	233	30·91	76·39	103·61	149·09
	301	82·45	97·55	82·45	97·55
	302	75·76	104·24	75·76	104·24
	310	76·89	76·89	103·11	103·11
	311	69·63	84·52	95·48	110·37
	312	63·53	91·66	88·34	116·47
	313	58·90	97·77	82·23	121·10
	320	65·99	65·99	114·01	114·01
	321	59·11	73·39	106·61	120·89
	322	53·41	80·59	99·41	126·59
	323	49·19	87·06	92·94	130·81
	331	51·51	64·93	115·07	128·49
	332	46·06	71·82	108·18	133·94
012	231	53·30	74·41	105·59	126·70
	232	43·85	84·34	95·66	136·15
	233	35·99	93·01	86·99	144·01
	301	79·52	100·48	79·52	100·48
	302	70·10	109·90	70·10	109·90
	310	80·97	80·97	99·03	99·03
	311	70·84	91·22	88·78	109·16
	312	61·79	100·62	79·38	118·21
	313	54·34	108·60	71·40	125·66
	320	73·65	73·65	106·35	106·35
	321	64·28	83·16	96·84	115·72
	322	55·80	92·09	87·91	124·20
	323	48·66	99·95	80·05	131·34
	331	60·04	77·03	102·97	119·96
	332	52·25	85·20	94·80	127·75
013	231	60·18	82·16	97·84	119·82
	232	50·10	92·39	87·61	129·90
	233	41·45	101·31	78·69	138·55
	301	78·49	101·51	78·49	101·51
	302	68·07	111·93	68·07	111·93
	310	83·41	83·41	96·59	96·59
	311	72·30	94·56	85·44	107·70
	312	62·23	104·77	75·23	117·77
	313	53·75	113·47	66·53	126·25
	320	78·11	78·11	101·89	101·89
	321	67·97	88·31	91·69	112·03
	322	58·62	97·83	82·17	121·38
	323	50·53	106·19	73·81	129·47
	331	65·45	83·46	96·54	114·55
	332	57·00	92·03	87·97	123·00

HKL	hkl	θ_{hkl}	$\theta_{hk\bar{l}}$	$\theta_{h\bar{k}l}$	$\theta_{h\bar{k}\bar{l}}$
021	021	00.00	41.35	138.65	180.00
	023	27.87	69.22	110.78	152.13
	031	6.55	34.79	145.21	173.45
	032	6.03	47.38	132.62	173.97
	100	90.00	90.00	90.00	90.00
	101	78.66	101.34	78.66	101.34
	102	73.56	106.44	73.56	106.44
	103	71.57	108.43	71.57	108.43
	110	51.57	51.57	128.43	128.43
	111	44.45	66.58	113.42	135.55
	112	46.46	79.11	100.89	133.54
	113	50.35	87.13	92.87	129.65
	120	35.37	35.37	144.63	144.63
	121	27.76	48.37	131.63	152.24
	122	28.92	60.84	119.16	151.08
021	123	34.06	70.58	109.42	145.94
	130	28.83	28.83	151.17	151.17
	131	20.99	39.49	140.51	159.01
	132	19.45	50.05	129.95	160.55
	133	23.19	59.22	120.78	156.81
	201	83.55	96.45	83.55	96.45
	203	75.50	104.50	75.50	104.50
	210	67.67	67.67	112.33	112.33
	211	62.18	74.94	105.06	117.82
	212	59.44	81.98	98.02	120.56
	213	58.76	87.66	92.34	121.24
	221	46.47	58.86	121.14	133.53
	223	44.85	73.47	106.53	135.15
	230	41.54	41.54	138.46	138.46
	231	36.54	48.38	131.62	143.46
021	232	34.29	55.77	124.23	145.71
	233	34.59	62.72	117.28	145.41
	301	85.58	94.42	85.58	94.42
	302	81.71	98.29	81.71	98.29
	310	74.59	74.59	105.41	105.41
	311	70.49	79.29	100.71	109.51
	312	67.45	83.96	96.04	112.55
	313	65.51	88.13	91.87	114.49
	320	61.51	61.51	118.49	118.49
	321	57.65	66.32	113.68	122.35
	322	55.11	71.43	108.57	124.89
	323	53.87	76.31	103.69	126.13
	331	47.83	56.30	123.70	132.17
	332	45.46	61.47	118.53	134.54
023	023	00.00	97.09	82.91	180.00
	031	34.42	62.66	117.34	145.58
	032	21.84	75.25	104.75	158.16
	100	90.00	90.00	90.00	90.00
	101	65.32	114.68	65.32	114.68
	102	53.07	126.93	53.07	126.93
	103	47.84	132.16	47.84	132.16
	110	63.91	63.91	116.09	116.09
	111	43.19	86.72	93.28	136.81
	112	32.73	102.71	77.29	147.27
	113	29.82	112.37	67.63	150.18
	120	54.76	54.76	125.24	125.24
	121	38.53	71.70	108.30	141.47
	122	26.63	85.97	94.03	153.37
	123	20.43	96.64	83.36	159.57
023	130	51.69	51.69	128.31	128.31
	131	39.18	64.43	115.57	140.82
	132	28.34	76.03	103.97	151.66
	133	20.16	85.77	94.23	159.84
	201	76.21	103.79	76.21	103.79
	203	57.89	122.11	57.89	122.11
	210	74.40	74.40	105.60	105.60
	211	61.53	87.85	92.15	118.47
	212	51.62	99.35	80.65	128.38
	213	45.15	108.02	71.98	134.85
	221	52.49	75.85	104.15	127.51
	223	36.68	95.68	84.32	143.32
	230	58.02	58.02	121.98	121.98
	231	48.16	68.20	111.80	131.84
	232	39.54	77.79	102.21	140.46
023	233	32.78	86.21	93.79	147.22
	301	80.59	99.41	80.59	99.41
	302	72.18	107.82	72.18	107.82
	310	79.16	79.16	100.84	100.84
	311	70.06	88.46	91.54	109.94
	312	62.08	97.04	82.96	117.92
	313	55.69	104.31	75.69	124.31
	320	70.27	70.27	109.73	109.73
	321	61.77	79.05	100.95	118.23
	322	54.26	87.37	92.63	125.74
	323	48.17	94.72	85.28	131.83
	331	56.13	71.92	108.08	123.87
	332	49.10	79.65	100.35	130.90
031	031	00.00	28.24	151.76	180.00
	032	12.59	40.83	139.17	167.41
	100	90.00	90.00	90.00	90.00
	101	82.19	97.81	82.19	97.81
	102	78.72	101.28	78.72	101.28
	103	77.38	102.62	77.38	102.62
	110	49.89	49.89	130.11	130.11
	111	46.74	62.19	117.81	133.26
	112	51.12	73.61	106.39	128.88
	113	55.95	81.17	98.83	124.05
	120	32.30	32.30	147.70	147.70
	121	28.46	43.39	136.61	151.54
	122	32.83	55.10	124.90	147.17
	123	39.37	64.51	115.49	140.63
031	130	24.76	24.76	155.24	155.24
	131	19.98	34.11	145.89	160.02
	132	22.27	44.15	135.85	157.73
	133	28.07	53.07	126.93	151.93
	201	85.55	94.45	85.55	94.45
	203	80.04	99.96	80.04	99.96
	210	66.80	66.80	113.20	113.20
	211	63.38	72.24	107.76	116.62
	212	62.40	77.98	102.02	117.60
	213	62.93	82.83	97.17	117.07
	221	46.82	55.55	124.45	133.18
	223	48.58	68.39	111.61	131.42
	230	39.12	39.12	140.88	140.88
	231	36.03	44.57	135.43	143.97
031	232	35.82	51.05	128.95	144.18
	233	37.79	57.45	122.55	142.21
	301	86.95	93.05	86.95	93.05
	302	84.28	95.72	84.28	95.72
	310	74.01	74.01	105.99	105.99
	311	71.31	77.39	102.61	108.69
	312	69.55	80.96	99.04	110.45
	313	68.67	84.28	95.72	111.33
	320	60.37	60.37	119.63	119.63
	321	57.89	63.93	116.07	122.11
	322	56.70	68.05	111.95	123.30
	323	56.62	72.17	107.83	123.38
	331	47.49	53.47	126.53	132.51
	332	46.50	57.74	122.26	133.50

Table 12 (cont.)

HKL	hkl	θ_{hkl}	$\theta_{hk\bar{l}}$	$\theta_{h\bar{k}l}$	$\theta_{h\bar{k}\bar{l}}$
032	032	00·00	53·42	126·58	180·00
	100	90·00	90·00	90·00	90·00
	101	75·50	104·50	75·50	104·50
	102	68·88	111·12	68·88	111·12
	103	66·27	113·73	66·27	113·73
	110	53·60	53·60	126·40	126·40
	111	42·95	70·79	109·21	137·05
	112	42·50	84·21	95·79	137·50
	113	45·34	92·62	87·38	134·66
	120	38·87	38·87	141·13	141·13
	121	28·35	53·19	126·81	151·65
	122	26·17	66·20	113·80	153·83
	123	29·55	76·20	103·80	150·45
	130	33·23	33·23	146·77	146·77
100	100	00·00	00·00	00·00	00·00
	101	33·85	33·85	33·85	33·85
	102	53·30	53·30	53·30	53·30
	103	63·58	63·58	63·58	63·58
	110	41·63	41·63	41·63	41·63
	111	48·08	48·08	48·08	48·08
	112	58·15	58·15	58·15	58·15
	113	65·56	65·56	65·56	65·56
	120	60·64	60·64	60·64	60·64
	121	62·24	62·24	62·24	62·24
	122	65·82	65·82	65·82	65·82
	123	69·57	69·57	69·57	69·57
	130	69·44	69·44	69·44	69·44
	131	70·02	70·02	70·02	70·02
101	101	00·00	67·71	00·00	67·71
	102	19·45	87·16	19·45	87·16
	103	29·72	97·43	29·72	97·43
	110	51·63	51·63	51·63	51·63
	111	36·43	72·23	36·43	72·23
	112	33·62	87·49	33·62	87·49
	113	36·14	96·91	36·14	96·91
	120	65·97	65·97	65·97	65·97
	121	55·89	77·72	55·89	77·72
	122	49·74	88·05	49·74	88·05
	123	47·07	95·82	47·07	95·82
	130	73·05	73·05	73·05	73·05
	131	65·70	81·02	65·70	81·02

HKL	hkl	θ_{hkl}	$\theta_{hk\bar{l}}$	$\theta_{h\bar{k}l}$	$\theta_{h\bar{k}\bar{l}}$
032	131	23·48	44·67	135·33	156·52
	132	18·52	55·59	124·41	161·48
	133	19·53	64·93	115·07	160·47
	201	81·78	98·22	81·78	98·22
	203	71·41	108·59	71·41	108·59
	210	68·73	68·73	111·27	111·27
	211	61·41	77·57	102·43	118·59
	212	57·03	85·73	94·27	122·97
	213	55·16	92·13	87·87	124·84
	221	46·77	62·20	117·80	133·23
	223	41·89	78·22	101·78	138·11
	230	44·39	44·39	135·61	135·61
	231	37·88	52·27	127·73	142·12
	232	33·82	60·32	119·68	146·18
100	132	71·48	71·48	71·48	71·48
	133	73·34	73·34	73·34	73·34
	201	18·54	18·54	18·54	18·54
	203	45·18	45·18	45·18	45·18
	210	23·96	23·96	23·96	23·96
	211	29·11	29·11	29·11	29·11
	212	38·82	38·82	38·82	38·82
	213	47·73	47·73	47·73	47·73
	221	43·53	43·53	43·53	43·53
	223	53·32	53·32	53·32	53·32
	230	53·13	53·13	53·13	53·13
	231	53·97	53·97	53·97	53·97
	232	56·18	56·18	56·18	56·18
	233	59·09	59·09	59·09	59·09
101	132	59·92	88·49	59·92	88·49
	133	55·97	94·78	55·97	94·78
	201	15·31	52·40	15·31	52·40
	203	11·32	79·03	11·32	79·03
	210	40·63	40·63	40·63	40·63
	211	27·28	55·78	27·28	55·78
	212	20·26	69·15	20·26	69·15
	213	20·66	79·54	20·66	79·54
	221	42·48	62·19	42·48	62·19
	223	33·81	80·72	33·81	80·72
	230	60·11	60·11	60·11	60·11
	231	53·25	67·75	53·25	67·75
	232	47·91	75·27	47·91	75·27

HKL	hkl	θ_{hkl}	$\theta_{hk\bar{l}}$	$\theta_{h\bar{k}l}$	$\theta_{h\bar{k}\bar{l}}$
032	233	32·43	67·70	112·30	147·57
	301	84·37	95·63	84·37	95·63
	302	79·43	100·57	79·43	100·57
	310	75·30	75·30	104·70	104·70
	311	69·98	81·15	98·85	110·02
	312	65·77	86·78	93·22	114·23
	313	62·83	91·70	88·30	117·17
	320	62·91	62·91	117·09	117·09
	321	57·85	68·76	111·24	122·15
	322	54·10	74·71	105·29	125·90
	323	51·75	80·22	99·78	128·25
	331	48·74	59·25	120·75	131·26
	332	45·14	65·14	114·86	134·86
100	301	12·60	12·60	12·60	12·60
	302	24·09	24·09	24·09	24·09
	310	16·50	16·50	16·50	16·50
	311	20·37	20·37	20·37	20·37
	312	28·21	28·21	28·21	28·21
	313	36·25	36·25	36·25	36·25
	320	30·65	30·65	30·65	30·65
	321	32·35	32·35	32·35	32·35
	322	36·59	36·59	36·59	36·59
	323	41·83	41·83	41·83	41·83
	331	42·51	42·51	42·51	42·51
	332	44·86	44·86	44·86	44·86
101	233	44·40	82·03	44·40	82·03
	301	21·25	46·46	21·25	46·46
	302	9·76	57·95	9·76	57·95
	310	37·23	37·23	37·23	37·23
	311	26·45	48·57	26·45	48·57
	312	17·95	59·19	17·95	59·19
	313	13·82	68·39	13·82	68·39
	320	44·40	44·40	44·40	44·40
	321	36·22	53·39	36·22	53·39
	322	29·91	62·18	29·91	62·18
	323	26·20	70·10	26·20	70·10
	331	45·25	58·64	45·25	58·64
	332	40·07	65·66	40·07	65·66

HKL	hkl	θ_{hkl}	$\theta_{hk\bar{l}}$	$\theta_{h\bar{k}l}$	$\theta_{h\bar{k}\bar{l}}$
102	102	00·00	106·60	00·00	106·60
	103	10·28	116·88	10·28	116·88
	110	63·47	63·47	63·47	63·47
	111	40·66	87·71	40·66	87·71
	112	27·98	104·61	27·98	104·61
	113	23·80	114·86	23·80	114·86
	120	72·96	72·96	72·96	72·96
	121	58·07	88·41	58·07	88·41
	122	46·73	101·29	46·73	101·29
	123	39·50	110·77	39·50	110·77
	130	77·89	77·89	77·89	77·89
	131	67·17	88·83	67·17	88·83
	132	57·89	98·73	57·89	98·73
103	103	00·00	127·15	00·00	127·15
	110	70·57	70·57	70·57	70·57
	111	45·68	95·97	45·68	95·97
	112	29·66	113·53	29·66	113·53
	113	21·58	124·17	21·58	124·17
	120	77·40	77·40	77·40	77·40
	121	60·86	94·16	60·86	94·16
	122	47·59	108·05	47·59	108·05
	123	38·35	118·27	38·35	118·27
	130	81·01	81·01	81·01	81·01
	131	69·06	93·05	69·06	93·05
	132	58·47	103·90	58·47	103·90
	133	49·88	112·90	49·88	112·90
110	110	00·00	00·00	83·27	83·27
	111	26·63	26·63	83·98	83·98
	112	45·08	45·08	85·25	85·25
	113	56·38	56·38	86·28	86·28
	120	19·01	19·01	102·28	102·28
	121	26·08	26·08	101·65	101·65
	122	37·82	37·82	100·23	100·23
	123	47·70	47·70	98·71	98·71
	130	27·81	27·81	111·08	111·08
	131	30·58	30·58	110·49	110·49
	132	36·85	36·85	108·99	108·99
	133	43·75	43·75	107·08	107·08
111	111	00·00	53·26	72·87	96·15
	112	18·45	71·71	66·96	104·08
	113	29·75	83·01	64·45	108·37
	120	32·31	32·31	100·96	100·96
	121	19·45	48·48	92·32	108·70
	122	17·74	62·62	84·98	113·90
	123	23·59	73·33	79·66	116·75
	130	37·75	37·75	108·75	108·75
	131	29·26	48·17	102·13	114·56
	132	25·00	58·38	95·73	118·81
	133	25·26	67·23	90·22	121·41
	201	39·11	60·60	39·11	60·60

HKL	hkl	θ_{hkl}	$\theta_{hk\bar{l}}$	$\theta_{h\bar{k}l}$	$\theta_{h\bar{k}\bar{l}}$
102	133	50·65	106·94	50·65	106·94
	201	34·76	71·84	34·76	71·84
	203	8·12	98·48	8·12	98·48
	210	56·90	56·90	56·90	56·90
	211	40·79	73·31	40·79	73·31
	212	27·79	87·33	27·79	87·33
	213	19·15	98·09	19·15	98·09
	221	51·08	76·21	51·08	76·21
	223	32·98	97·18	32·98	97·18
	230	68·99	68·99	68·99	68·99
	231	59·36	78·85	59·36	78·85
	232	50·80	88·09	50·80	88·09
	233	43·83	96·17	43·83	96·17
103	201	45·04	82·12	45·04	82·12
	203	18·40	108·75	18·40	108·75
	210	66·01	66·01	66·01	66·01
	211	49·37	82·74	49·37	82·74
	212	35·44	96·97	35·44	96·97
	213	25·11	107·87	25·11	107·87
	221	57·29	83·98	57·29	83·98
	223	36·48	105·81	36·48	105·81
	230	74·51	74·51	74·51	74·51
	231	64·00	85·12	64·00	85·12
	232	54·40	94·97	54·40	94·97
	233	46·26	103·55	46·26	103·55
	301	50·97	76·18	50·97	76·18
110	201	44·88	44·88	44·88	44·88
	203	58·21	58·21	58·21	58·21
	210	17·67	17·67	65·60	65·60
	211	24·36	24·36	66·73	66·73
	212	35·68	35·68	69·38	69·38
	213	45·46	45·46	72·29	72·29
	221	14·07	14·07	83·47	83·47
	223	36·95	36·95	84·62	84·62
	230	11·50	11·50	94·76	94·76
	231	16·12	16·12	94·67	94·67
	232	24·62	24·62	94·42	94·42
	233	32·98	32·98	94·08	94·08
111	203	37·92	81·20	37·92	81·20
	210	31·60	31·60	68·32	68·32
	211	18·97	46·92	61·02	77·19
	212	16·18	60·53	56·69	85·37
	213	21·51	71·12	54·88	91·80
	221	12·55	40·70	77·84	90·42
	223	10·32	63·57	69·32	100·70
	230	28·84	28·84	94·26	94·26
	231	18·70	39·61	89·10	99·28
	232	11·48	49·81	84·35	103·66
	233	11·02	58·78	80·32	107·17
	301	41·42	56·34	41·42	56·34

HKL	hkl	θ_{hkl}	$\theta_{hk\bar{l}}$	$\theta_{h\bar{k}l}$	$\theta_{h\bar{k}\bar{l}}$
102	301	40·70	65·91	40·70	65·91
	302	29·21	77·40	29·21	77·40
	310	55·04	55·04	55·04	55·04
	311	43·25	66·91	43·25	66·91
	312	32·58	77·84	32·58	77·84
	313	23·70	87·24	23·70	87·24
	320	59·06	59·06	59·06	59·06
	321	48·98	69·30	48·98	69·30
	322	39·85	78·93	39·85	78·93
	323	32·22	87·45	32·22	87·45
	331	55·06	72·04	55·06	72·04
	332	47·33	80·24	47·33	80·24
103	302	39·48	87·67	39·48	87·67
	310	64·74	64·74	64·74	64·74
	311	52·78	76·74	52·78	76·74
	312	41·84	87·75	41·84	87·75
	313	32·51	97·21	32·51	97·21
	320	67·49	67·49	67·49	67·49
	321	56·97	78·07	56·97	78·07
	322	47·25	87·95	47·25	87·95
	323	38·81	96·66	38·81	96·66
	331	61·60	79·61	61·60	79·61
	332	53·18	88·19	53·18	88·19
110	301	43·16	43·16	43·16	43·16
	302	46·98	46·98	46·98	46·98
	310	25·13	25·13	58·14	58·14
	311	27·72	27·72	58·93	58·93
	312	33·69	33·69	60·98	60·98
	313	40·41	40·41	63·64	63·64
	320	10·98	10·98	72·28	72·28
	321	15·42	15·42	72·61	72·61
	322	23·62	23·62	73·50	73·50
	323	31·76	31·76	74·72	74·72
	331	9·49	9·49	83·36	83·36
	332	18·48	18·48	83·62	83·62
111	302	37·54	64·73	37·54	64·73
	310	35·97	35·97	61·84	61·84
	311	27·71	45·78	56·26	68·44
	312	23·01	55·44	52·39	75·10
	313	22·61	64·01	50·26	81·12
	320	28·65	28·65	74·22	74·22
	321	18·84	39·00	69·40	79·49
	322	11·49	48·84	65·49	84·67
	323	10·23	57·58	62·64	89·34
	331	17·14	36·12	79·79	88·31
	332	8·15	45·11	76·03	92·45

Table 12 (cont.)

HKL	hkl	θ_{hkl}	$\theta_{hk\bar{l}}$	$\theta_{h\bar{k}l}$	$\theta_{h\bar{k}\bar{l}}$
112	112	00·00	90·16	55·96	116·29
	113	11·30	101·46	50·54	122·94
	120	48·12	48·12	98·63	98·63
	121	31·19	65·61	85·49	111·34
	122	18·76	80·29	74·71	120·97
	123	13·45	91·26	67·02	127·17
	130	51·35	51·35	104·71	104·71
	131	39·62	63·54	94·87	114·17
	132	29·91	74·74	85·87	122·11
	133	23·26	84·18	78·39	128·03
	201	43·49	74·03	43·49	74·03
	203	29·04	97·48	29·04	97·48
113	113	00·00	112·77	43·16	131·11
	120	58·44	58·44	96·76	96·76
	121	40·76	76·29	81·47	111·84
	122	26·50	91·16	68·94	123·78
	123	16·77	102·25	59·93	131·96
	130	60·68	60·68	101·48	101·48
	131	48·12	73·40	90·17	112·63
	132	37·07	84·94	79·94	122·34
	133	28·30	94·63	71·46	130·03
	201	48·92	82·67	48·92	82·67
	203	28·07	107·40	28·07	107·40
120	120	00·00	00·00	121·28	121·28
	121	18·20	18·20	119·56	119·56
	122	33·33	33·33	115·71	115·71
	123	44·61	44·61	111·69	111·69
	130	8·80	8·80	130·09	130·09
	131	15·87	15·87	128·81	128·81
	132	26·62	26·62	125·63	125·63
	133	36·19	36·19	121·73	121·73
	201	62·30	62·30	62·30	62·30
	203	69·78	69·78	69·78	69·78
	210	36·68	36·68	84·60	84·60
121	121	00·00	36·41	111·78	124·49
	122	15·13	51·54	103·91	125·72
	123	26·41	62·82	97·57	124·79
	130	20·16	20·16	127·71	127·71
	131	9·81	32·63	121·59	131·84
	132	10·77	44·27	114·85	133·36
	133	18·74	54·10	108·62	132·83
	201	57·26	69·99	57·26	69·99
	203	56·64	83·88	56·64	83·88
	210	40·37	40·37	84·88	84·88
	211	34·92	50·44	79·81	90·35

HKL	hkl	θ_{hkl}	$\theta_{hk\bar{l}}$	$\theta_{h\bar{k}l}$	$\theta_{h\bar{k}\bar{l}}$
112	210	47·72	47·72	73·04	73·04
	211	31·70	64·17	60·89	85·90
	212	19·32	78·25	51·77	96·97
	213	12·96	89·08	46·05	105·34
	221	31·01	59·15	75·38	95·27
	223	8·13	82·02	60·54	111·07
	230	46·21	46·21	93·36	93·36
	231	35·11	57·41	85·28	101·37
	232	25·00	67·82	77·88	108·59
	233	16·60	76·91	71·59	114·59
	301	47·97	68·87	47·97	68·87
	302	39·57	78·89	39·57	78·89
113	210	58·16	58·16	76·78	76·78
	211	41·55	74·91	62·44	91·45
	212	27·75	89·17	50·93	103·90
	213	17·83	100·10	42·94	113·28
	221	42·31	70·46	74·61	98·02
	223	19·44	93·33	56·47	116·66
	230	57·14	57·14	92·64	92·64
	231	45·88	68·43	83·15	102·09
	232	35·49	78·91	74·44	110·71
	233	26·51	88·05	66·98	118·02
	301	54·16	77·17	54·16	77·17
120	211	39·93	39·93	84·84	84·84
	212	46·86	46·86	85·40	85·40
	213	53·82	53·82	86·03	86·03
	221	23·50	23·50	101·90	101·90
	223	40·92	40·92	99·78	99·78
	230	7·51	7·51	113·77	113·77
	231	13·61	13·61	113·28	113·28
	232	23·12	23·12	111·96	111·96
	233	31·93	31·93	110·19	110·19
	301	61·42	61·42	61·42	61·42
	302	63·41	63·41	63·41	63·41
121	212	35·63	60·91	76·15	95·00
	213	39·44	69·55	73·91	98·38
	221	18·71	37·32	96·89	105·78
	223	25·10	57·99	88·49	110·44
	230	19·65	19·65	112·52	112·52
	231	9·96	30·50	108·29	115·92
	232	7·97	40·80	103·80	118·16
	233	14·60	49·85	99·58	119·29
	301	58·49	67·27	58·49	67·27
	302	56·45	72·69	56·45	72·69
	310	47·02	47·02	77·80	77·80

HKL	hkl	θ_{hkl}	$\theta_{hk\bar{l}}$	$\theta_{h\bar{k}l}$	$\theta_{h\bar{k}\bar{l}}$
112	310	50·26	50·26	68·11	68·11
	311	39·33	61·53	59·14	77·52
	312	29·93	72·03	51·56	86·36
	313	22·97	81·10	45·85	93·99
	320	46·12	46·12	77·59	77·59
	321	35·46	56·84	69·83	85·57
	322	25·68	66·88	62·95	93·08
	323	17·39	75·73	57·31	99·66
	331	35·59	54·57	78·56	92·01
	332	26·60	63·56	72·36	98·39
113	302	44·13	87·84	44·13	87·84
	310	59·92	59·92	73·01	73·01
	311	48·34	71·61	62·59	83·62
	312	37·92	82·39	53·36	93·42
	313	29·30	91·66	45·87	101·81
	320	57·08	57·08	80·30	80·30
	321	46·29	67·89	71·17	89·53
	322	36·27	77·98	62·85	98·15
	323	27·50	86·88	55·79	105·68
	331	46·89	65·87	78·39	94·20
	332	37·90	74·86	71·00	101·68
120	310	44·14	44·14	77·15	77·15
	311	45·43	45·43	77·44	77·44
	312	48·73	48·73	78·20	78·20
	313	52·87	52·87	79·22	79·22
	320	29·99	29·99	91·29	91·29
	321	31·73	31·73	91·27	91·27
	322	36·07	36·07	91·21	91·21
	323	41·40	41·40	91·12	91·12
	331	21·17	21·17	102·11	102·11
	332	26·27	26·27	101·63	101·63
121	311	42·94	53·05	74·21	81·89
	312	41·44	59·77	71·50	85·92
	313	42·07	66·15	69·71	89·50
	320	34·64	34·64	91·23	91·23
	321	29·89	41·50	87·82	94·59
	322	28·35	49·03	84·71	97·60
	323	29·69	56·19	82·09	100·06
	331	20·39	33·45	98·49	104·52
	332	18·04	41·17	95·31	106·89

HKL	hkl	θ_{hkl}	$\theta_{hk\bar{l}}$	$\theta_{h\bar{k}l}$	$\theta_{h\bar{k}\bar{l}}$
122	122	00·00	66·67	93·47	131·64
	123	11·28	77·95	85·58	134·01
	130	34·35	34·35	122·55	122·55
	131	21·62	47·34	113·43	130·52
	132	11·16	59·15	104·63	136·13
	133	7·52	69·07	97·02	139·16
	201	55·73	77·67	55·73	77·67
	203	47·27	95·80	47·27	95·80
	210	47·93	47·93	85·49	85·49
	211	36·71	61·34	76·34	94·93
123	123	00·00	89·23	76·70	139·15
	130	45·29	45·29	117·28	117·28
	131	32·25	58·42	106·57	127·39
	132	20·66	70·30	96·63	135·56
	133	11·53	80·26	88·23	141·23
	201	56·34	83·83	56·34	83·83
	203	41·91	104·61	41·91	104·61
	210	55·19	55·19	86·16	86·16
	211	41·27	70·12	74·35	98·15
	212	31·37	83·13	64·91	108·06
130	130	00·00	00·00	138·89	138·89
	131	13·25	13·25	137·17	137·17
	132	25·22	25·22	132·97	132·97
	133	35·25	35·25	127·97	127·97
	201	70·56	70·56	70·56	70·56
	203	75·67	75·67	75·67	75·67
	210	45·48	45·48	93·41	93·41
	211	47·91	47·91	93·26	93·26
	212	53·29	53·29	92·90	92·90
	213	58·93	58·93	92·51	92·51
131	131	00·00	26·51	131·40	140·03
	132	11·97	38·48	124·45	139·56
	133	21·99	48·50	117·81	136·99
	201	66·61	75·46	66·61	75·46
	203	66·20	85·51	66·20	85·51
	210	46·96	46·96	93·32	93·32
	211	43·97	54·17	89·32	97·04
	212	45·44	62·48	85·96	99·74
	213	48·89	69·68	83·53	101·41
132	132	00·00	50·45	115·78	142·96
	133	10·02	60·47	108·10	143·38
	201	64·11	80·47	64·11	80·47
	203	58·25	94·49	58·25	94·49
	210	50·63	50·63	93·08	93·08
	211	43·00	61·21	85·79	100·15
	212	40·23	71·46	79·81	105·58
	213	40·94	79·72	75·59	109·15
	221	28·39	47·74	102·24	114·78

HKL	hkl	θ_{hkl}	$\theta_{hk\bar{l}}$	$\theta_{h\bar{k}l}$	$\theta_{h\bar{k}\bar{l}}$
122	212	30·86	73·50	69·26	102·72
	213	30·10	83·03	64·55	108·31
	221	25·86	50·76	92·22	107·81
	223	15·93	72·48	79·14	118·18
	230	34·08	34·08	109·68	109·68
	231	23·01	45·28	102·81	116·01
	232	13·20	55·72	96·16	121·17
	233	6·73	64·85	90·25	124·91
	301	58·69	73·75	58·69	73·75
	302	53·25	81·40	53·25	81·40
123	213	26·41	93·16	58·36	115·22
	221	34·55	61·18	88·63	108·52
	223	16·25	83·35	72·47	122·90
	230	45·11	45·11	106·68	106·68
	231	33·86	56·41	98·21	114·83
	232	23·52	66·89	90·23	121·90
	233	14·72	76·04	83·26	127·49
	301	60·41	79·20	60·41	79·20
	302	52·75	88·17	52·75	88·17
	310	59·28	59·28	80·89	80·89
130	221	30·91	30·91	110·42	110·42
	223	45·02	45·02	106·70	106·70
	230	16·31	16·31	122·57	122·57
	231	19·80	19·80	121·86	121·86
	232	27·09	27·09	119·96	119·96
	233	34·75	34·75	117·45	117·45
	301	69·96	69·96	69·96	69·96
	302	71·31	71·31	71·31	71·31
	310	52·94	52·94	85·95	85·95
	311	53·90	53·90	86·04	86·04
131	221	27·02	38·80	106·49	113·28
	223	34·32	56·61	98·16	114·68
	230	20·90	20·90	121·60	121·60
	231	16·05	29·47	117·94	123·99
	232	17·78	38·65	113·61	124·87
	233	23·33	47·06	109·28	124·55
	301	67·45	73·53	67·45	73·53
	302	66·07	77·39	66·07	77·39
	310	54·08	54·08	86·06	86·06
132	223	26·41	67·46	90·22	121·08
	230	29·75	29·75	119·15	119·15
	231	20·73	39·91	113·17	124·16
	232	15·30	49·74	107·02	127·66
	233	15·52	58·47	101·34	129·58
	301	66·24	77·47	66·24	77·47
	302	62·36	83·34	62·36	83·34
	310	56·96	56·96	86·34	86·34
	311	51·51	63·66	81·27	91·54

HKL	hkl	θ_{hkl}	$\theta_{hk\bar{l}}$	$\theta_{h\bar{k}l}$	$\theta_{h\bar{k}\bar{l}}$
122	310	53·16	53·16	79·29	79·29
	311	45·46	61·90	72·73	86·19
	312	39·86	70·46	67·21	92·62
	313	36·72	78·05	63·03	98·10
	320	43·65	43·65	91·08	91·08
	321	35·47	52·64	85·11	97·03
	322	29·23	61·44	79·65	102·41
	323	25·66	69·39	75·03	106·92
	331	29·58	46·49	94·85	105·41
	332	22·57	54·90	89·67	110·04
123	311	49·70	69·37	72·42	89·56
	312	41·73	78·89	65·02	97·54
	313	35·94	87·15	59·13	104·28
	320	51·94	51·94	90·92	90·92
	321	42·43	61·79	83·29	98·54
	322	34·15	71·14	76·28	105·50
	323	27·74	79·46	70·30	111·41
	331	38·78	56·77	91·92	105·37
	332	30·57	65·44	85·46	111·48
130	312	56·37	56·37	86·28	86·28
	313	59·54	59·54	86·59	86·59
	320	38·79	38·79	100·09	100·09
	321	40·06	40·06	99·91	99·91
	322	43·33	43·33	99·42	99·42
	323	47·54	47·54	98·73	98·73
	331	29·26	29·26	110·78	110·78
	332	32·98	32·98	109·94	109·94
131	311	51·56	58·30	83·38	88·90
	312	50·99	63·33	81·17	91·56
	313	51·87	68·32	79·52	93·79
	320	40·65	40·65	99·82	99·82
	321	37·97	45·44	97·14	102·17
	322	37·78	51·26	94·41	103·98
	323	39·50	57·15	91·90	105·21
	331	27·51	35·77	107·91	112·52
	332	27·23	41·94	105·03	113·87
132	312	48·01	70·54	76·90	96·27
	313	46·44	76·82	73·48	100·18
	320	45·16	45·16	99·12	99·12
	321	39·38	52·27	94·31	103·66
	322	35·80	59·67	89·71	107·52
	323	34·54	66·57	85·66	110·51
	331	30·74	44·03	104·52	113·03
	332	26·62	51·41	99·99	116·34

Table 12 (cont.)

HKL	*hkl*	θ_{hkl}	$\theta_{hk\bar{l}}$	$\theta_{\bar{h}\bar{k}l}$	$\theta_{\bar{h}\bar{k}\bar{l}}$
133	133	00·00	70·49	99·76	146·67
	201	62·91	84·93	62·91	84·93
	203	52·30	101·96	52·30	101·96
	210	55·07	55·07	92·78	92·78
	211	44·23	67·77	82·95	102·45
	212	37·83	79·25	74·92	110·06
	213	35·70	88·23	69·21	115·23
	221	32·75	55·92	98·31	115·16
	223	22·46	76·68	83·56	125·56
201	201	00·00	37·08	00·00	37·08
	203	26·64	63·72	26·64	63·72
	210	29·96	29·96	29·96	29·96
	211	22·85	42·68	22·85	42·68
	212	25·20	55·08	25·20	55·08
	213	31·46	65·01	31·46	65·01
	221	40·12	52·41	40·12	52·41
	223	40·76	67·96	40·76	67·96
203	203	00·00	90·36	00·00	90·36
	210	49·90	49·90	49·90	49·90
	211	34·54	65·92	34·54	65·92
	212	23·09	79·72	23·09	79·72
	213	17·40	90·34	17·40	90·34
	221	46·88	70·21	46·88	70·21
	223	32·07	90·30	32·07	90·30
	230	64·98	64·98	64·98	64·98
210	210	00·00	00·00	47·93	47·93
	211	17·04	17·04	50·16	50·16
	212	31·51	31·51	55·16	55·16
	213	42·60	42·60	60·44	60·44
	221	22·45	22·45	66·37	66·37
	223	40·40	40·40	70·72	70·72
	230	29·17	29·17	77·09	77·09
	231	31·13	31·13	77·35	77·35
211	211	00·00	34·08	45·70	58·22
	212	14·47	48·55	45·62	66·85
	213	25·56	59·64	47·94	74·14
	221	17·27	35·67	62·97	71·83
	223	25·28	56·50	60·54	81·98
	230	33·40	33·40	77·67	77·67
	231	28·80	40·48	74·50	81·28
212	212	00·00	63·02	40·52	77·65
	213	11·09	74·11	39·26	86·17
	221	23·79	48·63	62·05	77·61
	223	15·56	70·42	53·44	91·87
	230	41·89	41·89	79·02	79·02
	231	33·60	51·19	73·15	85·21
	232	27·66	60·30	68·17	91·06

HKL	*hkl*	θ_{hkl}	$\theta_{hk\bar{l}}$	$\theta_{\bar{h}\bar{k}l}$	$\theta_{\bar{h}\bar{k}\bar{l}}$
133	230	38·39	38·39	116·08	116·08
	231	28·09	49·11	108·49	123·02
	232	19·51	59·23	101·09	128·56
	233	14·24	68·12	94·48	132·43
	301	66·06	81·15	66·06	81·15
	302	60·17	88·50	60·17	88·50
	310	60·52	60·52	86·69	86·69
	311	52·97	68·88	79·78	93·70
	312	47·17	77·00	73·71	100·04
201	230	55·33	55·33	55·33	55·33
	231	51·65	60·33	51·65	60·33
	232	49·73	65·86	49·73	65·86
	233	49·36	71·18	49·36	71·18
	301	5·94	31·15	5·94	31·15
	302	5·55	42·64	5·55	42·64
	310	24·63	24·63	24·63	24·63
	311	17·16	34·70	17·16	34·70
203	231	56·32	74·06	56·32	74·06
	232	48·91	82·67	48·91	82·67
	233	43·22	90·26	43·22	90·26
	301	32·57	57·78	32·57	57·78
	302	21·08	69·27	21·08	69·27
	310	47·48	47·48	47·48	47·48
	311	35·95	59·19	35·95	59·19
	312	25·75	70·02	25·75	70·02
210	232	35·90	35·90	78·04	78·04
	233	41·62	41·62	78·98	78·98
	301	26·90	26·90	26·90	26·90
	302	33·47	33·47	33·47	33·47
	310	7·46	7·46	40·47	40·47
	311	14·19	14·19	41·94	41·94
	312	24·31	24·31	45·64	45·64
	313	33·49	33·49	50·22	50·22
211	232	27·88	48·31	72·09	84·91
	233	29·98	55·72	70·47	88·20
	301	23·57	37·94	23·57	37·94
	302	23·48	47·32	23·48	47·32
	310	18·56	18·56	43·33	43·33
	311	8·74	30·06	39·41	49·47
	312	9·33	40·90	38·37	56·42
212	233	24·85	68·46	64·33	96·15
	301	29·03	49·74	29·03	49·74
	302	22·40	60·14	22·40	60·14
	310	32·29	32·29	49·56	49·56
	311	20·59	44·19	41·95	58·36
	312	10·61	55·18	36·67	67·04
	313	6·44	64·64	34·08	74·76

HKL	*hkl*	θ_{hkl}	$\theta_{hk\bar{l}}$	$\theta_{\bar{h}\bar{k}l}$	$\theta_{\bar{h}\bar{k}\bar{l}}$
133	313	43·44	84·16	68·86	105·29
	320	50·47	50·47	98·23	98·23
	321	42·77	58·93	91·81	104·45
	322	36·75	67·24	85·78	109·93
	323	32·89	74·76	80·53	114·36
	331	36·14	51·88	101·22	112·63
	332	29·77	59·86	95·49	117·48
201	312	16·10	44·75	16·10	44·75
	313	20·52	53·66	20·52	53·66
	320	35·35	35·35	35·35	35·35
	321	30·57	42·19	30·57	42·19
	322	28·90	49·68	28·90	49·68
	323	30·07	56·81	30·07	56·81
	331	41·29	49·72	41·29	49·72
	332	39·39	55·16	39·39	55·16
203	313	17·80	79·35	17·80	79·35
	320	52·67	52·67	52·67	52·67
	321	43·16	62·51	43·16	62·51
	322	34·85	71·86	34·85	71·86
	323	28·39	80·17	28·39	80·17
	331	50·46	66·25	50·46	66·25
	332	43·57	74·05	43·57	74·05
210	320	6·69	6·69	54·61	54·61
	321	12·76	12·76	55·34	55·34
	322	22·03	22·03	57·28	57·28
	323	30·66	30·66	59·90	59·90
	331	19·99	19·99	65·95	65·95
	332	25·36	25·36	66·93	66·93
211	313	17·09	50·30	39·62	63·05
	320	18·27	18·27	56·38	56·38
	321	8·94	28·70	53·20	60·77
	322	7·48	38·64	51·54	65·70
	323	14·32	47·47	51·24	70·54
	331	18·77	31·76	64·03	70·04
	332	16·88	39·55	62·12	73·64
212	320	32·14	32·14	60·42	60·42
	321	21·53	42·88	54·30	67·29
	322	12·05	52·94	49·58	74·15
	323	5·94	61·82	46·46	80·42
	331	27·46	44·35	64·31	74·85
	332	20·59	52·79	60·02	80·30

HKL	hkl	θ_{hkl}	$\theta_{hk\bar{l}}$	$\theta_{h\bar{k}l}$	$\theta_{h\bar{k}\bar{l}}$
213	213	00·00	85·20	34·79	95·45
	221	32·34	58·95	62·64	82·51
	223	14·68	81·16	49·47	99·43
	230	50·00	50·00	80·54	80·54
	231	40·21	60·22	72·86	88·42
	232	31·90	69·91	66·09	95·75
	233	25·83	78·44	60·62	102·07
221	221	00·00	28·15	80·25	87·07
	223	22·87	51·02	76·29	93·17
	230	18·10	18·10	94·62	94·62
	231	11·53	27·89	91·78	97·29
	232	13·45	37·72	89·08	99·53
	233	20·05	46·52	86·75	101·22
223	223	00·00	73·89	64·14	106·64
	230	38·45	38·45	93·81	93·81
	231	27·58	49·52	86·93	100·58
	232	18·02	59·86	80·63	106·62
	233	11·15	68·91	75·30	111·56
	301	44·43	63·14	44·43	63·14
230	230	00·00	00·00	106·26	106·26
	231	11·38	11·38	105·93	105·93
	232	21·92	21·92	105·06	105·06
	233	31·12	31·12	103·87	103·87
	301	54·16	54·16	54·16	54·16
	302	56·79	56·79	56·79	56·79
231	231	00·00	22·76	103·31	107·94
	232	10·55	33·30	100·43	109·17
	233	19·74	42·50	97·64	109·69
	301	51·90	57·93	51·90	57·93
	302	51·87	62·84	51·87	62·84
232	232	00·00	43·85	95·83	112·36
	233	9·20	53·05	91·68	114·54
	301	51·34	62·50	51·34	62·50
	302	48·66	69·17	48·66	69·17
	310	41·88	41·88	71·17	71·17
233	233	00·00	62·24	86·45	118·19
	301	52·12	67·14	52·12	67·14
	302	47·16	75·05	47·16	75·05
	310	46·60	46·60	72·67	72·67
	311	38·73	55·71	66·44	79·46

HKL	hkl	θ_{hkl}	$\theta_{hk\bar{l}}$	$\theta_{h\bar{k}l}$	$\theta_{h\bar{k}\bar{l}}$
213	301	36·47	59·42	36·47	59·42
	302	27·08	70·26	27·08	70·26
	310	43·12	43·12	55·94	55·94
	311	31·18	55·13	46·41	66·07
	312	20·35	66·17	38·61	75·64
	313	11·47	75·66	33·16	83·98
	320	43·02	43·02	64·77	64·77
221	301	40·49	49·12	40·49	49·12
	302	40·44	55·77	40·44	55·77
	310	28·58	28·58	59·20	59·20
	311	24·54	36·13	56·52	63·28
	312	25·45	44·66	55·50	67·99
	313	29·53	52·62	55·79	72·60
223	302	37·75	72·55	37·75	72·55
	310	43·65	43·65	65·05	65·05
	311	33·54	54·44	57·42	73·35
	312	25·59	64·65	51·35	81·32
	313	20·98	73·54	47·16	88·30
	320	38·32	38·32	75·92	75·92
230	310	36·63	36·63	69·63	69·63
	311	38·31	38·31	70·11	70·11
	312	42·47	42·47	71·35	71·35
	313	47·55	47·55	72·98	72·98
	320	22·48	22·48	83·78	83·78
	321	24·85	24·85	83·89	83·89
231	310	38·11	38·11	70·05	70·05
	311	35·84	43·29	67·98	73·01
	312	36·79	49·81	66·96	76·36
	313	39·79	56·29	66·82	79·62
	320	25·06	25·06	83·90	83·90
232	311	36·27	49·48	66·80	76·27
	312	33·76	57·51	63·65	81·40
	313	34·09	64·90	61·74	86·01
	320	31·00	31·00	84·23	84·23
	321	24·18	39·54	80·26	88·39
233	312	33·37	64·67	61·48	85·98
	313	30·97	72·65	57·98	91·66
	320	37·72	37·72	84·68	84·68
	321	29·02	47·22	79·12	90·38
	322	22·50	56·45	74·21	95·68

HKL	hkl	θ_{hkl}	$\theta_{hk\bar{l}}$	$\theta_{h\bar{k}l}$	$\theta_{h\bar{k}\bar{l}}$
213	321	32·24	53·84	56·88	73·10
	322	22·27	63·94	50·14	81·09
	323	13·70	72·85	44·97	88·23
	331	36·55	54·53	65·70	79·14
	332	28·38	63·22	59·80	85·76
221	320	17·78	17·78	72·83	72·83
	321	11·18	27·23	70·38	75·88
	322	12·57	36·74	68·73	79·15
	323	18·87	45·32	67·84	82·29
	331	4·59	23·56	81·24	85·87
	332	4·41	32·56	79·34	88·24
223	321	27·87	48·94	69·37	82·80
	322	18·55	58·91	63·72	89·36
	323	11·49	67·72	59·27	95·15
	331	27·46	46·43	78·96	90·38
	332	18·46	55·43	73·78	95·84
230	322	30·41	30·41	84·20	84·20
	323	36·84	36·84	84·62	84·62
	331	14·87	14·87	94·70	94·70
	332	21·66	21·66	94·52	94·52
231	321	22·06	31·54	81·86	86·16
	322	23·61	39·23	80·21	88·38
	323	27·97	46·69	79·01	90·38
	331	11·46	23·79	92·74	96·48
	332	13·18	31·94	90·84	98·03
232	322	20·92	48·24	76·83	92·31
	323	21·71	56·23	74·12	95·72
	331	16·63	33·38	90·83	97·91
	332	11·32	41·94	87·40	101·04
233	323	19·36	64·74	70·20	100·25
	331	24·13	42·08	89·14	98·94
	332	16·37	50·82	84·47	103·37

Table 12 (cont.)

HKL	hkl	θ_{hkl}	$\theta_{hk\bar{l}}$	$\theta_{h\bar{k}l}$	$\theta_{h\bar{k}\bar{l}}$
301	301	00·00	25·21	00·00	25·21
	302	11·49	36·70	11·49	36·70
	310	20·66	20·66	20·66	20·66
	311	16·13	29·64	16·13	29·64
302	302	00·00	48·19	00·00	48·19
	310	28·92	28·92	28·92	28·92
	311	19·71	39·63	19·71	39·63
	312	15·14	49·94	15·14	49·94
310	310	00·00	00·00	33·01	33·01
	311	12·10	12·10	34·92	34·92
	312	23·21	23·21	39·58	39·58
	313	32·75	32·75	45·15	45·15
311	311	00·00	24·20	32·26	40·73
	312	11·11	35·31	33·26	47·86
	313	20·65	44·85	36·58	54·81
312	312	00·00	46·42	30·27	56·42
	313	9·54	55·96	30·53	64·21
	320	26·98	26·98	51·32	51·32
313	313	00·00	65·50	27·65	72·51
	320	35·36	35·36	55·12	55·12
	321	25·44	45·68	48·41	62·65
320	320	00·00	00·00	61·30	61·30
	321	10·89	10·89	61·86	61·86
321	321	00·00	21·78	60·08	64·70
	322	10·15	31·93	59·47	68·14
322	322	00·00	42·09	56·82	73·18
	323	8·95	51·03	55·41	77·95
323	323	00·00	59·98	52·41	83·66
331	331	00·00	18·98	81·88	85·02
332	332	00·00	36·96	78·11	89·72

HKL	hkl	θ_{hkl}	$\theta_{hk\bar{l}}$	$\theta_{h\bar{k}l}$	$\theta_{h\bar{k}\bar{l}}$
301	312	18·92	39·29	18·92	39·29
	313	25·18	48·02	25·18	48·02
	320	32·91	32·91	32·91	32·91
	321	30·04	38·44	30·04	38·44
302	313	16·87	58·98	16·87	58·98
	320	38·25	38·25	38·25	38·25
	321	31·97	46·05	31·97	46·05
	322	28·41	54·10	28·41	54·10
310	320	14·15	14·15	47·16	47·16
	321	17·78	17·78	48·10	48·10
	322	25·18	25·18	50·60	50·60
	323	32·88	32·88	53·92	53·92
311	320	18·54	18·54	48·32	48·32
	321	13·91	26·94	46·17	52·17
	322	16·23	35·94	45·91	56·96
312	321	18·27	36·80	46·51	57·36
	322	13·28	46·35	43·55	63·78
	323	14·33	54·91	42·41	69·86
313	322	17·18	55·46	43·28	70·15
	323	12·38	64·15	40·03	77·00
	331	32·84	48·56	58·19	69·59
320	322	21·04	21·04	63·37	63·37
	323	29·99	29·99	65·42	65·42
321	323	19·10	40·88	59·81	71·70
	331	10·90	23·12	70·98	74·72
322	331	15·66	32·38	70·17	77·24
323	331	22·92	40·87	69·98	79·77
331	332	8·99	27·97	80·68	86·71

HKL	hkl	θ_{hkl}	$\theta_{hk\bar{l}}$	$\theta_{h\bar{k}l}$	$\theta_{h\bar{k}\bar{l}}$
301	322	30·47	45·15	30·47	45·15
	323	33·26	51·82	33·26	51·82
	331	40·94	46·89	40·94	46·89
	332	40·45	51·49	40·45	51·49
302	323	27·84	61·57	27·84	61·57
	331	42·25	52·72	42·25	52·72
	332	39·06	58·82	39·06	58·82
310	331	26·75	26·75	58·62	58·62
	332	30·83	30·83	59·96	59·96
311	323	22·18	44·24	47·10	61·89
	331	24·81	33·01	57·07	61·67
	332	25·04	39·36	56·22	64·97
312	331	27·67	40·91	57·08	65·57
	332	23·92	48·38	54·19	70·42
313	332	26·66	56·59	53·66	75·55
320	331	14·48	14·48	72·53	72·53
	332	21·40	21·40	73·23	73·23
321	332	13·05	31·31	69·92	77·08
322	332	10·65	40·96	67·47	81·05
323	332	15·24	49·61	65·89	84·75

TABLE 13. Angles between crystallographic zones in cementite (in degrees)

$u_1v_1w_1$	$u_2v_2w_2$	ρ
100	401	20·44
	301	26·42
	201	36·70
	302	44·82
	101	56·14
	203	65·90
	102	71·46
	103	77·39
	104	80·48
	414	56·61
	314	63·69
	214	71·76
	114	80·64
	014	90·00
	413	49·06
	313	56·95
	213	66·55
	113	77·76
	013	90·00
	412	38·54
	312	46·73
	212	57·88
	324	64·79
	112	72·58
	236	78·18
	124	81·08
	012	90·00
	423	51·37
	323	59·07
	223	68·22
	123	78·70
	023	90·00
	411	25·03
	311	31·90
	211	43·04
	322	51·23
	111	61·83
	233	70·35
	122	75·01
	133	79·88
	144	82·37
	011	90·00
	432	48·39
	332	56·33
	232	66·05
	132	77·48
	032	90·00
	421	34·01
	321	41·98
	221	53·46
	342	60·93
	121	69·67
	263	76·12
	142	79·50
	021	90·00
	431	42·69
	331	50·88
	231	61·54
	131	74·83
	031	90·00
	441	49·84
	341	57·67
	241	67·12
	141	78·09
	041	90·00
	410	15·71
	310	20·55
	210	29·36
	320	36·87
	110	48·36
	230	59·35
	120	66·04
	130	73·49
	140	77·47
	172	83·23
	176	85·20
010	401	90·00
	301	90·00
	201	90·00
	101	90·00
	203	90·00
	102	90·00
	103	90·00
	104	90·00
	414	81·10
	314	80·43
	214	79·86
	114	79·46
	014	79·32
	413	79·38
	313	78·20
	213	77·07
	113	76·21
	013	75·88
	412	77·29
	312	75·11
	212	72·60
	324	71·37
	112	70·31
	236	69·78
	124	69·59
	012	69·33
	423	69·44
	323	67·32
	223	65·33
	123	63·85
	023	63·29
	411	75·24
	311	71·44
	211	65·72
	322	61·99
	111	57·92
	233	55·44
	122	54·42
	133	53·63
	144	53·34
	011	52·96
	432	55·92
	332	51·41
	232	46·77
	132	42·98
	032	41·46
	421	62·21
	321	56·11
	221	47·95
	342	43·23
	121	38·58
	263	35·97
	142	34·95
	021	33·53
	431	51·67
	331	44·79
	231	36·47
	131	28·01
	031	23·83
	441	43·49
	341	36·67
	241	29·00
	141	21·75
	041	18·33
	410	74·29
	310	69·45
	210	60·64
	320	53·13
	110	41·64
	230	30·65
	120	23·96
	130	16·51
	140	12·53
	172	21·77
	176	48·81
	302	90·00
001	401	69·56
	301	63·58
	201	53·30
	302	45·18
	101	33·86
	203	24·10
	102	18·54
	103	12·61
	104	9·52
	414	34·87
	314	28·25
	214	21·05
	114	14·17
	014	10·68
	413	42·90
	313	35·62
	213	27·16
	113	18·60
	013	14·12
	412	54·34
	312	47·06
	212	37·59
	324	32·17
	112	26·79
	236	23·68
	124	22·44
	012	20·67
	423	45·74
	323	39·98
	223	33·95
	123	28·84
	023	26·71
	411	70·27
	311	65·05
	211	56·99
	322	51·52
	111	45·28
	233	41·26
	122	39·55
	133	38·21
	144	37·71
	011	37·04
	432	60·33
	332	56·57
	232	52·77
	132	49·74
	032	48·54
	421	72·01
	321	68·32
	221	63·66
	342	61·13
	121	58·81
	263	57·58
	142	57·11
	021	56·47
	431	74·10
	331	71·73
	231	69·19
	131	67·05
	031	66·17
	441	76·09
	341	74·59
	241	73·16
	141	72·08
	041	71·67
	410	90·00
	310	90·00
	210	90·00
	320	90·00
	110	90·00
	230	90·00
	120	90·00
	130	90·00
	140	90·00
	172	69·42
	176	41·59

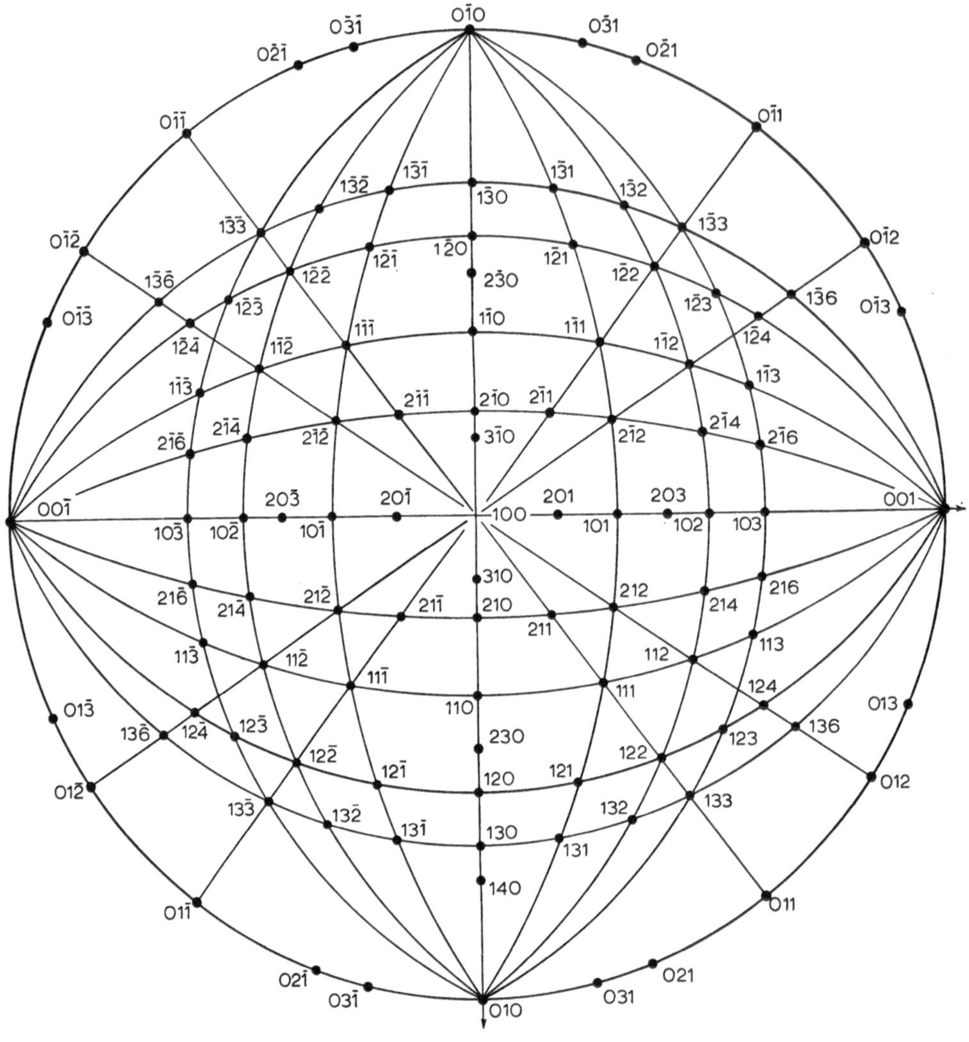

FIG. 46. Stereographic projection of Fe_3C on (100).

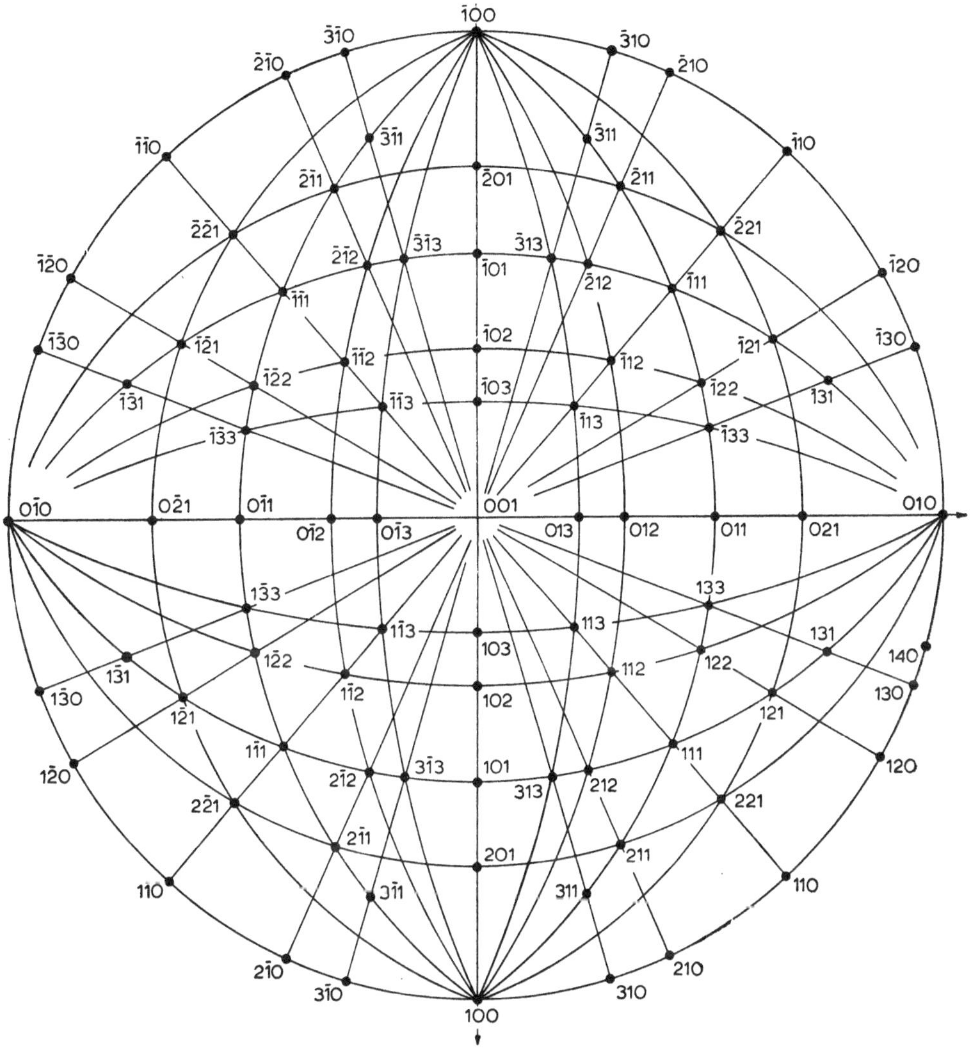

FIG. 47. Stereographic projection of Fe_3C on (001).

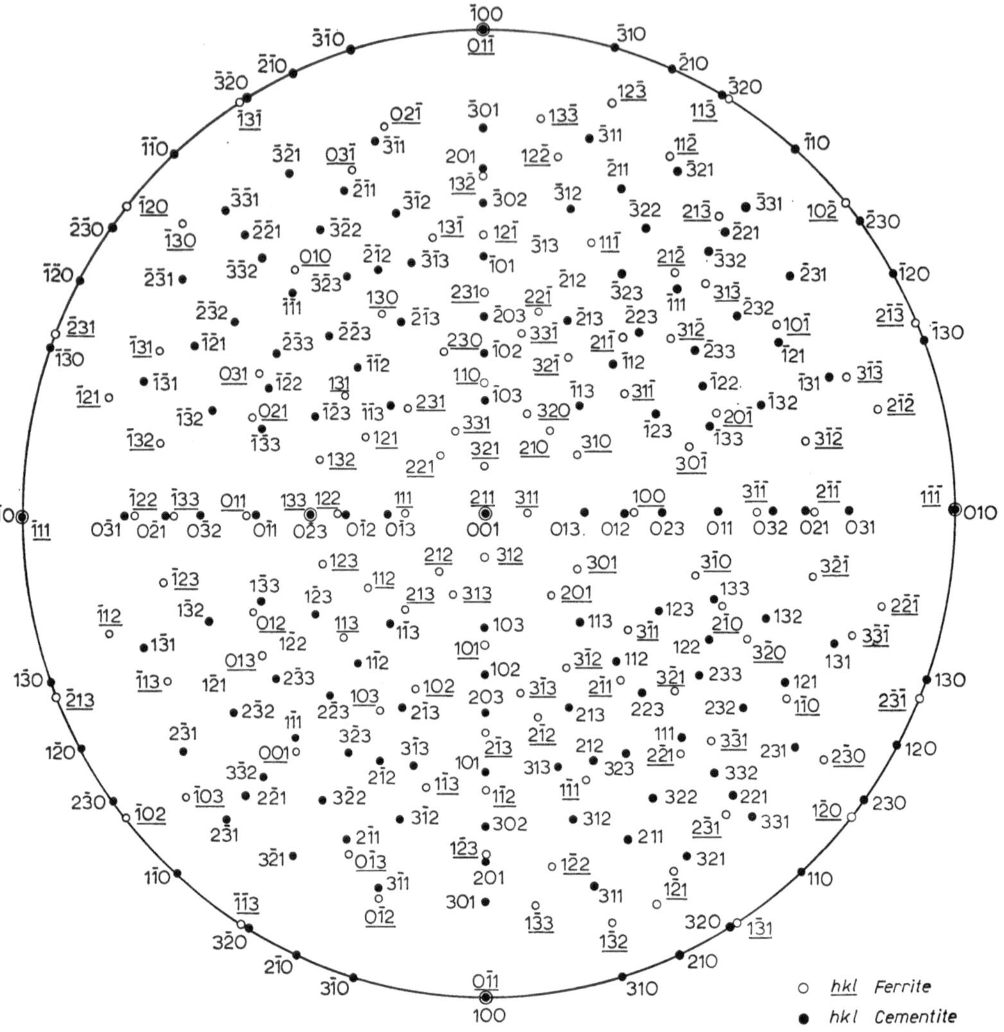

FIG. 48. Stereographic projection representing the Bagaryatskii orientation relationship between cementite and ferrite.
(100) cementite //$(0\bar{1}1)$ ferrite
(010) cementite //$(1\bar{1}\bar{1})$ ferrite
(001) cementite //(211) ferrite

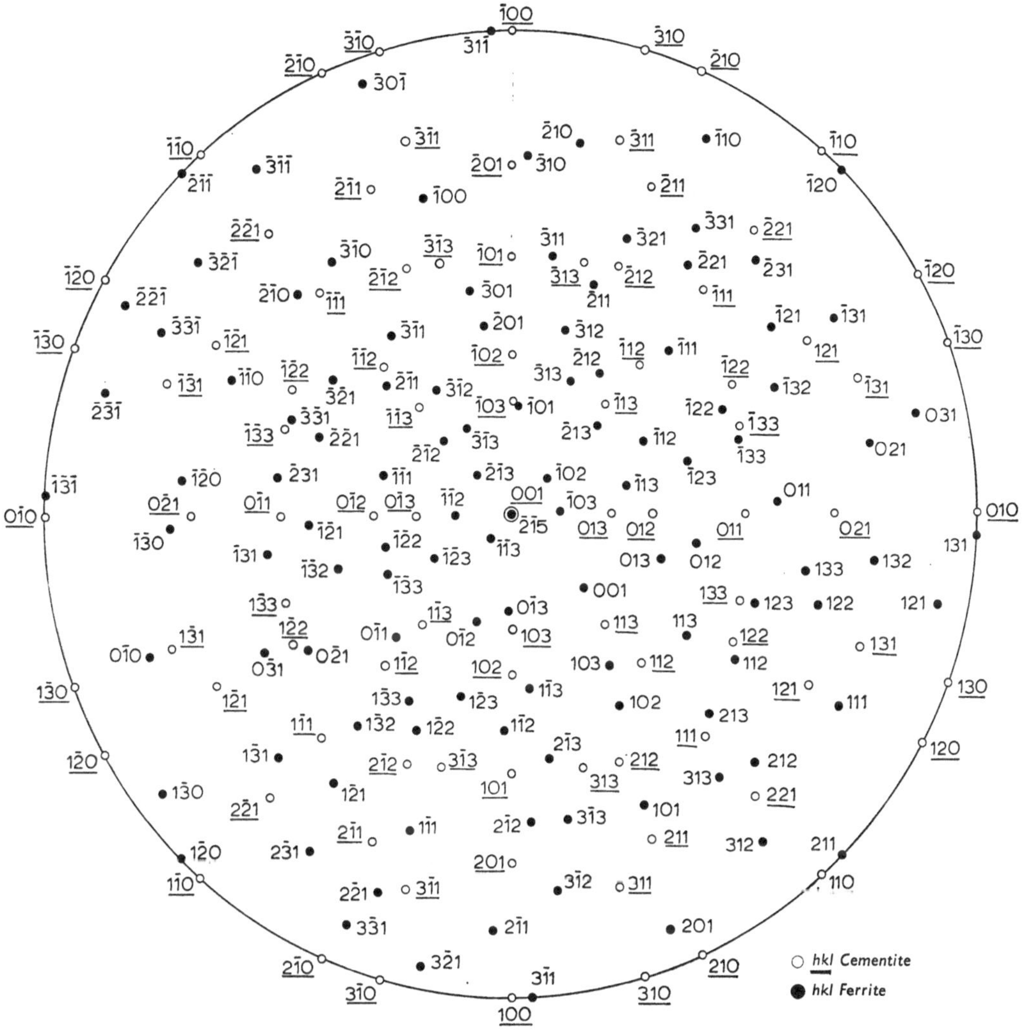

FIG. 49. Stereographic projection representing the **Pitsch** orientation relationship between cementite and ferrite.
(100) cementite //$(3\bar{1}1)$ ferrite
(010) cementite //(131) ferrite
(001) cementite //$(\bar{2}\bar{1}5)$ ferrite

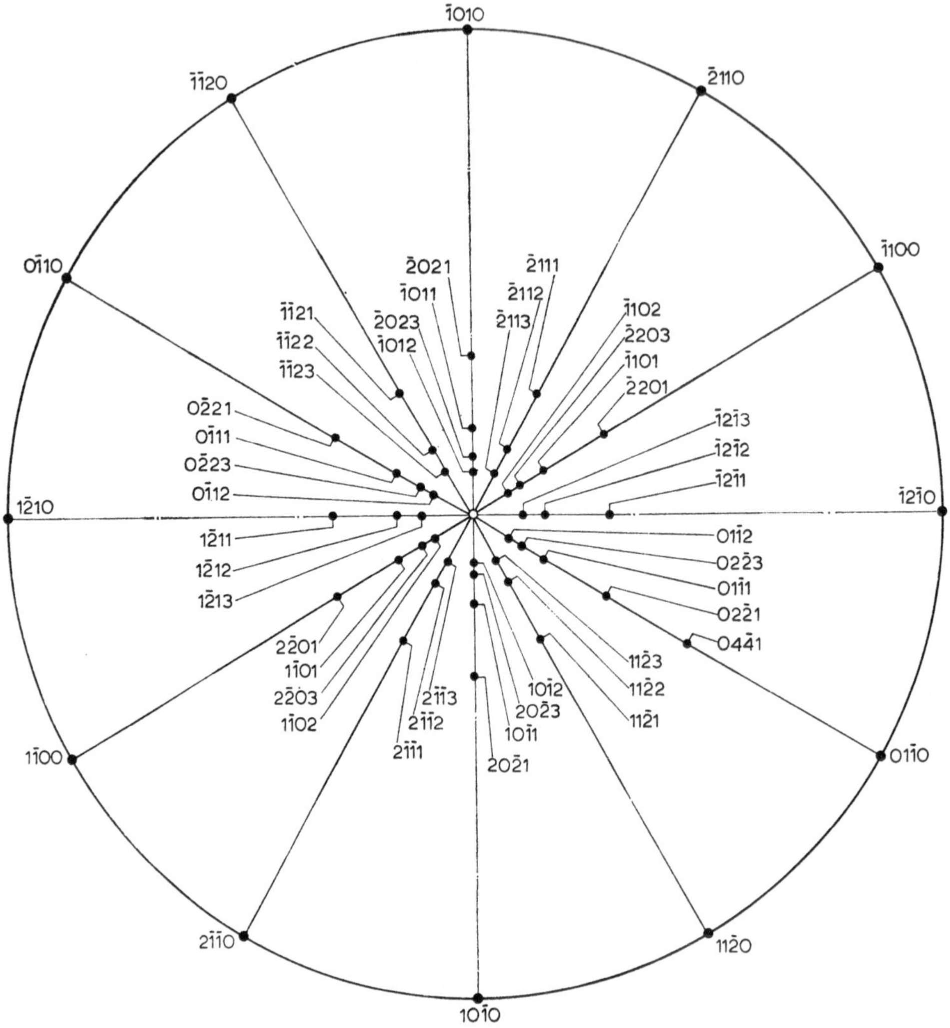

FIG. 50. Stereographic projection of M_7C_3 on (0001) with $c/a \sim 0.32$ (see §15.2).

Table 14. Interplanar spacings of M_7C_3=$(Cr,Fe)_7C_3$ (see §15.2)

Crystal structure: Pseudo-hexagonal

Unit cell: a_0 = 13·982$_0$ Å, c_0 = 4·506$_5$ Å

hk.l	*d*-spacing (Å)
00.1	4·5065
00.2	2·2532
00.3	1·5022
00.4	1·1266
01.0	12·1088
01.1	4·2235
01.2	2·2152
01.3	1·4907
01.4	1·1218
02.0	6·0544
02.1	3·6150
02.2	2·1117
02.3	1·4580
02.4	1·1076
03.0	4·0363
03.1	3·0066
03·2	1·9674
03.3	1·4078
03.4	1·0851
04.0	3·0272
04.1	2·5129
04.2	1·8075
04.3	1·3456
04.4	1·0559
05.0	2·4218
05.1	2·1332
05.2	1·6496
05.3	1·2765
05.4	1·0215
06.0	2·0181
06.1	1·8419
06.2	1·5033
06.3	1·2050
06.4	0·9837
07.0	1·7298
07.1	1·6149
07.2	1·3721
07.3	1·1342
08.0	1·5136
08.1	1·4348
08.2	1·2564
08.3	1·0662
09.0	1·3454
09.1	1·2892
09.2	1·1552
09.3	1·0022

hk.l	*d*-spacing (Å)
0,10.0	1·2109
0,10.1	1·1694
0,10.2	1·0666
0,11.0	1·1008
0,11·1	1·0694
0,11.2	0·9891
0,12.0	1·0091
0,12.1	0·9847
11.0	6·9910
11.1	3·7877
11.2	2·1446
11.3	1·4686
11.4	1·1123
12.0	4·5767
12.1	3·2111
12.2	2·0215
12.3	1·4273
12.4	1·0940
13.0	3·3584
13.1	2·6929
13.2	1·8711
13.3	1·3712
13.4	1·0681
14.0	2·6423
14.1	2·2794
14.2	1·7145
14.3	1·3059
14.4	1·0364
15.0	2·1748
15.1	1·9586
15.2	1·5648
15.3	1·2360
15.4	1·0004
16.0	1·8466
16.1	1·7087
16.2	1·4282
16.3	1·1653
16.4	0·9618
17.0	1·6038
17.1	1·5110
17.2	1·3066
17.3	1·0964
18.0	1·4172
18.1	1·3519
18.2	1·1997
18.3	1·0309

hk.l	*d*-spacing (Å)
19.0	1·2693
19.1	1·2218
19.2	1·1059
19.3	0·9695
1,10.0	1·1493
1,10.1	1·1137
1,10.2	1·0238
1,11.0	1·0500
1,11.1	1·0226
1,11.2	0·9517
1,12.0	0·9664
22.0	3·4955
22.1	2·7620
22.2	1·8939
22.3	1·3801
22.4	1·0723
23.0	2·7779
23.1	2·3648
23.2	1·7500
23.3	1·3214
23.4	1·0440
24.0	2·2883
24.1	2·0404
24.2	1·6056
24.3	1·2558
24.4	1·0108
25.0	1·9390
25.1	1·7811
25.2	1·4697
25.3	1·1875
25.4	0·9741
26.0	1·6792
26.1	1·5735
26.2	1·3464
26.3	1·1196
27.0	1·4793
27.1	1·4055
27.2	1·2366
27.3	1·0540
28.0	1·3212
28.1	1·2678
28.2	1·1397
28.3	0·9921
29.0	1·1931
29.1	1·1534
29.2	1·0544

hk.l	*d*-spacing (Å)
2,10.0	1·0874
2,10.1	1·0571
2,10.2	0·9793
2,11.0	0·9987
2,11.1	0·9751
33·0	2·3303
33.1	2·0700
33.2	1·6199
33.3	1·2626
33.4	1·0143
34.0	1·9907
34.1	1·8209
34.2	1·4919
34.3	1·1991
34.4	0·9805
35.0	1·7298
35.1	1·6149
35.2	1·3721
35.3	1·1342
36.0	1·5256
36.1	1·4450
36.2	1·2633
36.3	1·0704
37.0	1·3623
37.1	1·3041
37.2	1·1658
37.3	1·0091
38.0	1·2295
38.1	1·1861
38.2	1·0793
38.3	0·9514
39.0	1·1195
39.1	1·0864
39.2	1·0025
3,10.0	1·0271
3,10.1	1·0014
44.0	1·7478
44.1	1·6295
44.2	1·3810
44.3	1·1392
45.0	1·5504
45.1	1·4660
45.2	1·2772
45.3	1·0788

hk.l	*d*-spacing (Å)
46.0	1·3890
46.1	1·3274
46.2	1·1824
46.3	1·0198
47.0	1·2556
47.1	1·2095
47.2	1·0968
47.3	0·9634
48.0	1·1442
48.1	1·1090
48.2	1·0202
49.0	1·0500
49.1	1·0226
49.2	0·9517
4,10.0	0·9695
55.0	1·3982
55.1	1·3354
55.2	1·1881
55.3	1·0235
56.0	1·2693
56.1	1·2218
56.2	1·1059
56.3	0·9695
57·0	1·1598
57.1	1·1232
57.2	1·0312
58.0	1·0661
58.1	1·0375
58.2	0·9637
59.0	0·9854
59.1	0·9627
66.0	1·1652
66.1	1·1281
66.2	1·0350
67.0	1·0745
67.1	1·0452
67.2	0·9699
68.0	0·9953
68.1	0·9719
77.0	0·9987
77.1	0·9751

15.2 INTERPLANAR SPACINGS OF THE CHROMIUM-BASE CARBIDE, M_7C_3

Interplanar spacings are derived from the lattice parameters given at the head of Table 14. These were obtained by extrapolation after measuring the lattice parameters of several samples with different iron–chromium ratios.[55] Because of the large number of spacings, the table has been restricted to those ≧ 1·00 Å, and these have been arranged in zones. Fig. 50 is a stereographic projection on (00·1) and illustrates how decreasing the c/a value tends to crowd the poles towards the centre of the diagram.

15.3 INTERPLANAR SPACINGS OF THE IRON–MOLYBDENUM CARBIDE, $M_aC_b = Fe_2MoC$

There are a large number of spacings with $d \geqq 1·00$ Å because of the large values of the three parameters of the orthorhombic cell.[18] Table 15 therefore only lists those observed by X-ray diffraction although a fuller table is available from the authors. Fig. 51 is a stereographic projection of this phase on (100).

TABLE 15. Interplanar spacings of $M_aC_b \doteq Fe_2MoC$ (observed by X-ray diffraction)

Crystal structure: Orthorhombic
Unit cell: $a_0 = 16·27_6$ Å, $b_0 = 10·03_4$ Å, $c_0 = 11·32_3$ Å

hkl	*d*-spacing (Å)	*hkl*	*d*-spacing (Å)	*hkl*	*d*-spacing (Å)	*hkl*	*d*-spacing (Å)
002	5·662	342	2·113	345	1·606	952	1·307
020	5·017	043; 234; 514	2·089	254	1·605	428; 860	1·292
012	4·931	025	2·064	062	1·604	10,43; 727; 861	1·284
030	3·345	125	2·048	207	1·587	763	1·278
402	3·304	315	2·046	840	1·580	465	1·277
420	3·160	623	2·017	227; 805	1·513	12,03	1·276
322	3·088	334; 050	2·007	263; 407	1·503	109; 547; 080	1·254
421	3·044	810	1·994	462	1·492	119; 257; 374; 618	1·245
023	3·016	150	1·992	11,10; 10,30	1·464	13,01	1·244
231	2·984	632	1·975	554	1·463	280	1·240
313	2·960	802	1·915	164; 517	1·434	573	1·239
032	2·880	006	1·887	237; 070	1·433	628; 129; 319	1·217
004	2·831	820	1·885	108	1·410	229	1·207
223	2·828	633	1·840	753; 455	1·409	10,53; 480	1·199
403	2·767	244	1·829	11,22	1·377	275	1·198
204; 114	2·687	641	1·818	617; 371	1·376	482; 13,30	1·173
610	2·619	534; 450	1·800	028; 745	1·362	927	1·172
214; 430	2·584	822	1·789	471	1·343	067; 383	1·163
140	2·479	910	1·780	736	1·342		
503; 024	2·465	830; 605	1·738	12,11	1·335		
241	2·345	145; 060	1·672	951; 826	1·334		
531	2·286	643	1·655	12,12	1·308		
701	2·278	061	1·654				
340	2·277	036	1·644				
622	2·200						
115	2·190						
613	2·152						

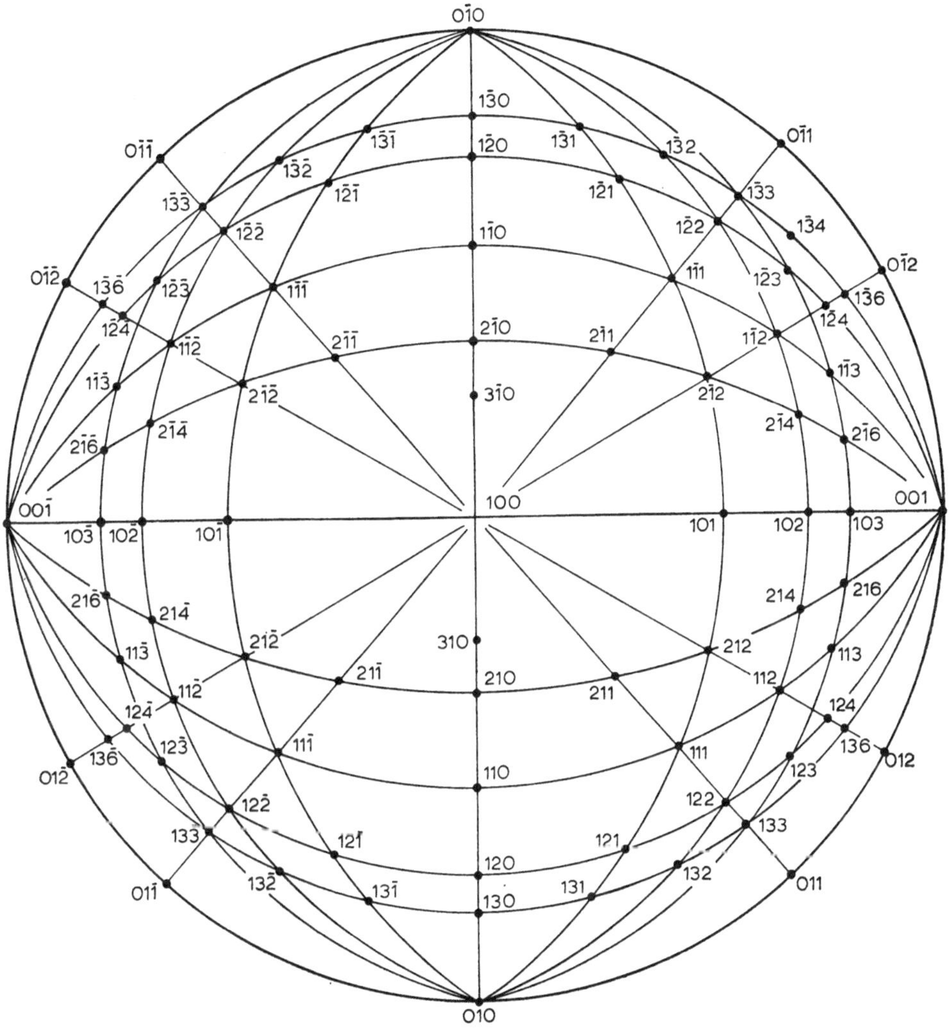

FIG. 51. Stereographic projection of M_aC_b on (100).

15.4 INTERPLANAR SPACINGS OF THE CUBIC CARBIDES, $M_{23}C_6$ AND M_6C

Tables 16 and 17 give *d*-spacings down to $d = 1{\cdot}00$ Å for the two face-centred carbides, $M_{23}C_6$ and M_6C. In order to minimize the length of the table, reflections forbidden by the crystal symmetry have been left out. Should these be needed for identifying an ordered form of either, they can quickly be determined using a slide rule. As can be seen, the parameters of these two carbides differ only slightly, and since they can occur together they can easily be confused if care is not taken. It is possible to distinguish between their spot patterns since the space groups are Fm3m and Fd3m respectively. Fig. 32 shows that in certain orientations double reflection does not fill in the M_6C reflections for which $\Sigma h^2 = 4n$, and the two patterns are then different and distinguishable.

TABLE 16. Interplanar spacings of an $M_{23}C_6$ carbide

Crystal structure: Face-centred cubic
Unit cell: $a_0 = 10{\cdot}621_4$ Å

hkl	*d*-spacing (Å)
111	6·132
002	5·311
022	3·755
113	3·203
222	3·066
004	2·655
133	2·437
024	2·375
224	2·168
333 115	2·044
044	1·878
135	1·795
006 244	1·770
026	1·679
335	1·620
226	1·601
444	1·533
117 155	1·487
046	1·473
246	1·419
137 355	1·383
008	1·328
337	1·298
028 446	1·288
228 066	1·252
157 555	1·227
266	1·218
048	1·188
119 357	1·166
248	1·159
466	1·132
139	1·113
448	1·084
557 177 339	1·068
00,10 068	1·062
02,10 268	1·042
159 377	1·023
22,10 666	1·022

TABLE 17. Interplanar spacings of an M_6C carbide

Crystal structure: Face-centred cubic (diamond type)
Unit cell: $a_0 = 11{\cdot}082_3$ Å

hkl	*d*-spacing (Å)
111	6·398
002	5·541
022	3·918
113	3·341
222	3·199
004	2·771
133	2·542
024	2·478
224	2·262
333 115	2·133
044	1·959
135	1·873
006 244	1·847
026	1·752
335	1·690
226	1·671
444	1·600
117 155	1·552
046	1·537
246	1·481
137 355	1·443
008	1·385
337	1·354
028 446	1·344
228 066	1·306
157 555	1·280
266	1·271
048	1·239
119 357	1·217
248	1·209
466	1·181
139	1·161
448	1·131
557 177 339	1·114
00,10 068	1·108
02,10 268	1·087
159 377	1·071
22,10 666	1·066
359	1·034
04,10 468	1·029
24,10	1·012

15.5 INTERPLANAR SPACINGS OF MOLYBDENUM CARBIDE, Mo_2C

Table 18 gives the *d*-spacings of the close-packed hexagonal carbide Mo_2C ($c_0/a_0 = 1{\cdot}573$) down to $d = 0{\cdot}70$ Å.

Fig. 52 represents the orientation relationship postulated by Pitsch and Schrader[56] for hexagonal close-packed ε-carbide with body-centred cubic ferrite. The same relationship also holds for Mo_2C (hexagonal) with ferrite.[57]

TABLE 18. Interplanar spacings of Mo_2C
Crystal structure: Hexagonal close-packed
Unit cell: $a_0 = 3{\cdot}002_0$ Å, $c_0 = 4{\cdot}724_0$ Å

hk.l	*d*-spacing (Å)	*hk.l*	*d*-spacing (Å)	*hk.l*	*d*-spacing (Å)	*hk.l*	*d*-spacing (Å)
00.1	4·724	11.2	1·267	11.4	0·928	02.5	0·764
01.0	2·600	02.1	1·253	12.2	0·907	03.3	0·759
00.2	2·362	00.4	1·181	01.5	0·888	12.4	0·755
01.1	2·278	02.2	1·139	02.4	0·874	01.6	0·754
01.2	1·748	11.3	1·087	03.0	0·867	22.0	0·751
00.3	1·575	01.4	1·075	03.1	0·852	22.1	0·741
11.0	1·501	02.3	1·003	12.3	0·834	13.0	0·721
11.1	1·431	12.0	0·983	03.2	0·814	22.2	0·715
01.3	1·347	12.1	0·962	11.5	0·800	13.1	0·713
02.0	1·300	00.5	0·945	00.6	0·787		

15.6 INTERPLANAR SPACINGS OF NIOBIUM CARBIDE, NbC

Niobium carbide is one of the family of compounds with the face-centred cubic B1 structure and as such represents this group of compounds. As this group only extends over a parameter range from $a_0 \simeq 4{\cdot}10$ Å to $4{\cdot}60$ Å, should two of these compounds form part of the same specimen it may be difficult to distinguish one from the other. Because the largest observable *d*-spacing for NbC is {111} = 2·581 Å, one never finds rows of spots of small separation appearing on the single-crystal patterns. However, if the identity of the diffracting medium is known, the indexing is very straightforward.

TABLE 19. Interplanar spacings of niobium carbide
Crystal structure: Face-centred cubic
Unit cell: $a_0 = 4{\cdot}470_0$ Å

hkl	*d*-spacing (Å)	*hkl*	*d*-spacing (Å)	*hkl*	*d*-spacing (Å)	*hkl*	*d*-spacing (Å)
111	2·581	333, 115	0·860	444	0·645	337	0·546
002	2·235	044	0·790	117, 155	0·626	028, 446	0·542
022	1·580	135	0·756	046	0·620	228, 066	0·527
113	1·348	006, 244	0·745	246	0·597	157, 555	0·516
222	1·290	026	0·707	137, 355	0·582	266	0·513
004	1·118	335	0·682	008	0·559		
133	1·026	226	0·674				
024	1·000						
224	0·912						

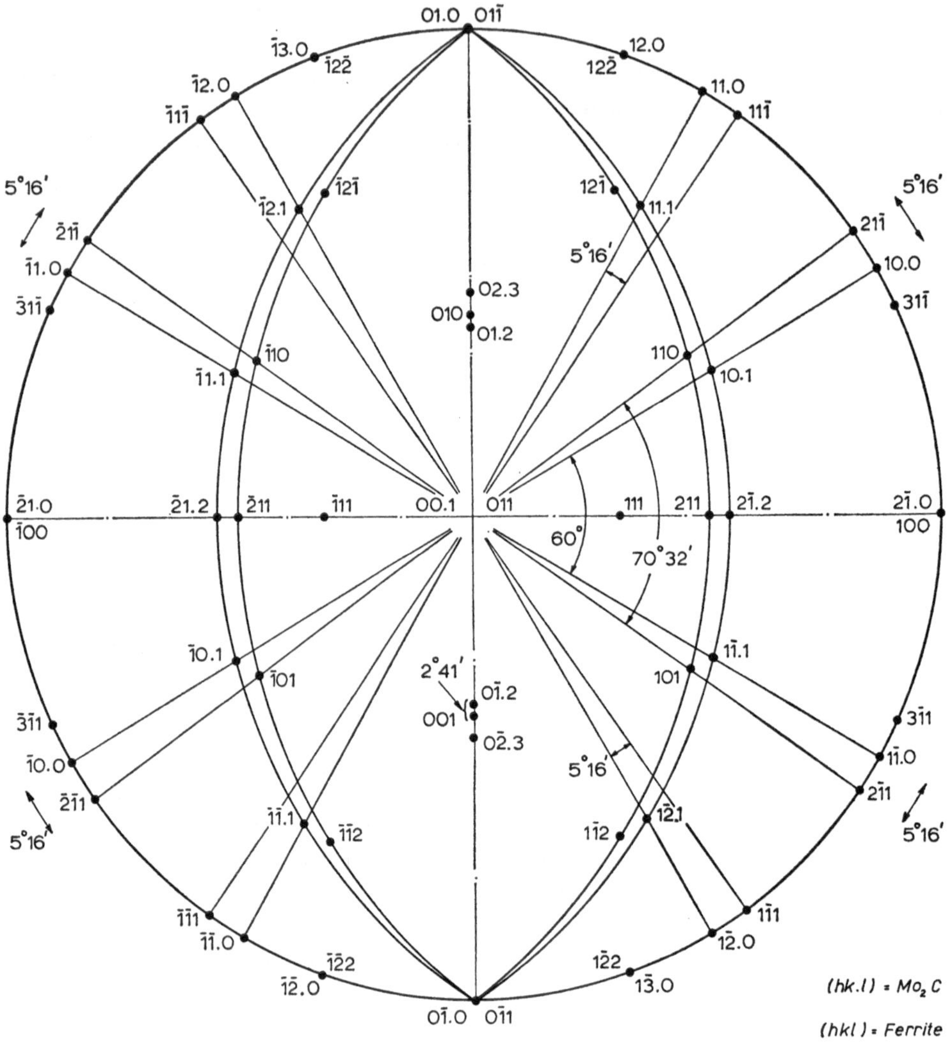

FIG. 52. Stereographic projection for Mo_2C with ferrite based on Pitsch and Schrader's ϵ/α orientation relationship.
($2\bar{1}.1$) Mo_2C // (100) ferrite
(01.0) Mo_2C // ($01\bar{1}$) ferrite
(00.1) Mo_2C // (011) ferrite

16 Metals

16.1 INTERPLANAR SPACINGS OF ALUMINIUM, COPPER and γ-IRON

Aluminium, copper and γ-iron each have the face-centred cubic A1-type structure. The difference in lattice parameter is fairly representative of the range over which this structure exists. Each metal has been included because of its use in either ferrous or non-ferrous metallurgy and as a standard.

TABLE 20. Interplanar spacings of aluminium
Crystal structure: Face-centred cubic
Unit cell: $a_0 = 4{\cdot}049_6$ Å

hkl	*d*-spacing (Å)	*hkl*	*d*-spacing (Å)	*hkl*	*d*-spacing (Å)	*hkl*	*d*-spacing (Å)
111	2·338	133	0·929	135	0·685	155, 117	0·567
002	2·025	024	0·906	006, 244	0·675	046	0·562
022	1·432	224	0·827	026	0·640	246	0·541
113	1·221	333, 115	0·779	335	0·618	137, 355	0·527
222	1·169	044	0·716	226	0·611	008	0·506
004	1·012			444	0·585		

TABLE 21. Interplanar spacings of copper
Crystal structure: Face-centred cubic
Unit cell: $a_0 = 3{\cdot}615_0$ Å

hkl	*d*-spacing (Å)	*hkl*	*d*-spacing (Å)	*hkl*	*d*-spacing (Å)	*hkl*	*d*-spacing (Å)
111	2·087	133	0·829	044	0·639	335	0·551
002	1·808	024	0·808	135	0·610	226	0·545
022	1·278	224	0·738	006, 244	0·603	444	0·522
113	1·090	333, 115	0·696	026	0·572	155, 117	0·506
222	1·044						
004	0·904						

TABLE 22. Interplanar spacings of γ-iron
Crystal structure: Face-centred cubic
Unit cell: $a_0 = 3{\cdot}585_2$ Å

hkl	*d*-spacing (Å)	*hkl*	*d*-spacing (Å)	*hkl*	*d*-spacing (Å)	*hkl*	*d*-spacing (Å)
111	2·070	133	0·823	044	0·634	335	0·547
002	1·793	024	0·802	135	0·606	226	0·541
022	1·268	224	0·732	006, 244	0·598	444	0·518
113	1·081	333, 115	0·690	026	0·567	117, 155	0·502
222	1·035						
004	0·896						

16.2 INTERPLANAR SPACINGS OF α-IRON

Ferrite is a member of the group of metals with the body-centred cubic A2-type structure. This phase is particularly common in low-alloy steels. In the examination of thin foils from such steels, the ferrite acts as a useful 'internal' standard for the electron diffraction pattern of the precipitated phase.

TABLE 23. Interplanar spacings of α-iron

Crystal structure: Body-centred cubic

Unit cell: $a_0 = 2{\cdot}866_4$ Å

hkl	*d*-spacing (Å)	*hkl*	*d*-spacing (Å)	*hkl*	*d*-spacing (Å)	*hkl*	*d*-spacing (Å)
011	2·027	013	0·906	114, 033	0·676	233	0·611
002	1·433	222	0·828	024	0·641	224	0·585
112	1·170	123	0·766			015, 134	0·562
022	1·013	004	0·717				

16.3 INTERPLANAR SPACING OF α-TITANIUM

α-titanium has a close-packed hexagonal A3-type structure. Even though this is different from the L′3-type structure of Mo_2C, Figs. 25 and 26 are approximately the stereographic projections for both.

TABLE 24. Interplanar spacings of α-titanium

Crystal structure: Hexagonal close-packed

Unit cell: $a_0 = 2{\cdot}950_4$ Å, $c_0 = 4{\cdot}683_3$ Å, $c/a = 1{\cdot}587_3$ Å

hk.l	*d*-spacing (Å)	*hk.l*	*d*-spacing (Å)	*hk.l*	*d*-spacing (Å)	*hk.l*	*d*-spacing (Å)
00.1	4·683	11.3	1·072	11.5	0·791	12.5	0·672
01.0	2·555	01.4	1·064	00.6	0·781	00.7	0·669
00.2	2·342	02.3	0·989	02.5	0·755	22.3	0·667
01.1	2·243	12.0	0·966	03.3	0·748	02.6	0·666
01.2	1·726	12.1	0·946	01.6	0·747	01.7	0·647
00.3	1·561	00.5	0·937	12.4	0·745	13.3	0·645
11.0	1·475	11.4	0·917	22.0	0·738	04.0	0·639
11.1	1·407	12.2	0·893	22.1	0·729	04.1	0·633
01.3	1·332	01.5	0·879	13.0	0·709	03.5	0·630
02.0	1·278	02.4	0·862	22.2	0·704	22.4	0·624
11.2	1·248	03.0	0·852	13.1	0·701	04.2	0·616
02.1	1·233	03.1	0·838	11.6	0·690	11.7	0·609
00.4	1·171	12.3	0·821	03.4	0·689	12.6	0·607
02.2	1·122	03.2	0·800	13.2	0·678	13.4	0·606

17 Intermetallics

17.1 INTERPLANAR SPACINGS OF Ni_3Ti

Ni_3Ti is found in some nickel-base alloys and austenitic stainless steels. It is representative of the group of compounds with the hexagonal DO_{24}-type structure. As the axial ratio is $\simeq$ 1·63, a separate stereographic projection is not required.

TABLE 25. Interplanar spacings of Ni_3Ti
Crystal structure: Hexagonal close-packed
Unit cell: $a_0 = 5{\cdot}101_0$ Å, $c_0 = 8{\cdot}306_7$ Å

hk.l	*d*-spacing (Å)	*hk.l*	*d*-spacing (Å)	*hk.l*	*d*-spacing (Å)	*hk.l*	*d*-spacing (Å)
00.1	8·307	02.3	1·727	12.4	1·301	04.0	1·104
01.0	4·418	12.0	1·670	03.3	1·300	03.5	1·102
00.2	4·153	00.5	1·661	22.0	1·275	04.1	1·095
01.1	3·900	12.1	1·637	22.1	1·261	22.4	1·087
01.2	3·026	11.4	1·610	13.0	1·225	11.7	1·076
00.3	2·769	01.5	1·555	22.2	1·219	04.2	1·067
11.0	2·551	12.2	1·549	11.6	1·217	12.6	1·066
11.1	2·438	02.4	1·513	13.1	1·212	13.4	1·055
01.3	2·346	03.0	1·473	03.4	1·201	02.7	1·045
02.0	2·209	03.1	1·450	00.7	1·187	00.8	1·038
11.2	2·173	12.3	1·430	12.5	1·178	04.3	1·026
02.1	2·135	11.5	1·392	13.2	1·175	23.0	1·014
00.4	2·077	03.2	1·388	02.6	1·173	22.5	1·012
02.2	1·950	00.6	1·385	22.3	1·158	01.8	1·011
01.4	1·879	02.5	1·323	01.7	1·146	03.6	1·009
11.3	1·876	01.6	1·321	13.3	1·120	23.1	1·006

17.2 SIGMA PHASES: INTERPLANAR SPACINGS OF σ Fe–Mo

It is well known that the sigma phases exist in a wide range of alloy systems and form extensive solid solutions. The phase is tetragonal with the $D8_6$-type structure. β-Ur has a similar unit cell. A short table showing the variation of the lattice parameters of this phase in different alloy systems has been compiled by Taylor and Kagle.[39]

	a	*c*	Axial ratio *c/a*
σ Fe–Cr	8·7995	4·5442	0·5164
σ Fe–Mo	9·188	4·812	0·524
σ Mo–Re	9·58	4·96	0·52
σ Pt–Ta	9·95	5·16	0·52
β-Ur	10·759	5·656	0·526

Fig. 53 shows a stereographic projection for this phase and is based on the values given for σ Fe–Mo. The interplanar spacings given in Table 26 should be sufficient for reasonable identification of σ Fe–Mo and the other related phases.

TABLE 26. Interplanar spacings of σ Fe–Mo
Crystal structure: Tetragonal
Unit cell: $a_0 = 9{\cdot}188_0$ Å, $c_0 = 4{\cdot}812_0$ Å

hkl	*d*-spacing (Å)	*hkl*	*d*-spacing (Å)	*hkl*	*d*-spacing (Å)	*hkl*	*d*-spacing (Å)
001	4·812	110	6·497	230	2·548	360	1·370
002	2·406	111	3·867	231	2·252	361	1·317
003	1·604	112	2·256	232	1·749	362	1·190
004	1·203	113	1·557	233	1·358	363	1·042
		114	1·183	234	1·088		
010	9·188					370	1·206
011	4·263	120	4·109	240	2·055	371	1·170
012	2·328	121	3·125	241	1·890	372	1·079
013	1·580	122	2·076	242	1·562		
014	1·193	123	1·494	243	1·264	380	1·075
		124	1·155	244	1·038	381	1·050
020	4·594						
021	3·323	130	2·906	250	1·706	440	1·624
022	2·131	131	2·487	251	1·608	441	1·539
023	1·514	132	1·853	252	1·392	442	1·346
024	1·164	133	1·404	253	1·169	443	1·141
		134	1·112				
030	3·063			260	1·453	450	1·435
031	2·584	140	2·228	261	1·391	451	1·375
032	1·892	141	2·022	262	1·244	452	1·232
033	1·421	142	1·635	263	1·077	453	1·069
034	1·120	143	1·302				
		144	1·059	270	1·262	460	1·274
040	2·297			271	1·221	461	1·232
041	2·073	150	1·802	272	1·118	462	1·126
042	1·661	151	1·688				
043	1·315	152	1·442	280	1·114	470	1·140
044	1·066	153	1·198	281	1·086	471	1·109
		154	1·001	282	1·011	472	1·030
050	1·838						
051	1·717	160	1·511	330	2·166	480	1·027
052	1·460	161	1·441	331	1·975	481	1·005
053	1·208	162	1·279	332	1·610		
054	1·007	163	1·100	333	1·289	550	1·299
				334	1·052	551	1·254
060	1·531	170	1·299			552	1·143
061	1·459	171	1·254	340	1·838	553	1·010
062	1·292	172	1·143	341	1·717		
063	1·108	173	1·010	342	1·460	560	1·176
				343	1·208	561	1·143
070	1·313	180	1·140	344	1·007	562	1·057
071	1·266	181	1·109				
072	1·152	182	1·030	350	1·576	570	1·068
073	1·016			351	1·498	571	1·043
		190	1·015	352	1·318		
080	1·149			353	1·124	660	1·083
081	1·117	220	3·248			661	1·056
082	1·037	221	2·692				
		222	1·933				
090	1·021	223	1·438				
		224	1·128				

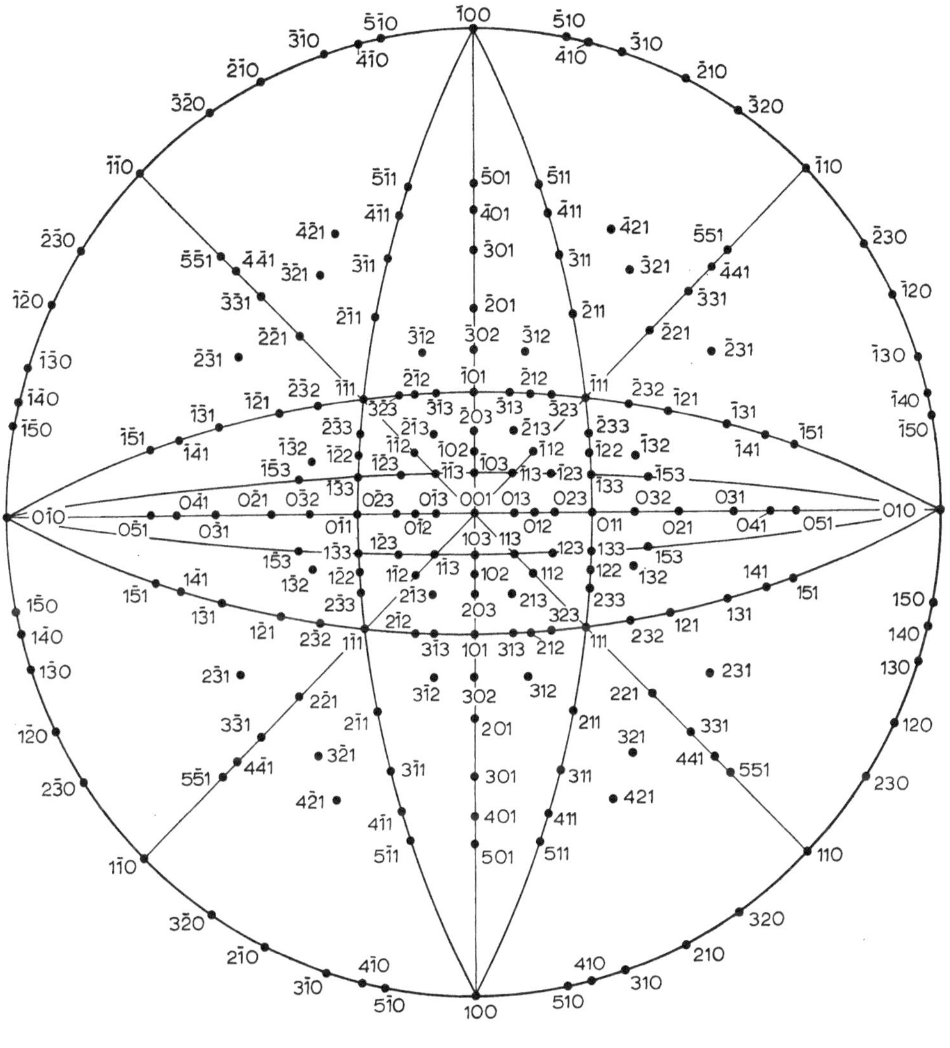

FIG. 53. Stereographic projection of σ iron-molybdenum (tetragonal) on (001), with c/a ≃ 0·524.

17.3 INTERPLANAR SPACINGS OF CHI PHASES

These cubic phases are found in certain alloy steels and other complex systems. The interplanar spacings are given for χ Re–Ta whose lattice parameter, together with that of the other phases of this group with the A12 α–Mn structure, has been compiled by Taylor and Kagle.[39]

System or formula	Lattice parameter
$Cr_6Fe_{18}Mo_5$	8·920 Å
Re_3W (~ 60 at % Re)	9·588 Å
χ Nb–Re (63 at % Re)	9·670 Å
χ Re–Ta (~ 30 at % Ta)	9·711 Å
χ Re–Hf (~ 17 at % Hf)	9·713 Å

Table 27. Interplanar spacings of chi phases
Crystal structure: Body-centred cubic
Unit cell: a_0 = 9·711 Å

hkl	*d*-spacing (Å)
011	6·867
002	4·856
112	3·965
022	3·433
013	3·071
222	2·803
123	2·595
004	2·428
033, 114	2·289
024	2·171
233	2·070
224	1·982
015, 134	1·905
125	1·773
044	1·717
035, 334	1·665
006, 244	1·619
116, 235	1·575
026	1·535
145	1·498
226	1·464
136	1·432
444	1·402
017, 055, 345	1·373
046	1·347
127, 336, 255	1·322
246	1·298
037	1·275
237, 156	1·233
008	1·214
118, 147, 455	1·195
028, 446	1·178
356	1·161
066, 228	1·145
057, 138, 347	1·129
266	1·114
257	1·100
048	1·086
019, 338	1·072
248	1·060
129, 167, 556	1·047
039, 158, 457	1·024
239, 367	1·002

Appendix: A simple stereographic plotting table

In §§3.4, 4.1 and elsewhere reference has been made to the use of stereographic projections. Patterns obtained from diffraction diagrams may have some quite arbitrary orientation which it will be necessary to relate to a standard projection, especially if orientation relationships are to be determined. It is useful therefore to have some means

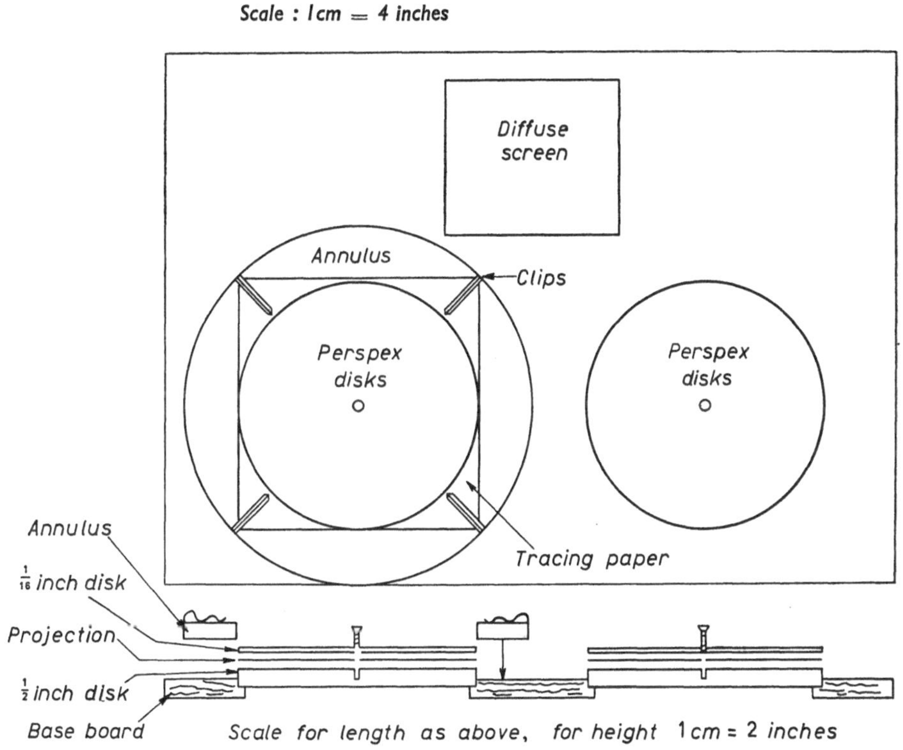

FIG. 54. Plan and sectional views of the plotting table.

of plotting projections and of rotating from standard to experimentally determined orientations or vice versa. The use of tracing paper suggests itself and there are advantages in being able to carry out some of the operations on an illuminated viewing screen. A further step is to use transparencies of the key projections in conjunction with the tracing paper plots of the experimental projections.

The handling of projections may be assisted by provision of a suitable plotting table. Opinions differ as to the most convenient arrangement. The plotting table described below was made from materials available in the authors' laboratory.[16] More complicated designs involving features such as adjustable set squares typical of draughting tables were rejected when it was realized that the three basic requirements could be met by the arrangement shown in Fig. 54. (Plate 5 also shows the arrangement.) These are:

(*a*) Provision for tracing over two separate projections although more may be desirable.

(*b*) An arrangement to transfer a tracing from one projection to the other and accurately superimpose.

(*c*) Provision for rotation of the tracing in both positions.

The basic details of construction are as follows. Two holes, 12½ inches in diameter, were bored in a sheet (28 × 40 inches) of ½ inch thick formica-faced plywood. The holes were counterbored to a depth of ¼ inch and a diameter of 13 inches. A transparent perspex disk, ½ inch thick and 13 inches in diameter, was fixed in each of these annular

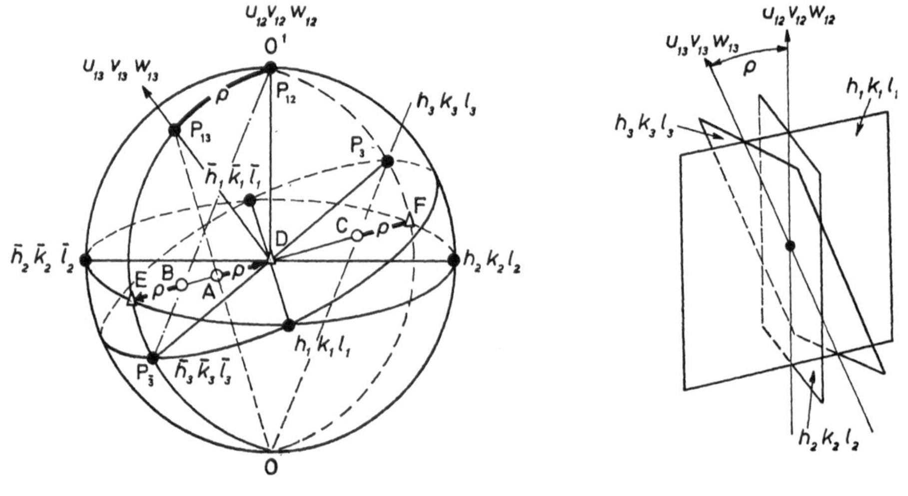

(*a*) (*b*)

FIG. 55. Rotation of stereographic projection and corresponding effect in crystal space. (*a*) Stereographic projection. (*b*) Crystal space.

recesses by a suitable adhesive. Another disk of transparent perspex, 13 inches in diameter and $\frac{1}{16}$ inch thick, was fastened to each of the thicker disks by countersunk screws through the common centre.

A transparent stereographic Wulff net 30 cm (= 11·811 inches) in diameter was fixed between one pair of disks. A transparent standard stereographic projection for the crystal under investigation was then placed between the other pair of disks. A small indentation was made at the centre of each screw head so that a compass point could be accurately located for drawing circles on the tracing. An annulus (internal diameter 13 inches and external diameter 19 inches), also made from perspex, was constructed so that it could be placed over either set of disks allowing freedom of rotary movement around 360° but no lateral movement. There was also very little friction between the annulus and the formica on which it rested. The thickness of the annulus was governed by the height of the disks and stereogram above the surface of the plywood, since it was required that a sheet of tracing paper placed over disks and annulus should be exactly level. The tracing paper was held on the annulus by four clips. A small diffuse screen was set in a recessed rectangular hole in the table. This acted as a convenient viewing screen and was illuminated, together with the disks, from below by a suitably placed light source.

All reflections which come from the zone whose axis is parallel to the incident beam are transferred to their relevant positions on the great circle, which defines the plane of the tracing. These are determined by taking an arbitrarily chosen zero line on both the plate and paper (usually taken as the horizontal through the centre) and using the angular scale, which the net itself provides, to measure the appropriate angle which the reflecting planes make with one another; for example, in a [111] zone of a cubic crystal the $(1\bar{1}0)$ and $(11\bar{2})$ which reflect at right angles to one another. These positions can be fixed on the net. Previously, however, the $(1\bar{1}0)$ may have been set at, say, 40° to the zero line; the intersection of this zero line with the great circle usually has no significance.

The annulus (plus tracing) is placed over the chosen standard projection. If the particle under investigation has the same orientation with respect to the electron beam as a standard projection, other poles can be immediately filled in by rotating the annulus until the poles already on the tracing coincide with those on the standard projection, and then transferring the other poles directly.

It is more frequently found that a stereogram in the required orientation is not available. It then becomes necessary to obtain this by choosing the projection which is nearest to the one needed and rotating the standard projection until the required one is obtained. The physical significance behind this is seen in Fig. 55.

Referring to Fig. 55, it is assumed that a projection is available with $u_{12}v_{12}w_{12}$ parallel to the incident beam. The planes $h_1k_1l_1$ and $h_2k_2l_2$, both in this zone, give points on the stereogram at $(h_1k_1l_1)$, $(\bar{h}_1\bar{k}_1\bar{l}_1)$, $(h_2k_2l_2)$ and $(\bar{h}_2\bar{k}_2\bar{l}_2)$, i.e. on the equatorial great circle shown. If, however, $u_{13}v_{13}w_{13}$ is the zone axis of the pattern, to obtain the stereographic projection for this condition $u_{13}v_{13}w_{13}$ must be rotated into $u_{12}v_{12}w_{12}$. The great circle which is perpendicular to $u_{13}v_{13}w_{13}$ is defined by $(h_1k_1l_1)$, $(\bar{h}_1\bar{k}_1\bar{l}_1)$, $(h_3k_3l_3)$ and $(\bar{h}_3\bar{k}_3\bar{l}_3)$ and is simultaneously rotated into the equatorial great circle. In this example the plane $(h_1k_1l_1)$ is common to both zones, so the easiest way to obtain the new projection is to rotate about the line $(h_1k_1l_1)-(\bar{h}_1\bar{k}_1\bar{l}_1)$ by an amount ρ.

A stereographic projection is obtained by joining the poles such as P_{13} or P_3 (these being where the sphere of reflection and the normal to the reflecting planes intersect) to either O or O′ depending on whether these poles are below or above the equatorial plane and then marking where these lines intersect the equatorial plane. Thus the planes $(h_3k_3l_3)$, $(\bar{h}_3\bar{k}_3\bar{l}_3)$ and $(h_{13}k_{13}l_{13})$ (perpendicular to $u_{13}v_{13}w_{13}$ in the cubic and certain other systems) will appear on the equatorial plane perpendicular to $u_{12}v_{12}w_{12}$ at C, B and A respectively. A rotation of ρ will change these points to F, E and D respectively and any other points accordingly.

The advantage of the rotating annulus is that it quickly enables one to discover the minimum amount of rotation which is required, and to make the rotation. Thus the stereographic projection on any plane of any known crystal structure can be quickly obtained.

For orientation relationship determination, after using the initial stereographic projection on, say, the product, unless the parent and product have the same crystal structure, the standard projection has to be changed and the above process repeated for the parent.

If the volume of work justifies it, a third pair of perspex disks could be provided for the second projection, thus facilitating the last process. A fourth, to take a polar net, could also be added.

In order to minimize the number of locations for the perspex disks, these could be held in the annular recesses by locating pins. In this way all projections could use one recess and the polar net and equatorial net could share the second. Thus the minimum number of useful locations is two.

REFERENCES

1. G. Pinsker, *Electron Diffraction* (Butterworth, London, 1953).
2. C. S. Barrett, *Structure of Metals* (Metallurgy and Metallurgical Engineering Series, McGraw-Hill, London, 2nd edition 1953).
3. A. Taylor, *X-ray Metallography* (J. Wiley & Son Inc., New York, 1961).
4. F. C. Phillips, *An Introduction to Crystallography* (Longmans, London, 3rd edition, 1963).
5. *International Tables for X-ray Crystallography*, Vol. I, 1952, Vol. II, 1959 and Vol. III, 1962 (The Kynoch Press, Birmingham).
6. S. R. Keown, and F. B. Pickering, *Iron & Steel, Lond.*, 1965, **38**, 600.
7. A. W. Agar, *Brit. J. App. Phys.*, 1960, **11**, 185.
8. R. Phillips, *Brit. J. App. Phys.*, 1960, **11**, 504.
9. W. K. Armitage, and A. MacConaill, *J. Sci. Instrum.*, 1964, **41**, 401.
10. R. H. Anderson, and J. S. Halliday, *Techniques for Electron Microscopy*, ed. D. Kay (Blackwell, Oxford, 2nd edition, 1965).
11. G. K. Richards, and G. K. Williamson, *J. Sci. Instrum.*, 1964, **41**, 174.
12. A. Howie, *Proc. Roy. Soc.*, 1963, **271A**, 268.
13. L. Sturkey, and L. K. Frevel, *Phys. Rev.*, 1945, **68**, 56.
14. H. M. Bendler, Proceedings of Fifth International Congress for Electron Microscopy, Philadelphia, 1962.
15. *Index to the Powder Diffraction File* (American Society for Testing and Materials, Philadelphia, U.S.A.).
16. D. J. Dyson, *Z. Kristallog.*, 1965, **122**, 307.
17. L. V. Azaroff, and M. J. Buerger, *The Powder Method in X-ray Crystallography* (McGraw-Hill, New York, 1958).
18. M. F. N. Henry, H. Lipson, and W. A. Wooster, *The Interpretation of X-ray Diffraction Photographs* (Macmillan, London, 1951).
19. T. Ito, *Nature*, 1949, **164**, 755.
20. H. J. Beattie, *Symposium on the Advances in Electron Metallography and Electron Probe Microanalysis* (A.S.T.M. Special Technical Publication Number 317, 1962), p.175.
21. D. J. Dyson, and K. W. Andrews, *J.I.S.I.*, 1964, **202**, 325.
22. A. H. Geisler, and J. K. Hill, *Acta Cryst.*, 1948, **1**, 238.
23. W. Cochrane, *Proc. Phys. Soc.*, 1936, **48**, 723.
24. E. Tekin, Ph.D. Thesis, University of Leeds, 1964.
25. G. Honjo, S. Kodera, and N. Kitamura, *J. Phys. Soc. Jap.*, 1964, **19**, 351.
26. S. Fujime, D. Watanabe, and S. Ogawa, *J. Phys. Soc. Jap.*, 1964, **19**, 711.
27. W. Johnson, and K. W. Andrews, *Brit. J. App. Phys.*, 1955, **6**, 92.
28. P. M. Kelly, *Trans. A.I.M.E.*, 1965, **233**, 264.
29. E. S. Meieran, and H. M. Richman, *Trans. A.I.M.E.*, 1963, **227**, 1044.
30. O. Johari, and G. Thomas, *Trans. A.I.M.E.*, 1964, **230**, 597.
31. M. J. Marcinkowski, and L. Zwell, *Acta Met.*, 1963, **11**, 373.
32. A. B. Glossop, and D. W. Pashley, *Proc. Roy. Soc.*, 1959, **250A**, 132.
33. A. Howie, and M. J. Whelan, *Proc. Roy. Soc.*, 1961, **263A**, 217; 1962, **267A**, 206.
34. G. W. Groves, and M. J. Whelan, *Phil. Mag.*, 1962, **7**, 1603.
35. M. von Heimendahl, W. Bell, and G. Thomas, *J. App. Phys.*, 1964, **35**, 3614.
36. G. Thomas, *Thin Films* (A.S.M., Cleveland, Ohio, 1964) p.227.
37. J. H. Donnay, G. Donnay, E. G. Cox, O. Kennard and M. V. King, *Crystal Data Determination Tables* (A.C.A. Monograph No. 5, 2nd edition, 1963).
38. W. B. Pearson, *A Handbook of Lattice Spacings and Structures of Metals and Alloys* (Pergamon Press, London, 1958 and 1967).
39. A. Taylor and B. J. Kagle, *Crystallographic Data on Metal and Alloy Structures* (Dover Publications Inc., New York, 1963).
40. T. B. Massalski, and H. W. King, *J. Inst. Met.*, 1961, **89**, 169.
41. H. W. King, and T. B. Massalski, *J. Inst. Met.*, 1962, **90**, 486.
42. *International Union of Crystallography World List of Crystallographic Computer Programs*, 1962.
43. G. Menzer, *Z. Kristallog.*, 1938, **99A**, 378.
44. G. Kurdjumov, and G. Sachs, *Z. Physik.*, 1930, **64**, 325.
45. Z. Nishiyama, *Scien. Rep., Tokohu University*, 1934, **23**, 638.
46. G. Wassermann, *Arch. fur das Eisenhutt.*, 1933, **16**, 647.
47. E. C. Bain, *Trans. A.I.M.E.*, 1924, **70**, 25.
48. J. B. Wagner, K. R. Lawless, and A. T. Gwathmey, *Trans. A.I.M.E.*, 1961, **221**, 257.
49. R. G. Baker, and J. Nutting, *I.S.I. Special Report*, 1959, p. 64.
50. H. Lipson, and N. J. Petch, *J.I.S.I.*, 1940, **142**, 95.
51. W. Hume-Rothery, G. V. Raynor, and A. T. Little, *J.I.S.I.*, 1942, **145**, 143.
52. K. W. Andrews, and D. J. Dyson, *Iron & Steel, Lond.*, 1967, **40**, 40 and 93.
53. Yu. A. Bagaryatskii, *Dokl. Akad. Nauk. S.S.S.R.*, 1950, **73**, 1161.
54. W. Pitsch, *Acta Met.*, 1962, **10**, 79.
55. D. J. Dyson, and K. W. Andrews. To be published.
56. W. Pitsch and A. Schrader, *Arch. fur das Eisenhutt.*, 1958, **29**, 715.
57. D. J. Dyson, S. R. Keown, D. Raynor, and J. A. Whiteman, *Acta Met.*, 1966, **14**, 867.

INDEX